FARM APPRAISAL AND VALUATION

5 TH EDITION

FARM APPRAISAL

THE IOWA STATE UNIVERSITY PRESS, AMES, IOWA, U.S.A.

AND VALUATION

5TH EDITION

William G. Murray

WILLIAM G. MURRAY, professor of economics and formerly head of the
Department of Economics and Sociology at Iowa State University, is a
recognized authority in the field of land appraisal. In addition to appraisals
of Indian lands for the U.S. Department of Justice he has made numerous
appraisals for insurance companies and various federal agencies. He has
served as chief economist of the Farm Credit Administration and research
director of the Iowa Taxation Study Committee and is a member of American Farm Economic Association, American Economics Association, National Tax Association, American Society of Farm Managers and Rural
Appraisers, and American Institute of Real Estate Appraisers. His writings
include many journal articles and he is coauthor of *Agricultural Finance*
(5 editions).

© 1940, 1947, 1954, 1961, 1969 The Iowa State University Press
Ames, Iowa 50010. All rights reserved

Composed and printed by The Iowa State University Press

First edition, 1940
Second edition, 1947
Third edition, 1954
Fourth edition, 1961
Second printing, 1963
Third printing, 1965
Fifth edition, 1969

Standard Book Number: 8138–0570–8
Library of Congress Catalog Card Number: 73–83317

CONTENTS

PREFACE

IN THIS NEW EDITION the emphasis has been shifted from income to sale value. Since modern appraisal requires more information on comparable sales, the contents of this volume have been re-arranged to give sale value analysis first attention and a much larger place than in previous editions. However, the importance of income values, especially as they support loan values, has not changed; the reader will find in the discussion of the income approach the primary underlying force determining long-range sale values.

This edition covers all phases of farm real estate valuation including loan appraisals, assessments, inheritance tax apprais-als, and condemnations. A comprehensive presentation is made to assist farm appraisers and assessors in difficult valuations such as those involving tracts on the rural-urban fringe where pressure from business and residential construction is raising market values. Another critical area for the appraiser is farm enlargement where small unimproved tracts are commanding a premium because adjoining farm owners are bidding for addi-tional acres to augment their farms. Other complex valuations included are highway takings and various types of easements.

It is strongly recommended that the reader select one or more farms in his community for use as problems while follow-ing this discussion. If possible, the farms selected should be readily accessible so that several trips can be made to obtain firsthand experience in appraising. Furthermore, it is important that the field trips dovetail with the reading assignments. In this manner the reader will see more clearly the purpose of the material in the text. For purposes of illustration, appraisals by the author of actual farms have been used at appropriate points throughout the book.

That more than one method of appraisal is now in use should not disturb the reader. In appraisal theory and practice, as in certain other fields, several schools of thought exist. Although this may seem undesirable and disillusioning to the practical minded, nevertheless it is an indication of a healthy condition. If these differences of opinion did not exist, there would be less hope for progress in solving the difficult problems involved in farm valuation.

The author has received ideas, suggestions, and assistance from many different individuals and organizations. I am especially indebted to Vern A. Englehorn, Charles H. Everett, William H. Scofield, Neill S. Thompson, and my colleagues in the College of Agriculture at Iowa State University. I am also indebted to Primrose Jackson for her efficient editing and preparation of the manuscript for publication. Finally acknowledgement is gratefully extended to officials in the Farmers Home Administration, federal land banks, insurance companies, highway departments, and other agencies for their assistance. The author, however, is alone responsible for the views expressed.

William G. Murray

1

INTRODUCTION

THE OBJECTIVE in this part is to give the reader an understanding of the whole field of farm appraisal and valuation before becoming immersed in the details. This overall view includes descriptions of the different kinds of farm appraisal, an actual appraisal, and discussions of the value that is sought, appraisal procedures, legal description, and maps.

1

FARM APPRAISAL

FARM APPRAISAL is the systematic process of classifying and evaluating the characteristics of a farm in order to make a well-reasoned judgment of its value. The making of a farm appraisal involves two major parts, the first devoted to classification and the second to valuation—the placing of a value on the property classified in the first part.

Before World War I, farm appraisal was little more than an estimate or guess made after driving past the farm or visiting briefly with the farmer in the farmyard. Such valuations have been frequently referred to as "horseback" or "windshield" appraisals. But with the coming of the professional appraiser, they have been largely replaced by detailed reports which make use of the latest scientific knowledge of soils, crops, farm structures, costs, price trends, sale value comparisons, and similar data. As a result, the reader of these up-to-date appraisal reports can study item by item the evidence presented in support of the appraised value of a farm.

Modern farm appraisal serves many purposes, as indicated by the different kinds of reports which the professional appraiser is called on to make, each of which reveals important aspects of the appraisal process and helps to explain the nature of an appraisal. Consequently in this chapter each of the important kinds of farm appraisal is briefly explained, and a complete appraisal example with common features of all types is presented at the end.

3

KINDS OF FARM APPRAISAL

Of the many kinds of farm appraisal the commonest types are:
1. Appraisals for farm loans.
2. Appraisals for the purchase or the sale of a farm.
3. Tax assessments of farm property.
4. Appraisals made for condemnation of farm property.
5. Other types including inheritance tax appraisals, scenic easements, and easements for power lines and pipelines.

Each of these appraisals has a different objective and consequently requires a different emphasis, but all are the same in requiring classification and valuation. The differences lie mainly in the amount of detail required. A brief explanation of the four principal types—the first four listed above—will not only indicate the similarities and differences but will also give an understanding of the wide area covered by the subject of farm appraisal.

FOR FARM LOANS

The main purpose of a farm loan appraisal is to provide an up-to-date valuation of a farm property. In the process the appraiser stresses long-range forecasts, because the loan agency or person interested in making the loan needs a long-range valuation. This emphasis on the future is based on the long period over which most farm mortgages are scheduled to run. When a loan is made for 20 or 30 years it is difficult to foresee developments, good and bad, which may affect the farm during the loan period. Developments such as the following may occur: prosperity, depression, changes in food preferences, new crops, new production techniques and equipment, government assistance programs, erosion, floods, insects, disease, accidents, and sale of the farm one or more times during the life of the loan.

Income, risks, and community morale are three long-range forecasts that are usually stressed in a farm loan appraisal. Income from the farm in the years ahead is especially important because it is from farm income that the borrower will most likely obtain the funds to pay the interest on the loan. Among the main weaknesses of farm loan appraising in the past have been overvaluation of low-income farms and undervaluation of high-income farms. Then there are the risks of loss during the loan period from such hazards as erosion and floods that may cut the income or reduce the value of the farm. Finally there is a question of community morale which indicates whether under depressed conditions landowners will make reasonable sacrifices to save their farms from foreclosure.

FOR PURCHASE AND SALE

A detailed market appraisal can serve both the buyer and the seller, because an appraisal by a competent appraiser includes all important factors affecting a farm's value—unfavorable as well as favorable. The appraiser in making his report is including not just the bad points in a farm which the buyer is particularly eager to have pointed out, nor is he including just the good points which a seller would like to have emphasized. Instead the appraiser is looking at the farm objectively to present all the facts which have a bearing on its value.

Consequently in a market appraisal the buyer should find information on weaknesses of the farm such as poor drainage, hard-pan layers in the subsoil, or alkali spots that may not be apparent on casual inspection. The report should show any deficiencies or bad features in the buildings and the water supply. Information will be provided on community developments that are unfavorable such as the loss of a nearby school, relocation of a highway, or absence of zoning restrictions that might lead to an undesirable roadside nuisance being located near the farm.

The buyer will also find in the appraisal an impartial listing and evaluation of the favorable features of the farm which the seller would want to see in the report. The appraisal would include positive aspects such as location with respect to markets and community institutions, areas of especially good soil, excellent building construction, and assured future community developments (paved roads, schools, and other attractive items) that might be overlooked. The farm in question might have a good physical layout in pasture, cropland, and buildings for cattle raising, and at the same time be conveniently located for selling purebred livestock. The seller in this instance may not be a breeder of livestock, but an appraiser should point out the possibility of attracting a cattle breeder to buy the farm. In fact the appraiser should specify that the highest and best use of the farm would be obtained from a buyer wanting it for this purpose.

FOR TAX ASSESSMENT

The appraisal of the tax assessor differs from the loan appraisal in that the tax assessor does not need to have the long-range forecasts needed by a lending agency. But the tax assessor or his appraisers do something just as difficult if not more so in the actual rating of all farms or tracts in a district, each tract above, below, or on a par with every other tract. This assessment valuation is usually public information for neighbors to compare and discuss. Since appraisal is not exact but composed of estimates and judgments, the assessor's position is obviously a difficult one.

Three tendencies are usually in evidence in valuing land for tax assessments. First, there is a tendency for any set of valuations to become fixed and to undergo little year-to-year change. Second, what change does occur in assessments (not taxes) is likely to be downward, with some properties being reduced more than others. Finally, there is definite evidence of an excessive concentration of valuations about the average.

In combatting these tendencies the assessor needs a systematic treatment of the value differences between various farm tracts in his district. His task is to establish the proper value relationship between Tract *A*, Tract *B*, and the other tracts he assesses. If Tract *A* is worth exactly twice as much as Tract *B*, this is the relationship that should show in the assessments.

The fact that tax assessment appraisals are usually made on a mass basis, with much less time and detail possible than for most other appraisals, sets them apart from other kinds.

FOR CONDEMNATION

Condemnation appraisal is highly specialized and especially significant because it may become the center of a controversy. Fair market value is the general standard used in providing compensation to owners in condemnation cases. This has proved to be the only satisfactory basis for rewarding the owner whose property is taken. The general principle is that fair market value for the property taken will make the owner "whole"; that is, it will enable him to replace what has been taken from him.

Examples cover cases of cities buying land for airports, of the federal government buying land for military purposes, and of highway authorities buying land for highways. If a new interstate highway is cutting through the center of a farm, separating it into two parts with a long roundabout access from one part to the other, the owner may want a special appraisal covering the market value of the land taken and also the damages to the farm as a whole. These damages will include added farm operating costs such as those arising from reduced farm size, irregular-shaped fields, and extra travel to fields on the other side of the highway. These added costs will be reflected in turn in a lower sale value for the divided farm. The appraiser in measuring the damages will find the difference between the appraised value of the farm before the taking and after the taking. In this same example, the highway authorities will want an appraisal which sets forth the fair market value of the taking and the damages—an appraisal which reflects the public's interest in compensating the owner in full but not in excess for his loss of value. When the two parties cannot get together, litigation may follow with the court determining the value after appraisers representing both sides have been called in to present and explain their appraisals.

Another type of condemnation which is becoming more common is that involving access to highways in rural areas. The problem is to appraise the damages to a farm from restrictions on the right of the owner to build a connecting drive from his land onto the public highway.

OTHER TYPES

Among the other types of appraisal are those for estate and inheritance taxes, book value, and easements. In the settlement of estates it is frequently necessary to appraise farm land for inheritance tax purposes, both state and federal. The object of the appraisal is to determine a value which meets the legal requirements and treats all cases on a uniform basis.

Book value appraisal, as the name implies, is for bookkeeping purposes. These appraisals usually require valuation of buildings for insurance coverage and itemized values for individual buildings and improvements so that detailed depreciation allowances can be figured for income tax purposes.

Easement appraisals include a group of valuations that have been growing rapidly in importance. An easement is a partial right to the use of land. The increase in restrictions on access to highways, resulting in some cases in tracts completely landlocked (without any road access) by interstate and similar type highways, has brought about a marked rise in the number of farmers needing easements through adjoining farms. Here the long-range loss to the owner giving the easement must be carefully evaluated.

Scenic easements, a new and promising development, are being obtained mainly by government bodies such as state highway and conservation departments to assure the public of attractive scenery. For example in buying a scenic easement from a farmer the highway department obtains the farmer's agreement not to cut down or destroy the timber that exists in a certain tract bordering a highway.

Other easements in which an owner gives up some of his rights include the right of a company to lay a pipeline across a farm or to build a power line over a farm.

A unique type of valuation similar to condemnation is the appraisal of lands involved in legal suits brought by various Indian tribes to recover compensation they claim due them in connection with cession of their lands to the United States. One of these cases, in which the author made the appraisal for the United States government, involved 5,000,000 acres of land in western Iowa and 900,000 acres in eastern Kansas—lands ceded to the United States by the Potawatomi Indians in 1846. In this valuation it was necessary to reconstruct the situation of 1846 to determine the market value of the land as of that date. The facts assembled in the appraisal included information on the land market situation in 1846 and in the years preceding in areas nearest to the cessions, statements of qualified contemporary observers, population trends, interest rates, transportation routes, and plats and field notes of the government surveyors who surveyed the land soon after the cessions.

This Indian land appraisal, even though it applied to a situation over a hundred years ago, had essentially the same characteristics as a modern appraisal. It contained an inventory or classification of the physical properties of the land and a well-reasoned judgment of land value based on the physical facts and sales of comparable tracts.

COMMON FEATURES

Although different kinds of appraisal have been stressed, based as they are on different purposes and uses, there are nevertheless some features that are common to practically all kinds of appraisal. These can be brought together in two main appraisal operations:

1. To inventory and classify the physical features including location and legal description, soil, topography, drainage, tillable and nontillable land, farm structures, irrigation facilities, and the like.

2. To establish a money valuation for the farm tract.

In all appraisals it is necessary to locate and legally describe the farm or tract. This is a universal requirement in a mortgage agreement or in a deed. So it is standard procedure in appraisals to start out with an accurate legal description and directions as to location, both of which will be explained in Chapter 4.

The physical inventory or classification of physical resources such as soil and topography is designed to give the user of an appraisal a mental picture of what is being valued. If it is a mortgage loan appraisal the physical classification should receive special attention with more detail on the long-term producing ability of the soil. If it is a tax assessment physical features are especially important because they establish differences between tracts that need to be reflected in the resulting assessments. But the assessor does not have the time or the money to spend on an individual farm that a mortgage loan appraiser would spend. So although there are differences in procedure in the two types of appraisal, the basic requirement of physical resources inventory is the same. This holds as well for purchase and sale appraisals, condemnation appraisals, and others.

The second major objective—placing a money value on the tract—is also a universal requirement. There are different kinds of money value, depending on the type of appraisal, but a dollar value has to be given. Assessments for example present a special problem because they are not generally regarded as current market value. In fact current market values and assessments are good illustrations of two entirely different kinds of money value.

Assessment values are frequently set for a period of several years. In some states real estate assessments are made every four years with no changes during this period unless there is a physical change in the property. During the period between assessments, however, the current market value of farms may change considerably. The basic objective in assessments is not current market value but a correct value *relationship* of one property tract to another as of a given point in time. This relationship then is supposed to last until the next assessment. If assessments were supposed to equal or coincide with current market value at all times, the assessor's task would be virtually impossible, because it would mean a constant changing of assessments to keep up with fluctuations in the real estate market.

Similarly a mortgage loan appraisal may differ in money value from current market value and from assessments. Loan officials are of course interested in current market values, but since they know these values are likely to change many times during the period of a mortgage loan, they may ask their appraisers either to give them a long-term loan value independent of current market value or to give them current market value and then they will decide what percentage of this value they will extend in a loan.

To provide an example of the common features of appraisal an actual case is presented. This appraisal was not made for any specific purpose—neither loan, purchase, sale, tax assessment, condemnation, inheritance tax, or any other purpose. However, it contains the basic elements which with appropriate additions and subtractions would make it fit any one of the specific purposes. The farm in this appraisal, designated as Farm A, was selected because it included many of the problems encountered in practical appraisal work.

Use of Farm A has several advantages. One is the inclusion of urban influence which appraisers are frequently required to handle. Another is cooperation of the owner, tenant, and farm management service in providing a complete set of yield, income, and expense figures over a ten-year period. Still another advantage is that this farm was appraised twice previously, in 1959 and 1963, in connection with farm appraisal schools. Thus Farm A provides the reader with an unusual opportunity to study valuation procedures at work in an actual case over a period of years.

The following comparison of important appraisal factors on Farm A in the three different appraisals may be helpful in showing how much has happened to physical yield and money value since the first appraisal was made in 1959.

Important Appraisal Items for Farm A per Acre

Appraisal Date	Corn Yield Estimate	Real Estate Property Taxes	Building Values	Net Income Value at 5%	Ap- praised Sale Value
1959, May 16	60 bu.	$3.20	$ 75	$240	$410
1963, Sept. 18	80	4.40	97	260	425
1968, Sept. 1	103	7.50	112.50	540	730

A glance at these figures indicates that all items increased between 1959 and 1963 and increased again, this time substantially, between 1963 and 1968. Before examining the forces responsible for these changes, we will take a look at the 1968 appraisal of Farm A. This will provide an opportunity to evaluate the evidence supporting the 1968 value of $730 an acre, a big jump over the 1963 value of $425 an acre. Then we will compare this 1968 appraisal with the two earlier ones and consider the forces which brought about these value changes.

APPRAISAL OF FARM A[1]

Owner: Mrs. Mary Coykendall.
Tenant: Mr. Vincent Hassebrock who has operated the farm on a
    crop-share lease since March 1958.

---

[1] Appraisal is presented without using a printed form to emphasize the material included. For printed appraisal forms of different lenders see Chapter 27.

Date of Appraisal: September 1, 1968.
I. Classification and Inventory
  A. Locational Factors
    Location: Three miles north of Ames, Iowa, on U.S. Highway 69, Story County.
    Legal Description: NW ¼ of Section 14, Township 84 North, Range 24 West of the Fifth Principal Meridian.
    Acres: 160 less 6.19 for road; taxable acreage 153.81.

Towns and Cities:	Name	Population 1960	Distance
	Gilbert	318	2½ miles
	Ames	27,003	3 miles
	Des Moines	208,982	36 miles

    Gilbert and Ames provide nearby grain markets. Ames has a market for hogs and Des Moines has packing plants providing markets for all types of livestock.
    Roads: Paved highway, U.S. 69, runs along west boundary of farm. Gravel road runs along north boundary. Road into farmstead from U.S. 69 is graveled. Location of farmstead back from highway is a favorable feature. High speed of cars passing farm on U.S. 69 is a hazard. Slight rise in highway north of farmstead driveway obscures approaching cars from the north.
    Facilities: Mail delivered on west side of U.S. 69 in front of farmstead. Telephone and electricity on the farm. Various pickup and delivery services from Ames are available.
    Schools: Farm is in Gilbert school district for elementary and high school grades. Gilbert district may merge with Ames district. Iowa State University located in Ames.
    Churches: Most church denominations represented in Ames.
    Community: A good neighborhood but not typical because of closeness to the city of Ames. Ames is growing rapidly with population in 1940 of 12,500, in 1950 of 22,900, in 1960 of 27,000, and in 1965 of 34,800. Estimated population in 1968 was 42,000.
      City of Ames is growing to the north in the direction of Farm A. A golf course and trailer court are located on U.S. 69 one mile to the south of the farm. A subdivision has started in the southwest corner of Section 14, the section in which Farm A is located.
    Zoning: Story County has a rural zoning law so Farm A is in area subject to zoning.
    Recreation: A flood control dam with supplementary recreational facilities has been approved by Congress for the Skunk River at a point 1½ miles directly southeast from the farmstead. Project is in planning stage.
      Present recreation opportunities in Ames and surrounding area include parks, golf courses, and swimming pools. Cultural events and activities, library facilities, and athletic programs are available.
    Health Services: Ames provides unusually fine health services with medical clinic, physicians and surgeons, den-

tists, mental health center, and municipal hospital.

Taxes: Real estate taxes have been rising over the years. Taxes on land and buildings on this farm in selected years follow:

1959—$511—$3.20 an acre   1965—$ 839—$5.45 an acre
1962—$708—$4.40 an acre   1967—$1,032—$6.45 an acre

Easements Given: None.

Buildings: Dwelling, barn, crib and granary, hog house, machine shed, and poultry house.

Typical Rental Rates: In this community the typical renting rate is a crop-share plan with 50 percent of the crop to the landlord and 50 percent to the tenant. Cash rent is paid for hay and pasture ground. Cash rent for buildings varies depending on the situation on each farm. Farm *A* is rented on a 50–50 crop-share basis with $10 an acre charged for hay and pasture, and $100 charged for the buildings.

B. Maps
C. Productivity

Tillable and Nontillable Land: The distribution of land follows:

Cropland	142	acres
Waste (ditch)	8	
Farmstead	4	
Roads	6	
Total	160	acres

A handicap on this farm is the open ditch which runs diagonally through part of the farm and along the southwest boundary. This ditch accounts for 8 acres of wasteland, land which is either nontillable or is kept in grassed waterway to prevent erosion.

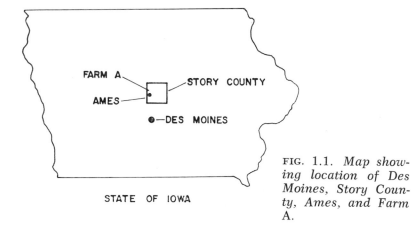

FARM A

STORY COUNTY

AMES

●—DES MOINES

STATE OF IOWA

FIG. 1.1. *Map showing location of Des Moines, Story County, Ames, and Farm A.*

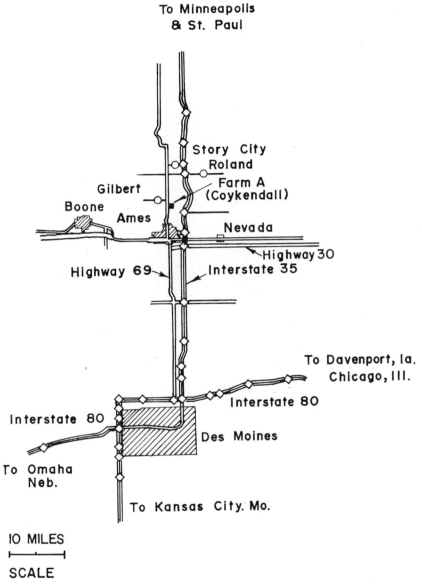

FIG. 1.2. *Map of central Iowa showing interstate and other major highways, cities, towns, and location of Farm A. State highway map used as base.*

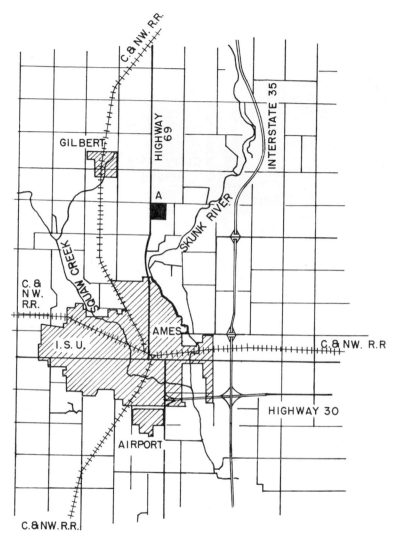

FIG. 1.3. *Map of Ames area showing all roads, railroads, Skunk River, and location of Farm A. County highway map used as base. In a regular appraisal either this map or Figure 1.2 could be omitted.*

FIG. 1.4. *Aerial map of Farm A. From ASCS, USDA.*

Topography: See soil map. Practically all of the tillable area is level to gently rolling (0 to 8 percent). The slopes along the ditch which vary from 9 to 13 percent are the serious erosion hazard on the farm.

Drainage: For drainage on this farm and surrounding area see the drainage maps, Figures 1.5 and 1.6. Drainage is good on the average but varies from only fair in the level southeast part to excessive along the ditch slopes.

Soil Productivity: See Soil Map, Figure 1.7, for location of different soils. Farm is made up of the following soil types with the appraiser's estimated yields for corn and soybeans per acre for each type:

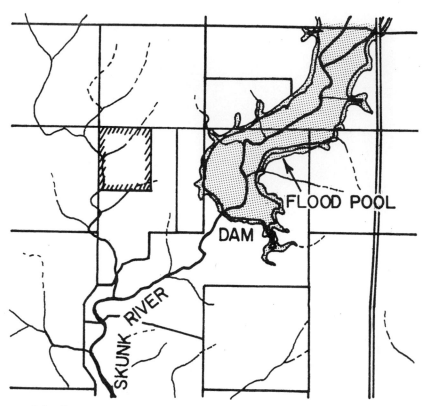

FIG. 1.5. *Drainage map of Farm* A *and surrounding area. Proposed Skunk River Dam and flood control pool are indicated.*

Soil Productivity (contd.)

Soil Number	Soil Type	Acres	Corn Yield	Soybean Yield
55	Nicollet loam	22	114 bu.	40 bu.
107	Webster silty clay loam	33	109	37
95	Harpster loam	3	90	30
6	Glencoe silty clay loam	1	85	25
138	Clarion loam (0–5%)	54	104	37
138	Clarion loam (6–9%)	21	92	33
62	Storden loam	5	75	27
133	Colo-Terril	3	100	35
	Total or average	142	103 bu.	36 bu.

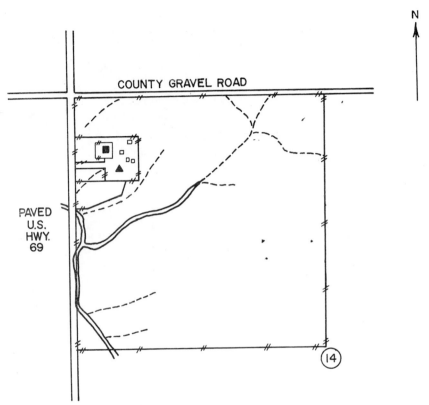

FIG. 1.6. *Drainage map of Farm* A *showing common type of appraisal map for drainage.*

The corn and soybean yields are in terms of bushels per acre, and the averages of 103 and 33 are obtained by multiplying each soil type in acres by the yield for that type, adding all types, and dividing by the 142 acres of tillable land.

Productivity Rating: Productivity of the 142 crop acres plus the 8 acres of wasteland is above the average for the county and the state. With the wasteland included the corn average is 97 bushels per acre while the Story County average for 1965–67 was 91 bushels. Farm A's overall productivity rating is B+.

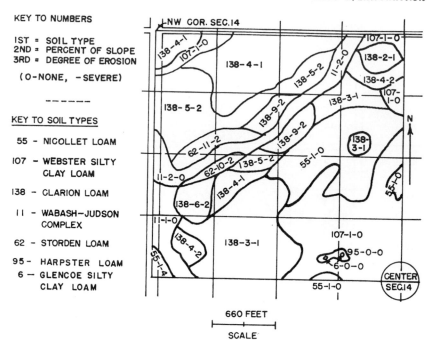

KEY TO NUMBERS

IST = SOIL TYPE
2ND = PERCENT OF SLOPE
3RD = DEGREE OF EROSION

(O-NONE, -SEVERE)

— — — — —

KEY TO SOIL TYPES

55 - NICOLLET LOAM

107 - WEBSTER SILTY
     CLAY LOAM

138 - CLARION LOAM

11 - WABASH-JUDSON
     COMPLEX

62 - STORDEN LOAM

95 - HARPSTER LOAM

6 - GLENCOE SILTY
    CLAY LOAM

660 FEET

SCALE

FIG. 1.7. *Soil map of Farm* A. *A map prepared by the Soil Conservation Service was used for a major portion of this map.*

D. Building Inventory and Value

Building	Dimensions[a]	Construction	Foundation	Roof[b]	Age	Replacement Value	Present Value
Dwelling	16x26x16 22x18x10	Frame	Stone	w.& c.sh.	70	$19,000	$ 8,000
Crib	27x32x14	Frame	Concrete	c.sh.	20	6,800	4,500
Barn	68x40x8 22x24x8	Frame	Concrete	c.sh.	60	10,000	3,000
Poultry	20x40x6	Frame	Concrete	w.sh.	35	2,500	800
Hog	18x32x7	Frame	Concrete	w.sh.	50	2,200	600
Machine	20x32x8	Frame	Concrete	c.sh.	50	1,400	600
Total value						$41,900	$17,500

[a] Dimensions are listed in order of length, width, and height to the eaves.

[b] Roof is either wood or composition shingles.

Notes: Dwelling is a typical older "added-to" house. Has a basement with a warm-air furnace, also has a bathroom with modern plumbing. Is in good condition.

Double crib and granary has slatted cribs for ear corn and overhead granary for small grain and shelled corn. Is partially obsolete under new picker-sheller harvesting method. In very good condition. No inside elevator.

Barn is an obsolete type but has been remodeled inside for pig farrowing. In fair condition.

Poultry house is partially obsolete under present poultry raising methods. In fair condition.

Hog house is practically obsolete except for use as hog shelter. In fair condition.

Machine shed is not adequate for operator's equipment but has been altered to accommodate combine. In fair condition.

Water system includes 160′ well. Water is pumped to supply tank.

Classification of buildings: An old but average set of buildings with good maintenance including painting, repairs, and some remodeling to bring buildings up to date. Overall rating: C +

II. Market Value Estimate

A. Sale History of Farm

Date	Grantor	Grantee	Book Page	Sale Price per Acre
1854	United States	Hunter	Orig. entry	$ 1.25
1900	Hunter	Johnson	50–55	$ 47.50
1913	Johnson	Wike	60–150	$160
1935	Wike	Coykendall	78–130	$100

B. Comparable Sale List

Farm Number	Sale Date	Grantor	Grantee	Book Page	Instrument	Sale Price per Acre	Verified
B	Sept. 1968	O'Neil	Hassebrock		Con.	$718	Grantee
C	Sept. 1967	Christensen	Jensen	117–91	Con.	$700	Grantee
C	Sept. 1967	Christensen	Jensen	100–611	Deed	$700	Grantee
D	Apr. 1967	Wirtz	Nelson	116–372	Con.	$637	Grantee
D	Feb. 1968	Nelson	I.S.U. Res.	101–204	Deed	$675	Grantee

Farmsteads Sold Separately						Total Sale Price	
E	Sept. 1964	Wirtz	Freel	114–56	Con.	$23,000	
F	Sept. 1963	Whattoff	Stahlman	95–571	Deed	$18,000	
G	Feb. 1963	Fatka	Cosgrove	114–577	Con.	$14,000	
G	Aug. 1965	Smith	Smith	95–115	Deed	$18,000	

C. Map Showing Location of Comparable Sales. See Figure 1.10.

D. Description of Comparable Sales

Farm B. O'Neil-Hassebrock sale. Sale was handled by a real estate broker who had several potential buyers. Most of the prospects were investors; the eventual buyer was the tenant on Farm A from 1958 through 1968. Buyer's father owns a farm adjoining Farm B and Farm A on the south. Farm B was bought on contract with interest at 6% and annual principal payments of $3,000. It has 22 acres of nontillable land and a fine set of buildings. There are actually more buildings than are needed. The permanent pasture is located in the southeast part of the farm close to the proposed flood control dam and recreational area on the Skunk River. The farm is not as well located as Farm A for suburban development. The sale price of $718 an acre was considered by this appraiser to be full market price for this farm.

FIG. 1.8. *Dwelling, barn,
and crib and granary
on Farm A.*

20

FIG. 1.9. *Machine shed, hog house, and poultry house on Farm A.*

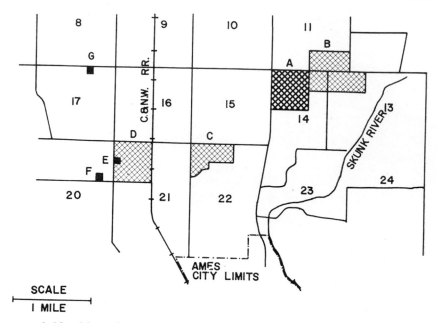

FIG. 1.10. *Map showing location of Farm* A *and comparable sales used.*

Farm *C.* Christensen-Jensen sale. Sale was made in two parts, 29 acres sold with warranty deed and 60 acres sold on contract. Buyer agreed to pay $1,500 a year on contract over a 15-year period, and agreed to pay 5½% interest on the outstanding balance. Also in the contract are provisions regarding the sale of parcels of not less than 5 acres. Jensen, the buyer, is an owner-operator with land and buildings adjoining on the east. He plans to sell acreages for suburban residential development. The farm is located ½ mile south and ½ mile west of Farm *A.* It has slightly higher productivity, is closer to town but on a side road, and its highest and best use is suburban residential eventually—not commercial use. With higher productivity but with only one small building and with no frontage on a paved road, this tract has a market value slightly less than Farm *A.* The price paid of $700 an acre was considered by this appraiser to be full market price at the time of the sale.

Farm *D.* Nelson-I.S.U. Research Foundation. Farm was purchased by Nelson in 1967 for $637 an acre on contract. Nelson sold his contract the following year at an increase in price of $38 an acre or a total of $6,000. Nelson paid a total of $100,000, the Foundation $106,000. Nelson was a farm operator; the Iowa State University Research Foundation acquired the land for the use of the University at some future time. When the Foundation bought the contract from Nelson they paid the contract in full and obtained a deed from Wirtz, the title owner. The farm is located 2 miles west and ½ mile south of Farm A. It is all tillable. There is one building which goes with the land which was assessed at $970. There are also two irrigation wells and irrigation equipment which were included in the sales of 1967 and 1968. The location near the University gives the tract added value. The farmstead of 3 acres which originally went with this land was sold separately in 1964. The price of $675 an acre paid by the Foundation was considered by this appraiser as the current market value of this tract at the time it was sold.

E. Comparison of Sales with Farm *A*

Farm Number	Acres	Est. Yields Corn	Soybeans	Buildings	Location	Road	Percent Tillable	Land Ass't.	Estimated Sale Value
B	190	101	30	V. good	V. good	Gravel	88	$62	$718
C	89	110	35	No house	V. good	Gravel	97	$73	$710
D	157	108	34	No house	V. good	Gravel[a]	100	$70	$700
A	160	103	33	Good	Excellent	Paved	95	$73	

Farmsteads

No.	Acres	Location	Road	Assessment	Present Estimated Sale Value
E	3	V. good	Gravel[a]	$2,525	$25,000
F	5	V. good	Gravel		
G	7	Good	Gravel	$2,085	$19,000
A	3	Excellent	Paved	$2,340	$18,000

[a] Gravel road is scheduled to be paved with asphalt in 1969.

F. Bench Mark Value Chart for Farm *A*.

Year as of Nov. 1	State Average	District Average[a]	County Average[b]	Farm A Estimate	Farm A above County Average
1959	$252	$427	$351	$410	17%
1963	250	413	363	425	
1964	265	443	385	460	
1965	293	481	415	525	
1966	331	550	468	600	
1967	362	602	513	675	
1968	375	636	531	730	37%

[a] District average refers to "High Grade Land" in "Central" district as reported in Annual Survey by Iowa State University Agricultural Experiment Station.
[b] County average for 1959 and 1964 represents federal census figures; average for other years is estimated from Annual Survey.

G. Market Value Estimate

Farm B, which sold in September 1968 for $718 an acre, was the comparable sale which came the closest to being similar to Farm A. Farm B has better buildings but poorer location for suburban development. Farm B, however, is closer to the Skunk River development. Farm A's location and smaller amount of permanent pasture gave it an edge over Farm B and also over Farms C and D which were not on paved highways and did not have farmsteads.

Another factor which gave added value to Farm *A* was the farmstead which was well located for renting or being sold separately. Three farmsteads sold north of Ames, Nos. *E*, *F*, and *G*, show the value that exists near Ames for rural farmsteads. The farmstead on Farm *A* when compared with the sales of the other farmsteads had an estimated market value of $18,000.

Farm *A*, according to the bench mark table, was considerably above the Story County average and above the district average for high value farms. The main reason for the higher value for Farm *A* was its location just north of Ames and the urban development which is moving toward Farm *A*.

On the basis of the comparable sales and the special features which Farm *A* has, I appraise the market value of Farm *A* as of September 1, 1968, at $116,800 or $730 an acre. If the farmstead were sold separately it would bring $18,000, with the remainder of the land bringing $98,800 or $617.50 an acre.

III. Income Value Estimate
      Valuation of Farm *A* on a Rental Income Basis
      A. Income

Crop Acreage	Average Yield (bu.)	Total Yield (bu.)	Landlord Share (bu.)	Estimated Price per Bu.	Landlord Income Per acre	Total
Corn 90	103	9,270	4,635	$1.05	$54.05	$4,867
Soybeans 45	36	1,620	810	$2.40	$43.20	$1,943
Pasture 15					$20.00	$300
Farmstead and roads 10						$600[a]
Total 160						$7,710
Average					$ 48.20	

B. Expense of Landlord

	Per Acre	Total		
Probable future property tax	$ 7.70	$1,230		
Maintenance of improvements	4.00	640		
Seed, fertilizer, etc.	6.50	1,040		
Management and miscellaneous	3.00	480		
Total	$21.20	$3,390	$21.20	$3,390

C. Valuation

Net return to landlord			$27	$4,320
Income capitalized at 3.7%[b]	$27	$4,320	$730	$116,800
Income capitalized at 5%	$27	$4,320	$540	$ 86,400
Income capitalized at 6%	$27	$4,320	$450	$ 72,000

[a] Note: The $600 of farmstead rent would be high for a tenant who lives in the dwelling and operates the farm, but it would be low for the renting of the farmstead to a separate party and the renting of the land to an operator living on a nearby farm.

[b] This 3.7% rate is the estimated rate of return for farms in the same location as Farm *A*.

IV. Final Appraisal Value
   A. *Market Value.* Comparable sales indicate a market value of $730 an acre. A sale value of $18,000 for the farmstead is one of the important factors supporting the $730 figure. The movement of Ames out in the direction of the farm plus the Skunk River dam and recreational area development are other important factors responsible for the $730 value. This value is well above the bench mark figure for comparable land without the special location advantages Farm *A* has.
   B. *Income Value.* Estimated income for Farm *A* is $27 an acre. This assumes a gross return of $600 a year from the buildings which is more than is being received at this time but is realistic in terms of the current and prospective demand for a dwelling so close to Ames. With a 3.7% rate of return, which is reasonable for well-located land like this, the capitalized market value is $730 an acre, the same as the market value obtained through the comparable sale approach.
   C. *Loan Value.* If a 60 percent loan is figured on a value obtained by capitalizing at 6%, the value would be $450 an acre and the loan would be $270 an acre. If a 50 percent loan is figured on a value obtained by capitalizing at 5%, the value would be $540 an acre and the loan would be $270 an acre. Other percentages can of course be used, as for example, a loan policy of 50 percent and capitalizing at 6% which would result in a loan recommendation of $225 an acre. On the other hand, a 60 percent loan policy on a 5% capitalization would result in a loan recommendation of $324 an acre.

Date of Appraisal   WILLIAM G. MURRAY
Sept. 1, 1968   Appraiser

APPRAISALS OF FARM *A* COMPARED

The 1968 appraisal of Farm *A* at $730 an acre is 78 percent above the $410 value for this same farm in 1959. Since high-grade farms in the central district of Iowa where Farm *A* is located rose only 41 percent in the 1959–67 period, it is evident that there were additional factors causing the increase in the value of Farm *A*. These additional factors were the expansion of the city of Ames toward Farm *A*, increased value in the farmstead as a separate unit, and approval of the Skunk River dam improvement.

One of the major reasons for the big value increase in Farm A between 1959 and 1968 was the estimated increase in per acre net income. This is shown in Table 1.1 where estimated net income of $13 an acre in 1963 rose to $27 an acre in 1968. Part of this rise is due to increased yields and part is due to the

TABLE 1.1. Comparative Appraisals of Farm *A*

Factors	1959 May 16	1963 Sept. 18	1968 Sept. 1
Corn yield	60 bu.	80 bu.	103 bu.
Soybean yield	26 bu.	30 bu.	36 bu.
Gross income, total	$3,670	$4,220	$7,600
Expenses, total	1,750	2,150	3,280
Net total	$1,920	$2,070	$4,320
Gross per acre	$ 22.95	$ 26.40	$ 47.50
Expenses per acre	10.95	13.40	20.50
Net per acre	$ 12.00	$ 13.00	$ 27.00
Capitalizing at 5%	$ 240.00	$ 260.00	$ 540.00
Adjustments	170.00	165.00	190.00
Market value	$ 410.00	$ 425.00	$ 730.00
Rate of capitalization for market value	2.9%	3.1%	3.7%

added income estimated from the farmstead. Expenses, it will be noted, rose during this period but not as much as estimated gross income. Actually there was not as much increase in net income between 1963 and 1968 as the appraisal estimates indicate. The income portion of appraisals, it should be remembered, are estimates of future income, not averages of past or present incomes. In 1963 appraisers were estimating yields and income conservatively because they were not sure that the current increases were going to persist; by 1968 these same appraisers felt sure the higher yields and incomes were justified and here to stay.

The small rise in values between 1959 and 1963, from $410 to $425, can be explained by the recession in land values which occurred at the same time that net incomes were rising. A glance at the bench mark portion of the Farm *A* appraisal, Section II F, shows this recession in values. According to USDA figures the percentage of net return on Iowa farm value rose from 3.9 in 1959 to 4.3 in 1963, and then fell again to 3.9 by 1968.

In the appraisal of Farm *A* many problems which deserve consideration and explanation have been necessarily omitted. In concentrating on an individual farm we have purposefully avoided general principles. However, Farm *A* has indicated definitely and specifically what an appraisal is, and it will provide many useful illustrations in later chapters to clarify problems discussed. With the background of Farm *A* and the explanation of different kinds of appraisal presented earlier in the chapter, we are ready now to consider principles and problems basic to an understanding of farm appraisal and valuation.

QUESTIONS

1. What is a farm appraisal? What are the main parts of an appraisal report?
2. Explain the important differences between appraisals for loans, for farm purchases, for farm sales, for tax assessments, and for condemnations. What elements in common do these appraisals have?
3. The market value of Farm *A* was appraised at $730 an acre. Outline the steps taken which resulted in this market value appraisal and explain the logical connection between the steps.
4. The income value of Farm *A* was appraised at $540 an acre. Outline the steps taken which resulted in this income appraisal and explain the logical connection between the steps.
5. Explain the difference between the $540 income value and the $730 sale value per acre of Farm *A*.
6. Describe the economic forces which caused the market value of Farm *A* to rise from $410 an acre in 1959 to $730 an acre in 1968.
7. Appraisal is referred to sometimes as a logical process of arriving at value. Do you agree? Explain.

REFERENCES

Appraisal of Real Estate, 5th ed., American Institute of Real Estate Appraisers, Chicago, 1967.

Rural Appraisal Manual, 2nd ed., American Society of Farm Managers and Rural Appraisers, DeKalb, Ill., 1967.

Wendt, Paul F., *Real Estate Appraisal,* Holt, Rinehart, & Winston, New York, 1956.

See also the following periodicals:

Appraisal Institute Magazine. Published quarterly by the Appraisal Institute of Canada, Winnipeg, Manitoba, Canada.

Appraisal Journal. Published quarterly by the American Institute of Real Estate Appraisers, 155 Superior St., Chicago, Ill. 60611.

Journal of the American Society of Farm Managers and Rural Appraisers. Published semiannually by the Society, Box 295, DeKalb, Ill. 60115.

2

APPRAISAL VALUE

WHAT IS appraisal value? A simple yet meaningful answer is: Appraisal value is quality measured in dollars at a given moment of time. Unfortunately, however, this answer is not as simple as it appears, because there are several problems connected with this measurement of quality which have no ready answers.

First, there is the difference in value depending on who is doing the value measurement. If it is the owner of the property, this is a subjective valuation. If it is a buyer and seller, their final agreement, called market price, is an objective valuation. If it is an assessor valuing a farm, it is referred to as a tax assessment. If it is an appraiser valuing a tract for a client, it is called appraisal valuation. Second, there is need for an explanation of the differences if any between value and price. Third, there is the problem of measuring changes in value over time. Fourth, there is the question faced by the appraiser of which approach to use in his measurement of value—market, income, or cost, or some combination of the three.

SUBJECTIVE AND OBJECTIVE VALUE

To start with, the term "value" needs to be explained. *Value is quality of worth which has the power to satisfy a human want.* When we say that tract A is worth more than tract B we mean that A has a greater value than B. Thus value is measured by comparison—the worth of one thing in comparison with another.

Value can be either subjective or objective. *Subjective value is the worth of a thing in the mind of an individual.* John Smith considers his farm to be worth $50,000, but his neighbor thinks it is worth $30,000. The owner in this instance has a sentimental attachment to the farm which the neighbor does not have. Both of these values, which represent individual opinions, are subjective values.

Objective, or group judgment, value is the worth of a thing in the market. It is measured by the price in dollars in a transaction between a willing buyer and a willing seller. To find this market or objective value of the Smith farm it is necessary to sell it or, as a substitute, to estimate what it would bring on the market. Let us assume that Smith decides to sell the farm. After an extended period of negotiations the farm is sold for $40,000. This $40,000 figure is market or objective value.

Henceforth, when the word "value" is used without any qualifying adjective, it will have reference to market or objective value. This is the common use of the term. Any other kind of value will be labeled specifically to avoid misunderstanding.

Is appraisal value the subjective value of the appraiser? In one sense it is, but in a more generally accepted sense it is not. It is subjective in that it is a value estimate of an individual; there is no willing-seller–willing-buyer bargain setting an objective market price. On the other hand an appraisal value is usually an estimate either of market value or of some variation of market value. Since the appraiser usually is not personally interested in the property, it is reasonable to say that appraisal value is related more closely to objective than to subjective value.

Earlier we noted the two subjective values of $30,000 and $50,000 for the Smith farm and a market value of $40,000. If Smith had called in an appraiser to appraise the farm for sale, or if a prospective buyer had obtained an appraisal for use in buying the farm, these appraisals would be estimates of objective market value. It is obvious, therefore, that appraisal values are closely related to market values. However, in the discussion on appraisal types it was seen that some appraisal values (such as for tax assessments), although related to market values, are made for different purposes. Consequently it should be emphasized again that use of the word "value" in this book always will mean "objective market value." When any other types of values are mentioned (such as appraisal values for loans or tax assessments), full descriptive terms will be used to distinguish them from market values.

VALUE AND PRICE

Confusion may be encountered in distinguishing between value and price. Value is an overall general quality of worth which exists in a thing; price is the measure of this worth in terms of money. If we did not have money we could still express value by comparing one object with another. A cow might be valued at or exchanged for four sows; or a farm might be valued at or exchanged for 100 cows, a tractor, and a full line of machinery. Money, it is evident, is a very convenient tool for measuring value.

It is common to use the term value when speaking of things that are not actually for sale or when summing up the total worth of a number of individual units. Thus we say that the Smith farm has a value of $40,000, although it is not on the market for the good reason that the owner considers it worth $50,000 to him. Similarly it is common to speak of the value of all the farms in the United States; for example all farms in the country according to the Federal Census of November 1964 had a total value of $160 billion and according to the USDA a value of $191 billion in January 1968.

Price is reserved for those instances where a farm is or has been sold and the value can be quoted in terms of sale price registered in the market. Thus we say the Jones farm sold for a price of $25,000, or $125 an acre, or we might say that farms sold in a certain county in 1969 brought an average price of $270 an acre.

VALUE CHANGES

An appraisal once made usually becomes out of date in a relatively short time as far as the dollar figures are concerned. In fact one of the few unchanging factors in farm real estate values is change itself. Crop yields, prices of farm products, expenses, interest rates, and attitudes of farm buyers and sellers are continually undergoing change—more rapid in some items and at some times than others.

Farm product prices and farm real estate values in the United States have fluctuated widely since 1910, as shown in Table 2.1 and Figure 2.1. Most of the time these two—product prices and land values—have varied in the same direction. For example the rise in farm product prices and farm values from 1910 to 1920, the decline in these two from 1920 to 1940, and the rise again from 1940 to 1950 were all movements of the two factors in the same direction. Values, it will be noted, fluctuated most (but not all) of the time around one-third of the price level index. The major exceptions, including both extremes 26 and 67 percent, occurred after World War II. More will be said about these later. There is no reason why values should fluctuate around any particular percentage of the price

TABLE 2.1.   Price Level of Farm Products and Farm Real Estate Values per Acre
Average for the United States

Year[a]	Price Level Farm Products[b]	Value per Acre Farm Real Estate[c]	Value of Price Level
		(dollars)	(percent)
1909–10	97	40	41
1919–20	218	69	32
1929–30	148	49	33
1939–40	95	32	34
1949–50	249	65	26
1959–60	240	119	50
1967–68	252	170	67

[a] The first year in each instance refers to "Price Level of Farm Products" (in Column 2)—1909 price level index is 97, 1919 index is 218. The second year in each instance refers to "Value per Acre of Farm Real Estate" (in column 3)—1910 value is $40, 1920 value is $69, etc., except 1959–60 where value is for Nov. 1959.
[b] Index numbers from USDA. Base of 1910–14 equals 100.
[c] Values from federal census except 1968 which is an estimate by USDA for Jan. 1, 1968.

level, but it is important to note that in recent years values have risen much more than prices of farm products. This situation was evident in the 1959–68 appraisal value of Farm A.

Changes in prices of farm products generally cause changes in farm real estate values. The causation runs from prices to income to value of farm real estate. The reason for this is the

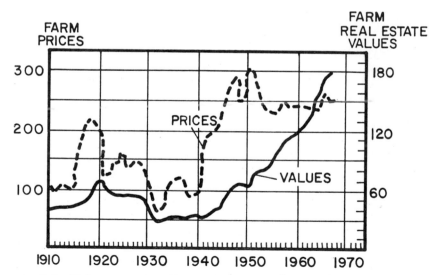

FIG. 2.1. *Value per acre of farm real estate and prices of farm products, United States average, 1910–1969. (Source of data, USDA.)*

relatively fixed supply of land. With a constant amount of land, changes in land prices are caused by changes in demand for the products produced on the land. For instance a continued rise in the price of cotton increases first the income from cotton land and then the demand and price of cotton land.

Farm values, as pointed out above, do not always move or follow exactly in the path of farm product prices. This is indicated clearly in Table 2.1 and Figure 2.1. In 1950 land values were unusually low relative to the price level, the percentage standing at only 26. In 1968 the reverse situation existed with values at an unusually high position relative to prices, the percentage standing at 67. In the 1950 period farm real estate values were relatively low because many believed a depression similar to that which followed World War I might occur again. In the years that followed 1950 on the other hand, not only did the depression fear disappear but many farmers caught in the cost-price squeeze with expensive labor-saving power and equipment bought additional land to enlarge their farms so they could operate more efficiently, thus exerting a strong upward pressure on farm land prices. This pressure, plus other factors such as higher yields and anticipated higher values, forced land values above the relationship with product prices that had existed previously.

APPROACHES TO APPRAISAL VALUE

In times past a great deal of controversy existed over the proper approach to value in a farm appraisal. There were those who held that market values provided the only safe and sound method of reaching a value in a farm appraisal. On the other hand there were just as many enthusiastic supporters of what was called the income method of farm valuation. These two approaches or methods are the two main avenues of logical reasoning leading to value in a farm appraisal. There is a third which, though of minor importance on farms, is of special significance in the valuation of buildings; this is the method of reaching value by means of costs.

As more and more effort and thought have gone into farm appraisal, it has become increasingly apparent that market value is the major approach where sales for comparison purposes are available. But for certain types such as loan appraisals the income approach may be stressed as much or more than market value. In any event, when an intensive analysis is made to determine what it is that makes the value of a producing farm, it is income—not sale value—which has to be emphasized. Of course where cost is a factor it must be included as an approach, but since cost values are limited principally to buildings and other improvements, the cost approach is a minor procedure in farm appraisal. However, where cost enters in, as it does where buildings are an important factor, it has a direct bearing on value. From this discussion it is evident that all three approaches have their place, with market value in the lead followed by income and cost.

MARKET APPROACH

The most common approach to value is to estimate what the property would bring if sold. This process of estimating sale value involves four steps:

1. A definition of the kind of sale to be used.
2. Selection and analysis of nearby sales.
3. Determination of the comparability of the farm tracts sold with the tract being appraised.
4. Adjustment for value changes between the sale of comparable properties and the time of the present appraisal.

Sales that can be used in establishing market value are defined as bona fide transactions. Bona fide sales are genuine market value transactions as contrasted with family transfers. For example bona fide sales are transactions in which the price is arrived at in free and open negotiations between a well-informed seller who is able, willing, and under no compulsion to dispose of his property and a well-informed buyer who is able, willing, and under no compulsion to buy the property in question.

Using the definition above, we would exclude foreclosure sales, sales between relatives, and sales for failure to pay property taxes. Likewise, we would usually exclude sales where an estate is sold to one of the heirs, where an owner is selling in a hurry because of sickness or pressure of debts, or where a city resident buys a farm as a hedge against inflation without making any inquiry—merely paying the price asked.

As in the appraisal of Farm *A* in Chapter 1, the market value approach involves selection of nearby farms that have sold recently. One of the difficulties in the selection is the shortage of sales of properties that are similar to the farm being appraised. An appraiser always hopes there will have been a number of such sales nearby so he can choose those where the property is similar to the appraised tract. But often it is necessary to use all the sales in recent years in the vicinity, and even in this group there may not be a farm sale that is strictly comparable. At other times it is necessary to go a considerable distance just to find a farm sale, let alone a comparable one.

Measurement of the quality of nearby sales in comparison with the farm being appraised is an essential but difficult task in the market value approach. The appraiser's knowledge of soils, crops, and yield variations is especially useful in arriving at differences between the farms sold and the farm being appraised. Later chapters contain material to help the appraiser with this measurement problem.

Finally the appraiser has to adjust sales to bring them up or down to the present level, because only in this way can the sales be considered as comparable. Even though a nearby farm similar to the one being appraised was sold only last year, this does not make it a comparable sale until an adjustment has been made in the sale price to compensate for whatever change has taken place in the local farm real estate market during the year that has elapsed. Fortunately there currently is a supply of farm real estate estimates available from the USDA and state experiment stations which the appraiser can use.

The appraiser who has followed the steps just outlined and has come up with a final value figure for the farm has been using the market value approach. In this process he has been studying carefully what others have arrived at as the going market price for farms. In short he is basing his appraisal value chiefly on what others have concluded in the bargaining process of a sale. His task has been mainly one of transferring the value from one farm to another. This of course is not a simple task because no two farms are exactly alike, and the skill required to measure the difference between two farms is a highly complex one.

At this point, the following question might be raised: Why did this comparable farm sell for what it did? The easy answer would be: Because some other farm sold previously for a certain price. But this is reasoning in a circle, basing the selling price of one farm on another without ever getting to the heart of the issue—What makes value? There must be basic factors that determine farm real estate values, factors that go deeper than the price at which the neighboring farm recently sold. The appraiser who has thought through this problem recognizes that values change and that there are factors which cause these changes. Two of the elemental factors that cause these changes and are basic to an understanding of farm value are income and cost—the other two approaches to value.

INCOME APPROACH

There are two questions the appraiser can ask when he is using the income approach: (1) What would the farm rent for? (2) What annual income would an average owner-operator make farming the tract? Most properties can be subjected to these two questions, whether they be farms, business buildings, or city dwellings. In the case of farms there are variations, with some areas being almost entirely devoted to owner operation and other areas where renting is common. In either case the appraiser can study the farm from the viewpoint of its income potential and come up with a figure representing the income value of the property.

An example of the rental income approach to value was provided in the net income statement for Farm *A* (Chapter 1). The procedure consisted of mapping the soil, estimating yields, determining a crop pattern, calculating total production and a rental share to the landlord, applying a price to this share, subtracting estimated landlord expenses to obtain an annual net landlord income, and capitalizing this net income to obtain an income value for the farm.

Each of the steps in the income valuation process outlined above has in it pitfalls and complexities that deserve and will receive special attention in later chapters. At this point it is important for the reader to note that the various steps fit together logically to give the income value estimates.

The same procedure as that followed in estimating income under rental operation can be applied to an owner-operator type of operation. In this case, however, there are many more expenses to estimate making the method more difficult to carry out with confidence in the results. Where renting is not practiced to any large extent, the owner-operator method may be the only reliable procedure to follow if an income value is desired.

One of the major difficulties in using the income appraisal is the selection of the capitalization rate. A slight change in the rate makes a large difference in the value. With Farm *A*, for example, reducing the rate from 5 to 4 percent raises the value from $540 to $675. What is needed is a clear-cut basis for choosing a certain rate, and this is not easy to establish. The problem of the capitalization rate is considered at length in Chapter 13.

COST APPROACH

Cost provides an approach to value. It is most easily seen in a subdivision in a growing city or town. With competition between builders it is likely that the price of new houses will not be much above or much below the costs incurred by the builders. These costs, in addition to the original price for the lots, are made up chiefly of materials and labor. The price of the completed houses will not be below these costs because if this happened there would be no new houses started until buyers were willing to pay these costs. On the other hand the price for the completed houses will not get far above the total costs, because this would bring a rush of builders into the area and with them an oversupply of houses which in turn would bring the price of houses down closer to cost.

This house-building example has been used because it gives a much clearer picture of cost as a factor in determining value than any example available on a farm. The cost of farm buildings does provide a partial approach to farm value, but it is so indirect and indistinct that it is difficult to recognize. A new set of attractive, well-adapted buildings on a dairy farm gives the best illustration of cost influencing value on a farm. A prospective purchaser may compare the opportunity of three large dairy farms available for sale. The first is the farm which has this new set of buildings, the second has a relatively old set of buildings, and the third is a tract without buildings. If we assume that buildings are essential to operating the farm, that the investment in buildings is not excessive, and that the land is about the same on all three farms, then the difference in value between the three units is largely one of cost—the first farm being worth more than the one without buildings roughly by the cost of the buildings, and worth more than the second by the amount of depreciation and obsolescence of the buildings on this second unit. The purchaser would not pay much more for the first unit, as compared to the third unit, than the cost of the buildings; to do so would be paying more than he would have to pay to build a group of buildings like the ones that are there. He probably would have to pay enough to cover the cost of these buildings, because if he were unwilling to do so, he might lose to a competitive purchaser who would realize that to get a set of well-adapted and attractive buildings like these, certain costs have to be incurred.

How much depreciation to figure in the case of farm buildings often arises as a troublesome problem in the cost approach.

Recently the pressure of farm enlargement has played havoc with the usual cost approach to building values because of this depreciation problem. With a premium being paid for land to add to existing farm units, many cases have developed where buildings in good physical condition have had to be valued at a very low figure or even at less than zero because the buildings were not wanted. A farmer purchasing an adjoining farm to add to his own may want to raise crops on the land where the buildings are located. He may have adequate buildings on his home farm and have no use for additional buildings and thus faces the expense of tearing down the buildings on the farm he is buying. Problem situations of this kind are considered later in Chapter 15.

Although cost is not an important factor in most farm appraisals, it is usually present in some form and should be recognized for the part it plays. This part which is related closely to buildings and other improvements will be discussed in the chapters on buildings.

THREE APPROACHES COMPARED

Market value, income, and cost are all approaches to value. Market value, however, is frequently and logically looked to as the final or absolute test, because there are elements of nonincome or noncost value in farms that do not show up in either income or cost analysis. Such items as a good view, a location close to school or church, or a desirable neighborhood belong to this grouping. Rental or owner-operator income does not measure these features clearly, and cost has no bearing on their value. Consequently the market approach has a primary position.

One method frequently suggested to equate market and income values is the use of a capitalization rate in the income approach which is low enough to include the so-called nonincome features in the resulting income value. To get this rate the procedure followed is to compare net return received annually with sale prices on farms that have sold. Figuring the net return as a percentage of the sale price gives a rate which reflects all features—income and nonincome, but to get this rate it is necessary to obtain sale prices which in themselves represent the market value approach.

The courts have looked with favor on the market value approach because of the conflicting methods of computation associated with income and cost compared to the relatively simple objectivity in actual sales. However, even the courts have not always found sale values adequate. In some areas bona fide sales may be so infrequent as to be of little help, and the sales that do occur may be unreliable as market value because they involve abnormal circumstances such as transfers between members of the same family. In situations of this kind the income approach is usually stressed. But even in cases of this kind it is necessary to estimate sale value in order to get a measurement of the nonincome features of the farm.

Sales by themselves are not enough. It is often desirable to go back of the sale to determine its reasonableness. In doing this we are likely to find that earning power is the chief factor. Why did Farm F sell for $300 an acre and Farm G sell for $200 an acre? In analyzing these sales we find differences in quality which cause differences in income, and these differences are reflected in the sale prices. It is the emphasis on income which has given us in this instance the explanation of the difference in sale value. Thus we see that income and sale value are not two unrelated factors but instead are two characteristics which reflect the same thing—the worth of the property.

Then we may run across two farms with similar income potential but with different building situations. One has a relatively new set of buildings, the other an old set. A cost less depreciation analysis of the two farms will clearly indicate a higher value for the farm with the new set. The difference will probably be reflected in market value, but it is the cost analysis which provides detailed evidence to explain the difference.

SUMMARY

To sum up, market value, income, and cost are all involved in farm appraisal. Market value is the final or absolute test but income and cost are important as means of explaining market value. Market value stands out as the principal approach because it includes not only the income and cost elements but also those nonincome features of a farm such as attractiveness or nearness to town which are not easily measured by either income or cost. Finally, market value has won the approval of the courts because of its objective nature. When all is said and done, however, it is still true that when there is a desire to get behind the market—to find out what makes farm value—income and cost analysis are necessary and helpful. In the chapters which follow all three approaches to value will be included for the contributions they make to our understanding of farm value.

QUESTIONS

1. Explain the difference between subjective, objective, and appraisal value.
2. If two sons both wanted to buy the family farm, and both were well fixed financially, it is conceivable that they might bid the price of the farm well above the going market value for comparable farms in the community before it was sold to one of them. Would the sale price in this instance be objective or subjective value?
3. What is the difference between the value of a farm and its price?
4. Explain changes in farm product prices and farm real estate values in your home community since 1945.
5. Discuss the application of the three appraisal approaches—market value, income, and cost—to a farm with which you are familiar.
6. There is an old rhyme that goes like this:

> The price of pig is something big;
> Because its corn, you'll understand,
> Is high-priced, too; because it grew
> Upon the high-priced farming land.
>
> If you'd know why that land is high
> Consider this: its price is big
> Because it pays thereon to raise
> The costly corn, the high-priced pig!

Is there a fallacy in this? If so, explain it.

REFERENCES

Appraisal of Real Estate, 5th ed., American Institute of Real Estate Appraisers, Chicago, 1967.

Babcock, Henry A., *Appraisal Principles and Procedures*, Richard D. Irwin, Inc., Homewood, Ill. 1968.

Barlowe, R., *Land Resource Economics*, Prentice-Hall, Englewood Cliffs, N.J., 1958.

Medici, Giuseppe, *Principles of Appraisal*, Iowa State University Press, Ames, 1953. See especially Book 1. (This book is an abridged English edition of *Principii di Estimo* by the same author, published in Italian in Bologna, Italy, in 1948.)

Ratcliff, Richard U., *Modern Real Estate Valuation, Theory and Application*, Democrat Press, Madison, Wis., 1965.

Renne, R., *Land Economics*, rev. ed., Harpers, New York, 1958.

Ring, Alfred A., *The Valuation of Real Estate*, Prentice-Hall, Englewood Cliffs, N.J., 1963.

Rural Appraisal Manual, 2nd ed., American Society of Farm Managers and Rural Appraisers, 1967.

Wendt, Paul F., *Real Estate Appraisal*, Holt, Rinehart, and Winston, New York, 1956.

3

MAKING AN APPRAISAL

APPRAISALS unfortunately require more work than is indicated by a glance at a finished report. An important part of the appraisal overview is understanding the basic procedures involved in making an appraisal. Chapter 1 considered the finished product—the various kinds of farm appraisals and an example of a complete report, and Chapter 2 dealt with the meaning of the value we are after. This chapter will go behind the scenes to see how a finished appraisal, such as that for Farm A in Chapter 1, was made. This will begin with the assignment to make the appraisal and follow through to the final report.

Three major stages are involved in the making of an appraisal:

1. Collection of materials, data, and equipment.
2. Inspection of the farm and comparison with other farms.
3. Final analysis, valuation, and preparation of the report.

Each of these three stages will be discussed at length with frequent reference to Farm A so the reader can follow closely the procedure to see how it relates to a finished appraisal.

COLLECTION OF MATERIALS, DATA, AND EQUIPMENT

Upon receipt of an appraisal assignment, the appraiser's first task is to check the location and legal description, using maps, ownership plat books, and other information which he

41

has in his files. Once the location and description have been checked, the appraiser will take from his files other maps, bulletins, and information which bear on the farm in question. Soil maps and reports, data on crop yields and prices of farm products, tax data, and published sale value estimates of farm real estate in the county or region are the types of materials needed. Many of them are published by the USDA in Washington, D.C., by state statisticians of the USDA, and by state agricultural experiment stations. Sources and uses of these materials will be discussed later.

The appraiser before going to the farm will usually stop at the county courthouse to obtain data about the farm being appraised and comparable farms. A courthouse records assignment appearing at the end of this chapter indicates the type of information available in the different offices. The official plat should be checked to make certain that no mistake is made in obtaining the exact legal description of the boundaries of the tract. Top priority should be given to information on recent sales in the vicinity which will indicate farms that can be used as "comparables." The previous sales history of the farm may be recorded. Tax data including assessment values, irrigation or drainage district charges, taxes levied, and related items should be tabulated carefully, not only for the current year but for one or more previous years. At the engineer's office information may be obtained on developments in the planning stage such as new highways, road changes, new drainage, flood control, or irrigation projects; also a record of any county tile or drainage ditches existing on or near the farm. At the appropriate office, information may be obtained on zoning regulations pertaining to the area in which the farm is located. These courthouse items and others of a similar nature are public information available without cost to anyone, and especially to the appraiser who can use the information in his appraisal report.

Access to an aerial map of the farm being appraised is important. If the appraiser does not have such a map he may be able to trace one in the county ASC (Agricultural Stabilization and Conservation) office. Aerial maps (like the one for Farm *A* in the preceding chapter) are frequently a great time saver; they provide accurate detail on irregular fence boundaries, separation of permanent pasture and tillable land, location of streams, trees, and the like.

The appraiser will need equipment appropriate for appraising in his area. Common items of equipment include a soil auger or probe, a level for estimating slope percentages, tapes or wheels for measurement, and a camera for taking pictures of buildings. Last but not least the appraiser will take his clip board, ruler, pencils (including colored pencils), erasers, appraisal forms, notebook, and other materials for use in making an appraisal map of the farm and in recording necessary information.

FARM INSPECTION AND COMPARISONS

Armed with his materials, data, and equipment, the appraiser is now ready to take to the field. It is a good plan to drive first around the community in which the farm is located in order to get a general idea of the setting. It usually is desirable to drive next around the accessible farm boundaries before driving into the farmstead. This provides an overall view that has much to commend it—enabling the appraiser to size up the whole before he starts his detailed examination of the parts. Besides, with the use of the car's mileage meter he may wish to make a rough check of boundary distances and of the total area to be appraised.

It is helpful to divide the inspection procedure into five important parts:

1. Community evaluation.
2. Appraisal map and crop yields.
3. Building and improvement inventory.
4. Talk with farm operator.
5. Sale value comparisons.

These items should be checked by the appraiser before he finishes his inspection.

COMMUNITY EVALUATION

Evaluation of the community is made by noting such factors as churches, schools, predominance of nationality groups, roads, attractiveness of farmsteads, and distance of the farm in question from market points and shopping centers. A large part of Farm *A*'s value, it was noted, was caused by its nearness to the city of Ames and to Iowa State University which uses farmland in its various agricultural experiment programs. A study of the development of Ames and the University showed that both had expanded rapidly in the last 10 years. However, the city of Ames was more a factor than the University.

PREPARATION OF THE APPRAISAL MAP

This second step is an absolute requirement for any detailed appraisal report. Before he starts out and after acquainting the farmer with his assignment, the appraiser makes an accurate outline map of the farm using the exact legal description which he has obtained from the courthouse; on this he will record what he observes as he walks over the farm. He may use the aerial map or a tracing of it in making this outline map. For instance the aerial map of Farm A (Fig. 1.4) was used as the base for the appraisal map. If it had not been for this aerial map, a large amount of time would have been spent in fixing the location and meanderings of the ditch and stream. In a case like this the appraiser without an aerial map is likely to spend a great deal of time locating various features on the map and a minimum of time on the essential task of evaluating what he sees as he walks over the farm.

Another valuable aid if available is a soil or land capability map prepared by Soil Conservation Service personnel. Maps based on aerial photographs showing the soils on the individual farm are usually available at the home of the operator on those farms where the operator is cooperating with the SCS on a conservation program. Use of these soil capability maps in preparing an appraisal map is explained in detail in Chapter 19. At this point, however, it should be emphasized that the capability map can save the appraiser considerable time and effort and provide him with more technical information on the soils than he would be able to obtain on his own.

Helpful as the aerial and soil maps are, they do not record all of the details needed in making a complete farm appraisal. For this reason the appraiser should provide plenty of time for walking over the farm. Figure 19.12 illustrates the kind of route that was required in obtaining the information for the appraisal map pictured in Figure 19.1. Since the nature and complexity of the soils will determine the route, and since this complexity cannot be determined entirely in advance, it is a good plan to allow extra time for unexpected soil conditions that may be encountered in the inspection trip over the farm.

The principal items noted on an appraisal map are the kinds of soil—including the slope and depth of surface soil. As brought out later, the appraiser is interested not in a technical soil map but in a map showing the farm divided into areas of uniform productivity. The soil appraisal map for Farm A (Fig.

1.7) gives a classification on this basis, with symbols designating soil type, depth of surface soil, and slope.

As he makes this map, the appraiser will be estimating in the back of his mind what the various soils will yield and selecting the kind of a cropping pattern the typical operator would follow.

One method of evaluating productivity is to make a yield estimate for the principal crop for each of the uniform soil areas. Another method is to determine a soil productivity index for each uniform soil area. Neither method is easy, but since this productivity evaluation is a fundamental part of the appraisal it cannot be ignored. To get the best results in these estimations it is desirable for the appraiser to take into consideration all the evidence he has collected in his trip over the farm, to study his appraisal map, and to compare the different productivity areas on this farm with similar areas on other farms he has appraised. If yields and a crop plan have been estimated for a farm by SCS personnel, this information can be helpful to the appraiser in his task, although it should not take the place of the appraiser's own judgment on yields and crop pattern for valuation purposes.

The depth factor or third dimension needs emphasis here. Too often values are arrived at through a superficial view of the surface—the length and breadth of a soil area. In reality, however, depth of surface soil is of great significance because root systems need depth to produce high-yielding crops. Since the depth factor is hidden from view, there is a tendency to overlook it. For this reason an appraiser should generally carry a spade or soil auger to examine the depth and quality of the soil beneath the surface. The phases of Clarion loam on Farm A— one deep and the other shallow, as shown in Figure 1.7—have estimated yields of 104 and 92 bushels of corn respectively. This difference in estimated yield means an important difference in the value of these two areas.

BUILDING AND IMPROVEMENT INVENTORY

This inventory includes not only measurements, pictures, and descriptions of the farm buildings but also an estimate of annual maintenance, adaptability, and usefulness to the farm. In addition, information on other improvements such as water system, fences, and tile is required—their importance depending upon their contribution to the productivity of the farm. The building inventory of Farm A follows the information on soil productivity.

A TALK WITH THE OPERATOR

This talk with the farm operator is not always possible. Sometimes there is no operator available, as when a farm has been sold or a tenant has just left. However, when the operator is available, the appraiser will want to acquaint him with the purpose of the visit and ask his permission to inspect the farm. It usually is better to save any extended discussion with the operator until the appraiser has finished his inspection of the farm, so that he will be in a better position to discuss the features of the farm intelligently or to ask about important points, such as broken tile lines, frequency of stream overflow, and alkali conditions, on which the operator may be able to supply useful information.

SALE VALUE COMPARISONS

Comparing the farm appraised with other farms in the vicinity that have sold recently is the last part of the inspection. It is not possible, of course, to give each of these comparable farms the same painstaking physical inspection given to the farm being appraised. But with his knowledge of soils, crops, yields, and related data for the area and the farm being appraised, the appraiser is able to make reasonably accurate estimates of the physical productivity of these comparable farms. Finally the appraiser can compare the nonincome features of these farms—farmstead attractiveness, distance to town, and road type—with the farm being appraised.

ANALYSIS AND VALUATION

The third and last stage of making an appraisal is the economic phase. It can be divided into four steps:

1. Market value estimate.
2. Income value estimate.
3. Building value estimate based on cost.
4. Final judgment of value.

These four steps, properly termed the appraiser's "homework," will ordinarily be done in the office after returning from the inspection of the farm and comparable farms.

MARKET VALUE ESTIMATE

By analyzing sales of comparable farms and using market value data, a reasonable market value estimate is obtained. In the appraisal of Farm *A* six transactions were used—three sales of land without farmsteads and three sales of farmsteads only. All of these tracts and farmsteads were within three miles of Farm *A* and all had been bona fide transactions. A bona fide transaction means a genuine market value sale, not a family transfer or some other less-than-full-value transaction. It would have been better, of course, if there had been farms available that were more nearly comparable and had sold more recently, but in this example as in many others an appraiser will encounter, he has to do the best he can with what is available. Each of the sales is briefly described and compared in tabular fashion or in some similar manner with the farm being appraised. In each case it is practically a requirement that the sale price be verified by someone who is familiar with the details of the sale.

County, district, and state market value data come next. These are Federal Census values, USDA indexes of farm real estate values, and state agricultural experiment station data on estimated values. As indicated for Farm *A*, this information gives the appraiser overall bench marks of market value that enable him to adjust individual sales to a current basis.

Another useful item in market value analysis is putting together the previous sale history of the farm appraised. An abstract of title of the farm can be especially helpful in this task by providing the leads to sales transactions which can then be run down in the courthouse records.

With all the various types of market value data at hand, the appraiser is in a position to set a market value estimate for the farm being appraised. In doing this he compares this farm with the farms for which he has adjusted sales price information and with available land market reports.

INCOME VALUE ESTIMATE

Stated briefly, the income or earning statement is a translation of physical data into dollars and cents. In making this translation either the rental or owner-operation form may be used. The rental form is frequently used in areas where tenancy is common because it provides a simple method of measuring market forces. Where tenancy is not common, an owner-operation estimate of income and expense is used. In some cases both forms are used.

The income statement requires several important decisions. It is assumed that the crop pattern and crop yields have been estimated. In the rental statement it is necessary to divide the crop between the landlord and the tenant, or show a cash rental rate for land on which no crop share division is provided. For both the rental and owner-operator farms, prices for farm products have to be selected and expenses have to be estimated. With this information an estimate of net income to the farmland and buildings can be figured.

Some appraisers and loan agencies using farm appraisals stop with the income statement, others proceed with the capitalization of net income to obtain an income value estimate. Capitalization is calculating the worth of a property by dividing a net income figure by an interest or capitalization rate. The capitalizing formula is:

$$\text{Value} = \frac{\text{Net income}}{\text{Interest rate}}$$

The capitalization rate expresses the relationship between an annual income and the value of the property. For example, with the interest rate at 5%, a property yielding an estimated return of $6.50 an acre indefinitely into the future would be worth $130 an acre:

$$\frac{\$6.50}{.05} = \$130$$

Looked at from another angle, farms selling for $130 an acre on an income basis netting $6.50 an acre in rent are providing their owners with a 5% return. Reasons for using a certain capitalization rate are given in Chapter 13.

BUILDING VALUE ESTIMATE BASED ON COST

Buildings on a farm present the most elusive and variable part of the economic valuation process. Although buildings are directly or indirectly included in the market and income value estimates, it is not easy to measure their contribution to value in these approaches. Consequently it is important to single out buildings for a cost-less-depreciation value estimate. Such an estimate can be useful also in figuring the amount of fire and windstorm insurance to carry. In an ordinary appraisal the separate cost estimate is especially important if buildings make up a large portion of farm value. A new house on a small acreage or a big dairy building layout on a farm where most of the feed is purchased are examples where building costs are particularly appropriate.

New buildings provide the easiest and old buildings the most difficult situation for a building cost report. With new buildings actual cost figures may be available, and if the buildings are well suited there may be little if any depreciation. Under these circumstances the actual cost figures are essentially the building value report.

With an old set of buildings the appraiser has many depreciation estimates to make and not much to use as a basis for the estimates. First, if the building is not going to be replaced with a similar structure, no comparison is possible with a new building. Second, depreciation on old buildings, even if they are to be replaced with similar structures, can involve considerable guesswork because there is no easy way of estimating the years of life remaining in the old buildings. Despite these difficulties, it is possible to compare an old set of buildings with a new set (as was done in the Farm A appraisal) and arrive at a reasonable cost-less-depreciation value estimate for the buildings.

FINAL JUDGMENT OF VALUE

After all the facts are in and evaluated, the final judgment process starts. First, value estimates based on sales, income, and building costs are compared. Nonincome features, which may account for the difference between the market and income value estimates, are given special consideration. Then the appraiser goes back over his appraisal, asking himself what item or items would attract a buyer, what the weak and strong features of the farm are, and what features if any have been missed. He analyzes his appraisal from the standpoint of sources of value—how much of the value is based on income, how much on nonincome factors, and how much on the buildings if any. He compares the farm as a unit with his comparable farm sales, with other sales, and with other farms he has appraised. After carefully considering all of the factors and weighing all of the evidence, the appraiser makes his final judgment of the appraised value of the property.

PROJECTS

ORIENTATION PROJECT

The instructor may invite an appraiser to the class to explain methods he follows in making appraisals. An alternative would be to use a film showing an appraisal being made, such as the film "What Is a Farm Worth?" produced by the USDA, Washington, D.C.

COURTHOUSE PROJECT

This assignment may be used either at this stage or after the student has finished Chapter 4. Since courthouse records vary by states, the outline below which was prepared for students visiting their home county courthouses in Iowa is not likely to fit other states. However, it does provide an example which can be amended to fit the situation in other states. This assignment has been designed to give the student an opportunity to get acquainted with officials and records in his own county courthouse and to familiarize himself with the sale transactions and financial history of his home farm or a farm in which he is interested.

Courthouse Records Assignment

The following records are available at most courthouses. This is your opportunity to get acquainted with these official sources of information and to ask questions on points that are not clear to you. In looking over these materials study them in the order in which they are listed below. Please be courteous to county officials and treat the record books with care.

A. 1. Before you go to the courthouse you will find it helpful to look at the abstract of title for the farm you are appraising. If you do not have access to this, one of the abstract offices at the county seat may have a copy. An abstract is useful in noting historical record of deeds, mortgages, and foreclosures on a farm.

B. Auditor's Office (Plat book and Transfer books)
   2. Official county plat book. Note boundaries of farm. Also note fractional acreages in sections along north and west of township.
   3. Transfer book. You will find dates of all sales listed by section, township, and range. Take down names and dates of recent sales of your farm and nearby farms.

C. Recorder's Office (Register of Deeds)
   4. Index books for Land Deeds and Land Mortgages. These are separate from Town Lot index books and should not be confused. Index books enable you to find exact book and page where deed or mortgage is recorded. Transfer book in Auditor's office gives you date and persons involved so you can find deed transaction listed in Index book.
   5. Deed book. Copy major items from a deed, preferably involving land you are appraising. Check legal description, consideration, date, and transfer stamp tax if any.
   6. Contracts. A special effort should be made to look up one or more recorded land contract sales. These are particularly useful because they give so much specific and accurate information including sale price, interest rate, and payment terms.

7. Farm mortgage book. Mortgages usually taken out at time of land sale can be looked up in Land Mortgage Index.
8. Original entry book. Note original sale by the United States of farm you noted on official county plat book. (Original entry book is not available in some counties.)

D. Assessor's Office
9. Tax Assessment book. Copy information on tax assessment for property you are appraising and other lands you are using in comparison.

E. Treasurer's Office
10. Tax book. Copy information on tax (millage times assessed valuation) for property on which assessment was noted.

F. Clerk of Court's Office (This is optional.)
11. Foreclosure file. List interesting points in the file of a mortgage foreclosure pertaining to your farm. If none occurred on your farm, a foreclosure for some other farm may be used in its place.

G. Outside Courthouse
12. ASC County office. Aerial maps are often available for tracing in these offices. Also visit local SCS office to see soil capability map if one is available.

H. Report
13. Prepare statement on what you have obtained from the records.
14. Comment on any questions which came up in this study of courthouse records. Note any items of interest or importance.

QUESTIONS

1. Name the three major stages in the making of an appraisal. Point out briefly what the appraiser does in each stage.
2. Explain in detail the different types of appraisal information that can be obtained from a county courthouse in your state.
3. Why is the appraisal map important?
4. What is meant by the "third dimension" in appraisal?
5. How would you make sale value comparisons?
6. What is an appraiser's "homework"?
7. Compare income capitalization and market values. Should these two valuation methods give the same answer on a given farm?
8. Of what use to the appraiser are indexes of real estate values and Federal Census values of farms?

REFERENCES

Murray, W. G., and Ackerman, J., "Appraisal of Farm Real Estate," *Land: The 1958 Yearbook of Agriculture*, USDA, pp. 190–97.
Rural Appraisal Manual, 2nd ed., American Society of Farm Managers and Rural Appraisers, DeKalb, Ill., 1967.

4

LEGAL DESCRIPTION
AND LOCATION

A LEGAL DESCRIPTION is to a farm what a name and address are to a person. Of the three million farms in the United States, no two have the same legal description. Consequently the first task in the appraisal of a farm is to note carefully the legal identification and location, thereby avoiding the error an appraiser should never be guilty of committing—appraising the wrong tract. Two errors in interpreting legal descriptions which occurred in central Iowa indicate how important it is for the appraiser to be absolutely certain and correct on the boundaries of the tract he is appraising. One of the errors resulted in a mistake amounting to over $28,000. Newspaper accounts of the two errors appear at the end of this chapter.

To simplify the procedure of locating and identifying a farm, this discussion is presented in three parts:

1. The form and interpretation of legal descriptions.
2. The common means of locating farms with respect to towns and other landmarks.
3. The use of maps.

Two methods are mainly used in describing land in the United States: (1) a system of section, township, and range for land that lies within the U.S. rectangular survey; (2) that of metes and bounds for land that lies outside the rectangular survey and for special cases such as small irregular tracts within the rectangular survey. The rectangular survey includes 30 states, principally the states from Ohio west and south, as shown in Figure 4.1. Alaska is included in the survey, Hawaii is not.

RECTANGULAR SURVEY DESCRIPTIONS

In May 1785 the Continental Congress passed an ordinance providing for the disposal of "lands in the western territory" and specifying a system of land survey which became known as the "Rectangular Survey." The 1785 ordinance which contained the basic elements of our present rectangular system includes these significant statements:

> The surveyors . . . shall proceed to divide the said territory into townships of six miles square, by lines running due north and south, and others crossing them at right angles. . . .
> The plats of the townships . . . shall be marked by subdivisions into lots one mile square, or 640 acres, in the same direction as the external lines, and numbered from 1 to 36. . . .[1]

This basic law has been amended several times, the last important amendment occurring in 1805. All of the changes since have been minor in nature.

If one compares the metes and bounds descriptions presented later in this chapter with rectangular survey descriptions, the simplicity and accuracy of the rectangular survey is apparent at once. We owe a great debt to our forefathers who used intelligence and foresight in establishing the rectangular survey system which eventually covered a major portion of the United States.

For example consider the following description of an actual farm:

NE¼ sec. 14, township 84 north, range 24 west of the fifth principal meridian.

At first sight this description may fail to make any sense whatever. With an understanding of the simple rules, however, it should present no difficulties.

A cardinal rule in reading rectangular survey descriptions is to work backward. The common practice is to place the general reference to location at the end and the specific one at

---

1. Lowell O. Stewart, *Public Land Surveys: History, Instructions, Methods,* Iowa State Univ. Press, Ames, 1935, p. 16.

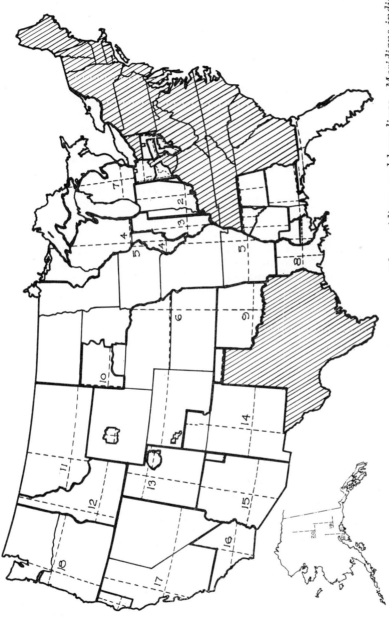

FIG. 4.1. Map showing area covered by rectangular survey with meridians and base lines. Meridians indicated on map by number are identified as follows: 1—First; 2—Second; 3—Third; 4—Fourth; 5—Fifth; 6—Sixth; 7—Michigan; 8—Louisiana; 9—Oklahoma; 10—Black Hills; 11—Principal; 12—Boise; 13—Salt Lake; 14—New Mexico; 15—Gila and Salt Rivers; 16—San Bernardino; 17—Mt. Diablo; 18—Willamette; 19—Copper River; 20—Fairbanks; 21—Seward. Names of other meridians are listed in Manual of Instructions for the Survey of the Public Lands of the United States, 1947, Bureau of Land Management, United States Department of the Interior, p. 168. (Courtesy United States Department of the Interior.)

the beginning; hence the necessity of starting at the end. In simple form, the procedure is similar to that of locating the residence of an individual from a mailing address; at the bottom or end is the state or nation, next above is the city, and above that the street address. Just as the street address is usually unintelligible without knowing first the city and state, the first part of a legal description is meaningless without the general references at the end.

PRINCIPAL MERIDIANS

The first objective in finding the tract of land described above is an explanation of "fifth principal meridian." The meridians used in legal descriptions of farms have no connection with the global geographical meridians—the lines measured from Greenwich, England, which run through the North and South Poles marking off the areas of the earth in longitude. Meridians used in legal descriptions of land (of which there are 34 in all of the area covered) represent arbitrarily chosen north and south lines from which measurements were made to the east or west. In Figure 4.1 it will be seen that the fifth principal meridian is a north and south line running through eastern Arkansas, Missouri, and Iowa; all land lying within the dark-bordered area will be designated east or west of the meridian line and north or south of the base line (see Fig. 4.2).

RANGES

It is clear up to this point that the farm in question is located within the rectangular survey somewhere in the area measured from the fifth principal meridian; in other words, it is in one of six states: Arkansas, Missouri, Iowa, Minnesota, North Dakota, or eastern South Dakota.

Continuing to read backward, the phrase "range 24 west" presents another clue to the location: it specifies that the farm is west of the fifth principal meridian, and also that it is 24 ranges west. According to the methods used in the survey, illustrated in Figure 4.2, land is divided into townships approximately 6 miles square; in measuring east and west from the meridian, the term "range" is used to represent the distance, one range equaling one township or approximately 6 miles. From this explanation it may be determined that the farm is located 24 townships or about 144 miles west of the fifth principal meridian.

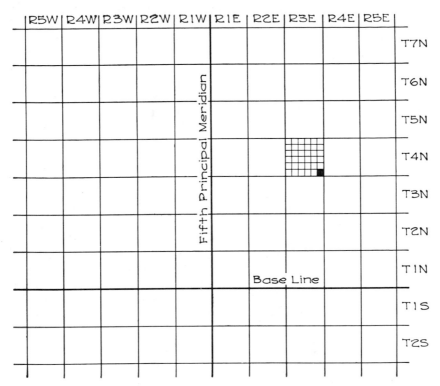

FIG. 4.2. *Illustration of meridian and base line which serve as chief guides in rectangular survey system. The spot shown is legally described as section 36, township 4 north, range 3 east of the fifth principal meridian. The basic unit of the United States Public Land System is the township. Distances are measured in terms of number of townships east or west, and north or south from the two reference lines: principal meridian and base line.*

TOWNSHIPS—BASE LINE

Next in the legal description, reading backward, appears "township 84 north." This is the additional information necessary to single out the township in which the farm is located. In addition to meridians, the General Land Office established base lines—east and west lines without descriptive designation—from which measurements are made to the north and south. The base line for the territory covered by the fifth principal meridian is an arbitrary line running east and west through Arkansas. Measurements from the base line, as from the meridian, are in units of one township. There is one difference, however, in that the unit distance is designated as "township" instead of "range." Sometimes the word "tier" is used to designate the number of townships north or south of the base line. Hence the farm is 84 townships or tiers north of the base line and 24 ranges, or townships west of the meridian. These directions, as a moment's reflection will confirm, indicate the only 6-mile square (township) in which this farm can be. Moreover, these directions are complete without any reference to state lines.

CONVERGENCE AND CORRECTION LINES

Inspection of a survey map of a large area like a state may show some irregular county boundary lines which are the result of correction lines. Correction lines are east and west lines run at intervals in order to take care of convergence caused by the shape of the earth. If the earth were a flat surface, as it was once considered to be, surveying would be relatively simple. As it is, allowance has to be made for the fact that north and south lines cannot be parallel and run through the North Pole.

In the early surveys, correction lines or parallels were not placed at regular intervals. In fact it was not until after the Ohio surveys were completed that the correction line procedure was adopted. Iowa for example has two correction lines approximately 60 miles apart. In 1855, however, a definite 24-mile spacing was adopted between correction or parallel lines and also between guide meridians running north and south (see Fig. 4.3). This gave a uniform grid of 24 miles each way including a total of 16 townships.

Convergence increases as one moves northward from the equator to the North Pole. If we take a township as an illustration, the convergence in the six miles between the south and north boundary will be as follows:

1. In latitude 33 degrees near the north boundary of Louisiana—31.1 feet.

2. In latitude 37 degrees near the south boundary of Missouri—36.1 feet.

3. In latitude 41 degrees near Iowa's south boundary—41.6 feet.

4. In latitude 45 degrees near Minneapolis—47.9 feet.

This means that near Minneapolis the two east and west lines bounding a township will be 47.9 feet closer together at the north end than at the south end, if allowance is made for convergence.

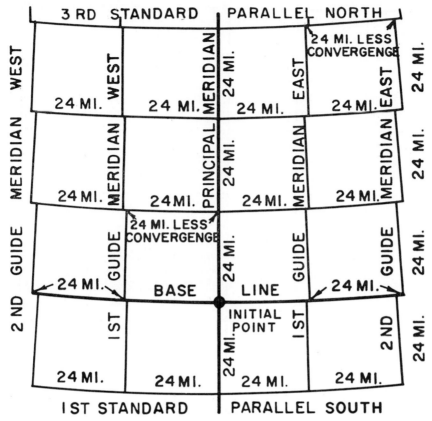

FIG. 4.3. *Example of guide meridians and standard parallels. As the rectangular survey system developed, instructions provided for the establishment of guide meridians at 24-mile intervals east and west of the principal meridian, and of standard parallels at 24-mile intervals north and south of the base lines. (Courtesy L. O. Stewart, Public Land Surveys, p. 93.)*

SECTIONS

The next objective is to determine where the farm is located within the township. The farm is located in section 14, according to the legal description given. Figure 4.4 shows the standard plat of a township. This township is called a "survey" or "government" township to distinguish it from the civil township, a local unit of government which is usually designated by a name such as Washington Township. The civil township may or may not have the same boundaries as the survey township. Section

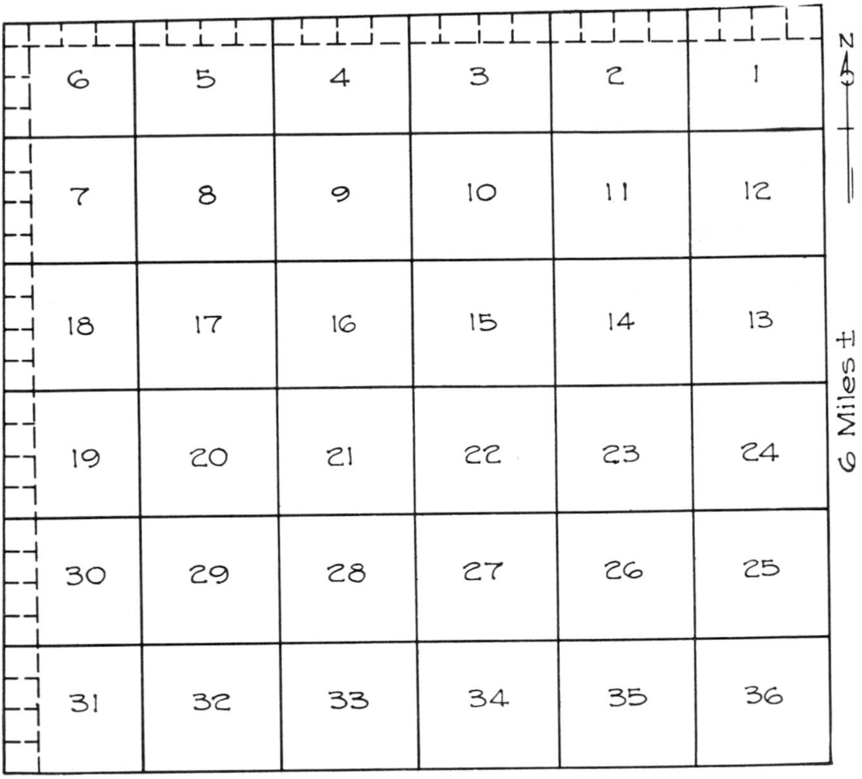

FIG. 4.4 *Numbering of sections in township in rectangular survey. The township is roughly six miles square. Inasmuch as the townships could not be rectangular and at the same time have their sides parallel to the cardinal directions (because meridians converge), it was necessary from the beginning to introduce corrective devices such as excess or deficit measurements and correction lines. It was decided early that, to provide the largest possible number of sections of uniform size, all excess or deficiency of measurement should be thrown northward and westward. For this reason the tracts outlined in the sections on the north and west boundaries are fractional; that is, they do not necessarily contain the regulation area of 40 acres.*

1 is always found in the northeast corner and 6 in the northwest corner; below section 6 is section 7, and the numbering continues back and forth across the township to the last section, 36, which occupies the southeast corner. Section 14 will be found in the third tier of sections from the north, the second section from the east boundary. In Canada the township is a 6-mile square of 36 sections as in the United States, but it is numbered

from south to north; section 1 is in the southeast corner, section 6 in the southwest corner, and section 36 in the northeast corner.

Each section of regulation size contains 640 acres. Not all sections are of regulation size, however, because of the curvature of the earth and slight errors in measurement. To take care of these discrepancies and still have as many uniform sections of 640 acres as possible, corrections are carried to the sections along the north and west boundaries of the township. Consequently sections 1 to 6 on the north boundary and 7, 18, 19, 30, and 31 on the west often do not contain 640 acres, but either more or less depending on the corrections which are applied to that part of the section lying close to the north or west line. If section 1 for example is 20 acres short, the deficiency will be found in the 40-acre tracts along the north boundary of the section as shown in Figure 4.4. These tracts, of which there are four, may each contain approximately 35 instead of 40 acres. The odd-sized units are referred to as "fractional." If the section in the description given earlier had been 4 instead of 14, the proper description would have read:

NE fractional ¼ sec. 4, or NE fr. ¼ sec. 4 . . . .

Here the location of the farm along the north boundary of the township as well as the term "fractional" clearly indicates that the farm is not of regulation size. Each unit in the description which is fractional should be designated as fractional, as for example the small tract in the northeast corner of section 1 in Figure 4.4 would be described as:

NE fr. ¼, NE fr. ¼ sec. 1 . . . .

The usual practice in subdividing the section is to use quarters or halves. Subdivision is a quick and simple process since the standard section of 640 acres is a square. The quarter section, or 160-acre unit, is the most common subdivision. The use of 160 acres as a unit of transfer as well as a farm unit was encouraged by the government through provisions in the Homestead Act and other acts by which the government disposed of the public domain. Quarter sections are designated as northeast, northwest, southwest, and southeast quarters of the section, in the order named; each quarter section thus being uniquely described by denoting its direction from the center of the section. The same procedure is followed for still smaller areas. Each quarter section is divided into quarters of approximately 40 acres each, often called quarter-quarter sections, and each forty is indicated by its direction from the center of the quarter section. Figure 4.5 illustrates the usual division of a section into quarters; and within each of these 160-acre units a second division into 40-acre units each being named respectively, northeast, northwest, southwest, and southeast quarter of the 160-acre unit. Thus if the description calls for the northeast quarter of the southwest quarter of section 8 (NE¼ SW¼ sec. 8), a specific 40-acre unit has been designated. A still finer division into 10-acre or 2½-acre tracts may be made in the same way.

A quarter section tract that is regulation in size and not fractional may be assumed to contain 160 acres. On the other hand a fractional tract, such as NE fractional ¼ of section 4, leaves a question as to the actual acreage. Often the legal description, particularly when a fractional tract is included, gives the acreage at the end, as "containing in all, 151 acres more or less according to government survey."

Descriptions may be written in different ways; for instance, the tract designated in shaded area Figure 4.5 may be described in three forms as follows:

> 1. NW¼SE¼ and NE¼SW¼SE¼ and W½SW¼SE¼ sec. 14, township 55 north, range 1 east of the fifth principal meridian.
> 2. The northwest quarter of the southeast quarter and the northeast quarter of the southwest quarter of the southeast quarter and the west half of the southwest quarter of the southeast quarter, all in section 14, township 55 north, range 1 east of the fifth principal meridian.
> 3. The northwest quarter of the southeast quarter (NW¼ SE¼) and the northeast quarter of the southwest quarter of the southeast quarter (NE¼SW¼SE¼) and the west half of the southwest quarter of the southeast quarter (W½SW¼SE¼), all in section 14, township 55 north, range 1 east of the fifth principal meridian.

Of the three forms given, Number 1 is recommended because of its brevity and clearness. Moreover, it is the form approved by the government for use in describing land belonging to the government. (A more extensive consideration of the forms used by the government is included in Appendix A.) A good rule to follow when descriptions are written out as in Number 2 above is to group as one unit all parts of the description linked together by the word "of," and to separate one group from another wherever the word "and" occurs. It is not a matter of great importance in what order the groupings are considered, but within each group it is usually easier to work backward—that is, from the large to the small unit.

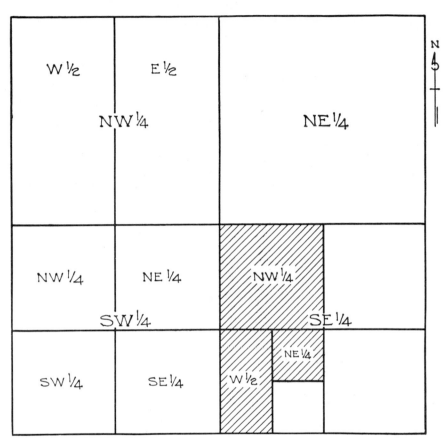

FIG. 4.5. *Division of section of 640 acres into quarters and smaller parts in rectangular survey descriptions.*

In the description just given for Figure 4.5, you already have located section 14. In the first grouping (NW¼SE¼) you locate the last unit—the southeast quarter section (160 acres)—and make sure that the first unit covers the whole northwest quarter of it (40 acres). The second grouping having three units (NE¼SW¼SE¼) indicates it will be less than a 40-acre area; the last unit indicates the location is still within the southeast quarter of section 14; the middle unit leads you to the southwest quarter of the quarter section; and the first unit designates or covers the northeast quarter of it (10 acres). The third grouping, with three units (W½SW¼SE¼), again indicates only a portion of a forty; the last two units describe it in the same 40-acre tract, and the first unit locates it as the west one-half (20 acres).

Blocked in on a section map, as in Figure 4.5, it is easy to visualize and to check the total: $40 + 10 + 20 = 70$ acres.

However, many farms do not have the precise regular boundaries indicated in Figure 4.5. Here the appraiser will often encounter winding roads, railroads, lakes, streams, rivers, and other natural features which make irregular boundaries even within the rectangular survey. These irregular areas and small parcels are usually divided into lots as indicated in Figure 4.6. These lots are numbered as Lot No. 1, Lot No. 2, etc. beginning in the northeast with Lot No. 1, going west to Lot No. 2, and ending with the last lot in the southeast, similar to the manner in which sections are numbered in the township. When

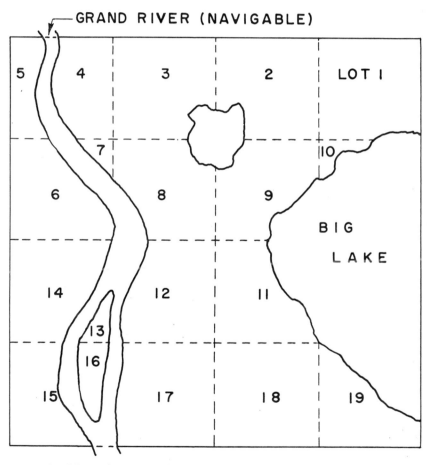

FIG. 4.6. *Map of section showing designation of lots.*

a tract is bounded by a railroad, the railroad right-of-way can become part of the legal description, as "all that part of the southwest one quarter of section ten lying south and west of the right-of-way of the Blank Railroad, etc."

Important rivers and lakes are bounded by meander lines. These lines, representing the normal high-water mark or mean high-water elevation, approximate the boundary between land and water. All navigable rivers and usually all other rivers with a width of 12 rods are meandered. The banks of the river are designated by assuming an individual facing downstream, with the shore on the left being the left bank of the river and that on the right, the right bank. Lakes with an area of 25 acres and more are meandered; that is, the shore line of the lake is officially outlined on the government survey plat. Islands are meandered at their mean high-water elevation if they are surrounded by a navigable stream or other meanderable body of water. All irregular land areas formed by the boundaries of meandered rivers, lakes, and islands are numbered as lots as shown in Figure 4.6.

The question may arise as to who owns the bottom of a river or lake. Unfortunately state laws vary so much that no exact general statement will apply. In most cases the bottom (or submerged land as it is sometimes called) is privately owned if the river or lake is not navigable. On the other hand if the river or lake is navigable the bottom will usually be public property with the owner's land rights ceasing at the water's edge.

One of the sad facts of the rectangular survey is that the regulation 640-acre section does not always contain the 640 acres it is supposed to contain and is frequently far from uniform in shape. The common ending to a rectangular survey description in a deed reading "160 acres more or less" has more meaning than is commonly supposed. An actual example, Figures 4.7 and 4.8, indicates how far reality can be from what a regulation section should be. The appraiser should be on the alert for just such situations. Before we place all the blame for such inaccuracies on the original government surveyors, we should pause to consider the rugged frontier conditions under which the surveys were run. We should keep in mind that in the early days large areas of the land being surveyed were regarded as unfit for settlement and hence of little value, and those areas regarded as fit for settlement were not expected in most cases to have a value of more than $1.25 an acre. Accuracy under such conditions was not considered important, and the United States government was not willing to pay to get it. Nevertheless

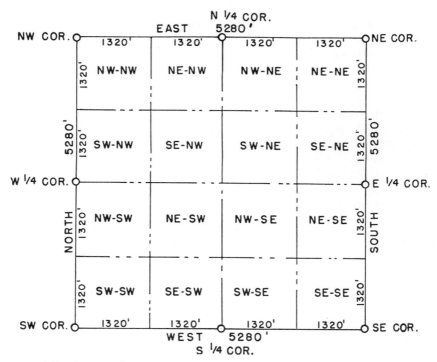

FIG. 4.7. *A typical section the way it is supposed to look.*

FIG. 4.8. *The same section the way the section actually measures.*

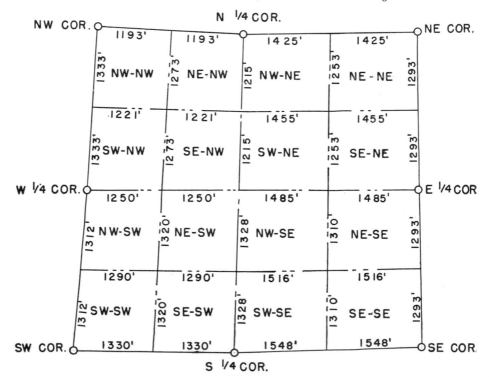

it may come as a shock today in some cases to see large measurement errors in supposedly regulation tracts. The answer is for the appraiser to check for any important discrepancy in what is described as a regulation tract.

The significance of the statement "containing (blank) acres more or less according to government survey," which has been standard wording in deeds to tracts in all sections, should not be overlooked. This use of "more or less" has undoubtedly prevented countless law suits that might have originated in those instances where the farm buyer measured his purchase and found it contained less than the stated acreage.

A handy aid for use in working with rectangular survey descriptions is presented in Figure 4.9. It can be helpful in locating sections in adjoining townships, and it also provides information on subdivisions within a section and includes various measurements such as the link and chain used in the original surveys. Also shown are different ways to figure the area of an acre.

METES AND BOUNDS

In those areas where the rectangular survey does not apply and in special cases within the survey, legal descriptions are usually given in metes and bounds or by reference to lot numbers. "Metes and bounds" means a description of the boundaries; literally, the word "mete" means measures, the whole phrase meaning a measure and identification of the boundaries. This was the system in use in eastern United States prior to the rectangular survey.

In a metes and bounds description a common practice is to use trees, stones, stakes, roads, and rivers as landmarks in bounding the tract of land. Since these marks are not permanent, some of the old descriptions appear to be out of date, inexact, and somewhat amusing in their phraseology. Yet these old descriptions are perpetuated in land transactions by the usual custom of copying verbatim the description as it appears in previous deeds and mortgages. An example of a metes and bounds description taken from a recent deed for a tract of 172 acres in an eastern state is given below. This description is pictured in Figure 4.10. The length and complexity of this description, including expressions like "about four chains and perhaps a few links," should be compared with the short and simple description of the farm in the rectangular survey presented at the beginning of this chapter.

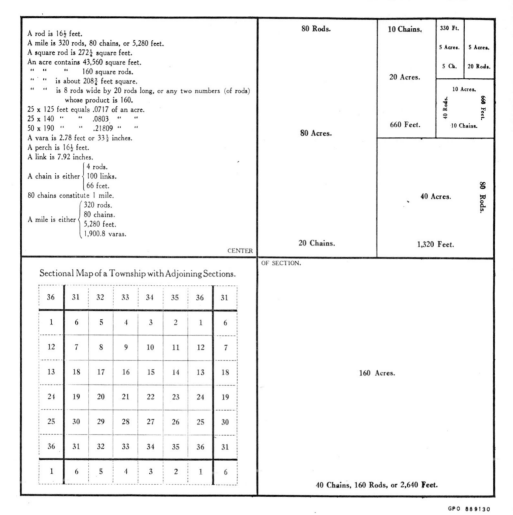

GPO 889130

FIG. 4.9. *An information guide for land descriptions in the rectangular survey.*

DESCRIPTION BY METES AND BOUNDS

All that Tract or Parcel of Land which was formerly known as the north part of the David D. Jones farm, and bounded and described as follows, viz.: Beginning in the middle of the highway running from Brown-hill easterly, down through Knickerbocker Hollow past the late Homestead residence of said Lorenzo M. Jones to the Cohocton-Valley Main road which runs from Cohocton village southerly to Bath, at the southwest corner of the lands lately owned by Benjamin Sawdy Hoig, now deceased,

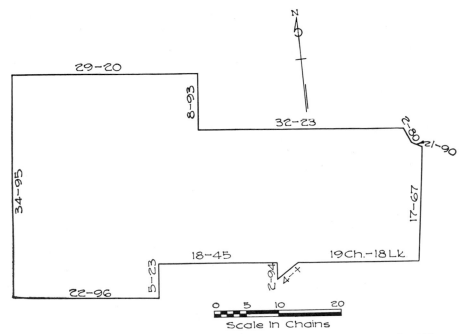

FIG. 4.10. *Plat of farm described in text by metes and bounds. The start is made at the southeast corner. (Description and plat furnished by Stanley W. Warren of Cornell University.)*

and running thence North 84¼° west 19 chains and 18 links along the middle of said highway first above named to its inter- section with the other highway known as the "Back-Road," which runs northerly past the twenty-five acre lot lately owned by F. B. Ireland; thence Southwesterly along the middle of said first mentioned highway which leads up through Knickerbocker Hollow to Brown Hill, about four chains, and perhaps a few links more than four chains to a stake in the center of said highway; thence north five and ¾° east two chains and ninety- four links to a stake and stones which are at a point straight with and in the course of said first described line, thence eighty-four and one-fourth° west (viz. North 84¼ degrees west), eighteen chains and 45 links to a stake and stones; thence south five and ¾° west five chains and twenty-three links to a corner near a watering-trough; thence north eighty-four and ¼ degrees west twenty-two chains and 96 links to a stake and stones in the east linc of the farm lately owned by Fonda Veeder, thence North five and ¾ degrees east thirty-four chains and ninety-five links to a stake and stones, thence south eighty-four and one-fourth degrees east twenty-nine chains and twenty links to a stake and stones, thence south five and ¾ degrees west eight chains and 93 links along the west line of said twenty-five acre lot lately owned by F. B. Ireland; thence south eighty-four and ¼ degrees east, thirty-two chains and twenty-three links to a stake on the westerly bank of the Cohocton River, thence along the same south twenty-three and three-fourths degrees east, two chains and eighty links; thence south along the said bank 64° east one chain and ninety links to a stake on the bank of said river, at the corner of said farm; thence south five degrees and 45 minutes west, seventeen chains and sixty-seven links to the place of beginning, and containing one hundred and seventy-two acres and 34/100 of an acre as surveyed by Robert M. Lyon in 1881, be the same more or less.

This description is long and involved but is short and simple compared to one brought to the attention of the author for a 550-acre farm in eastern New York state; this filled eighteen double-spaced typewritten pages in covering 21 separate parcels that made up the farm.

A more recent metes and bounds description is the following from the state of Hawaii; this example is taken from "Specifications for Description of Tracts of Land," issued by the federal government for use in executive orders and proclamations involving land descriptions:

Beginning at the northerly corner of Parcel No. 1 on the boundary of Land Court Application 900, the azimuth (measured from south) and distance to United States Military Reservation Monument No. 67 is 161°52′, 2,352.64 feet, the coordinates of monument No. 67, referred to Government Triangulation Station "Uka" being 5,263.18 feet north and 10,120.63 feet west.

From the initial point by azimuths and distances,

351°54′, 1,183.36 feet, along Land Court Application 900 to a point;

161°46′30″, 334.59 feet, along the new west side of Kamehameha Highway, to a point;

On a curve to the right, with a radius of 1,969.86 feet, long chord azimuth and distance being 166°39′20″; 335.5 feet, to a point on said highway;

187°44′08″, 123.07 feet, along the highway to a point;

On a curve to the left along the highway, with a radius of 1,472.50 feet, long chord azimuth and distance being 179°49′04″, 405.68 feet to the point of beginning.

The tract as shown on map No . . . . . . dated . . . . on file in the office of the Department Engineer, Fort . . . . , State of Hawaii, contains 1,195 acres.[2]

Appraisers interested in the mapping of farms by triangulation using data of the Coast and Geodetic Survey will find the

2. U.S. Department of the Interior, Revision of 1942, Washington, D.C., pp. 23–24.

following publication helpful: "Use of Coast and Geodetic Survey Data in the Survey of Farms and Other Properties."[3] Further discussion of metes and bounds descriptions, including up-to-date procedures recommended by the federal government, will be found in Appendix A.

LOCATION

Frequently a farm to be appraised is identified by name and location. Descriptions like the following are common: the Jones farm three miles west of Danbury on the south side of the ridge road; or the farm three and a half miles northeast of Medford in Buffalo County, Ohio; or the Smith place two miles south and one mile east of Lohrville, in Wheatland County, Kansas. Information like this does not outline the farm boundaries, but it does tell the appraiser where the farm is located. Once he finds the farm, he can use the legal description to check the boundaries. The short description by name and location is usually sufficient when referring to a farm, since every appraiser has or should have good local maps showing roads and other pertinent data.

The appraiser when writing up an appraisal should include both the legal description and the location with reference to roads, towns, and the like. The usual place for this information is at the beginning of the appraisal form because it represents the name or subject of the appraisal.

A location map is recommended for every detailed appraisal. Such a map gives the reader a clear picture of the farm setting in relation to market points, towns, cities, highways, railroads, and other important geographical features. The location map for Farm *A* (Fig. 1.3) shows the proximity of the farm to the city of Ames and to Iowa State University. The exact position of Farm *A*, which would be difficult to visualize without this location map, has an unusually important bearing on the farm's value; in fact, it points up the major value factor—the strong demand for this farm because of its potential for commercial and residential use.

Another location map is presented in Figure 4.11, this one of Farm *X* five miles north of Ames. It has been included to show a more typical farming area away from the suburbs of a city. Although suburban influence is not a vital factor with Farm *X*, the location map for Farm *X* is highly useful because

---

3. *Coast and Geodetic Surveys*, rev. ed., Washington, D.C. 1940.

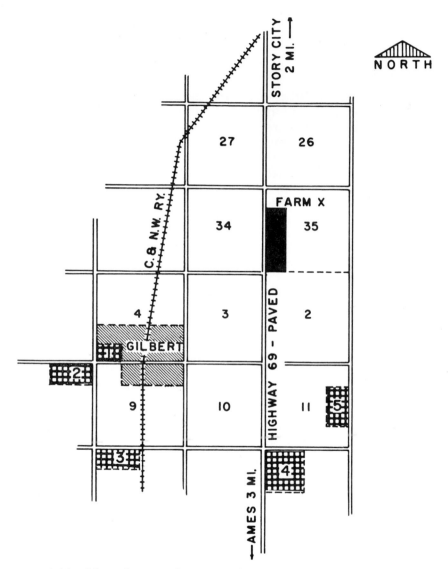

FIG. 4.11. *Map showing location of Farm* X *in relation to roads, market points, and comparable tracts sold or appraised in the years 1958–60. Scale .9 inch = 1 mile.*

it provides at a glance the general setting of the farm with respect to highways, railroad, market point at the town of Gilbert, and farms recently sold, all important factors as we will see later in the appraisal of this farm.

MAPS

Maps are essential in locating farms. Anyone making a business of appraisal should have an adequate set of maps covering his territory. Such a set would include several state maps showing highways and railroads, together with township and range lines of the rectangular survey. Also included would be county maps on which individual farms could be spotted both by road and by section number if in the rectangular survey. Often various county maps are available showing taxing districts, drainage districts, towns, highways, and railroads. Ownership plat maps, usually prepared for a county in book form with a page for each township, are useful because they indicate the ownership of individual tracts at the time the plat is prepared (see Fig. 4.12). An appraiser working intensively within a county may wish to purchase a recent plat map of the county; otherwise, he may borrow one or use a copy on file in the county courthouse.

Soil, aerial, and topographic maps should be used wherever they are available. One such map may save the appraiser several hours of field work. Soil maps are published either in detail for individual counties or as reconnaissance maps for areas as large as a state. These maps may be obtained from the state agricultural experiment station or the USDA. The county soil maps, besides giving information on soil, may also be helpful as base maps in locating farms, schools, churches, railroads, and towns. Enlargements of air photographs for many areas may be purchased from the USDA; directions for ordering may be secured from the ASC county or state offices. These offices also have copies of the aerial pictures on file. Aerial photographs save time in preparing an appraisal map of a farm, particularly if the farm has a meandering stream, irregular fences, and patches of timber. Topographic maps are available in areas which have been surveyed by the United States Geological Survey. This agency will furnish information on how and for what areas these topographic maps may be secured. Reference will be made in the chapters that follow not only to soil, aerial, and topographic maps but also to other maps, including land classification and land value maps which the appraiser will find useful.

It cannot be overemphasized that an adequate set of maps is a necessary part of the appraiser's equipment. Maps are

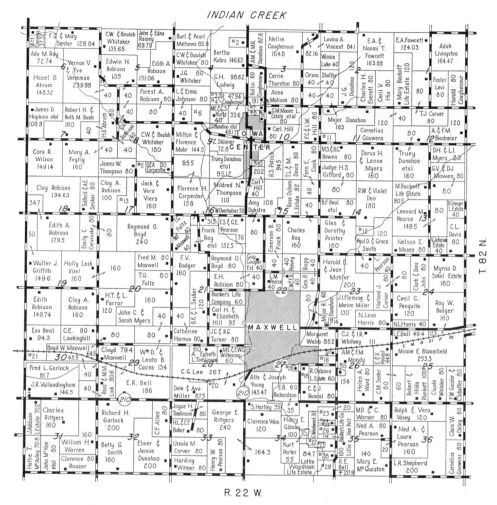

FIG. 4.12. *Reproduction of an ownership plat map for a township in Story County, Iowa. (Courtesy Ames Engineering and Testing Service, Ames, Iowa.)*

not only time savers but also an integral part of the appraisal in that they show the farm to be appraised in its proper setting with reference to surrounding farms and territory.

REPORT ON ERRORS IN LEGAL DESCRIPTION INTERPRETATION

1. Reject city's $28,000 bill for error
    A Des Moines land appraisal company has rejected,

although with respect, a request from the city of Ames for payment of $28,826.50.

The claim arose from the sale last Aug. 9 of the city-owned site on Pearle St. on which the inspection building stands.

As requested by law, an appraisal was obtained by the city before the tract was offered for sale. The appraisal, of a parcel of land with 138 feet on Main and Fifth St., 215.4 feet on Pearle St., was $100,000.

At the sale Aug. 9 there was only one bidder [who] offered $112,576.50 for the site. The bid was accepted.

However, it turns out that the legal description of the property was in error, by some 5,000 square feet. . . . Instead of 29,750.2 square feet, the lot actually is only 24,340.2 square feet in area.

A reappraisal established a value of $83,750 on the land and the [bidder] has offered to buy it at that figure. The city council last night accepted the offer and will give possession April 1.[4]

2.   Commission "partners" in restaurant

The Iowa Highway Commissioners Tuesday learned they are "partners" in a new Ames restaurant.

Right-of-way director Gordon Sweitzer told the commissioners the commission is technically part owner of the new . . . restaurant on U.S. 30 at the west edge of Ames.

Sweitzer said the restaurant has inadvertently been built partially on the highway right-of-way.

The southwest corner of the building, 10 by 30 feet, sits on commission property, he said, adding, "Technically it belongs to us."

The right-of-way director suggested the property be appraised by a private appraiser, then sold to the restaurant owner.[5]

QUESTIONS: PRACTICE IN READING DESCRIPTIONS

It is recommended that some time be spent in reading legal descriptions on mortgages and deeds, or records of these documents kept at the county courthouse. A survey of the various maps and plats kept at the courthouse may be made at the same time to familiarize the student with background material he should have on this subject. If legal descriptions are not available, or as a preliminary to such an assignment, the problems listed below provide an opportunity to apply the rules for reading legal descriptions discussed in this chapter.

1.   Outline the boundaries of the farms described legally as follows:

a.   NW¼ and W½NE¼ sec. 11, and S½SW¼ and SW¼SE¼ sec. 2, all in township 45 north, range 25 west of the fifth principal meridian.

4. *Ames Daily Tribune*, Ames, Iowa, Feb. 22, 1967.
5. *Ames Daily Tribune*, Ames, Iowa, Apr. 3, 1968.

    b.   West 25 acres of NW¼SW¼ sec. 10, excepting 1 acre thereof in the northwest corner of said tract reserved for school purposes, all land being in township 90 north, range 10 west of the second principal meridian.

    2.   Write out legal descriptions for one or more farms whose boundaries are shown in Figure 4.13.

    3.   If located in an area where metes and bounds descriptions are common, plat one or more farms from descriptions appearing on deeds or mortgages.

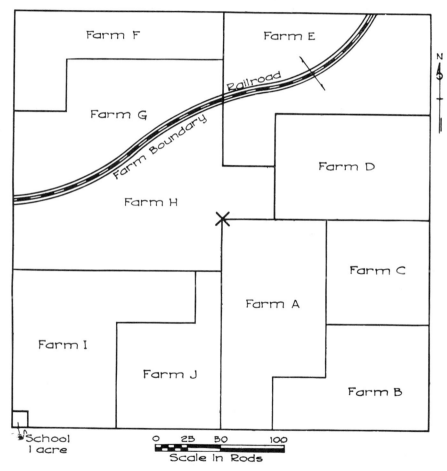

FIG. 4.13. *Diagram of section 3, township 6 north, range 2 west of the ———— meridian. Center of section is indicated by an "x."*

REFERENCES

Beuscher, Jacob H., *Law and the Farmer,* 3rd ed., Springer Pub. Co., New York, 1960.

Davis, Raymond E., Foote, Francis S., and Kelly, Joe W., *Surveying, Theory and Practice,* 5th ed., McGraw-Hill, New York, 1966.

Kratovil, Robert, *Real Estate Law,* 3rd ed., Prentice-Hall, Englewood Cliffs, N.J., 1958.

Manual of Instructions for the Survey of the Public Lands of the United States, 1947, Bureau of Land Management, U.S. Department of the Interior, Washington, D.C., 1947.

Morris, Fred C., *Your Land: Surveys, Maps and Titles,* Va. Poly. Inst. Eng. Exp. Sta., Bull. No. 71, Blacksburg, 1949.

Patton, Rufford G. and Patton, Carroll G., *Land Titles,* 2nd ed., West Publishing Co., St. Paul, Minn., 1957.

Specifications for Descriptions of Tracts of Land. For Use in Executive Orders and Proclamations. Prepared by the Committee on Cadastral Surveys, Board of Surveys and Maps of the Federal Government, Washington, D.C., 1943.

Stewart, Lowell O., *Public Land Surveys: History, Instructions, Methods,* Iowa State University Press, Ames, 1935.

Land Measure Aids: Pronto Land Measure Compass is a device for plotting metes and bounds descriptions. Pronto 4-Square Land Area Scale is a device for plotting and estimating areas of irregular shaped tracts. These two mapping tools along with a land description chart, map sheets on a scale of 660 and 330 feet to 1 inch, and a special scale rule can be obtained from Beach and Stull Printing Co., 512 Buckham, Flint 2, Mich. An acreage grid for use in estimating acreages is described at the end of Chapter 19.

2

COMPARABLE SALE APPROACH

THE PRINCIPAL OBJECT in appraisal is to provide a well-supported estimate of market or sale value. One method of obtaining this figure is by using the sales of comparable tracts. This would appear to be a simple listing of sales near the farm being appraised, with application to the appraised farm of an appropriate sale value estimate based on the comparable sales. In general this is the process, but it does not include the problems, dangers, and possibilities of error that may beset the appraiser as he carries out this "simple" task.

Appraisal of a farm's market value involves several important steps. First, sales of farmland comparable to the farm being appraised have to be located and the sale prices verified. Second, each sale, whether deed or contract, needs to be analyzed to determine whether it is a bona fide transaction and how close the price is to the current market. Such factors as interest rate, payment terms, who pays the current year's property taxes, and who gets the current year's crop are involved. In addition it is desirable to find out the main points in the minds of the buyer and seller as they reached their agreement on the price for the farm.

The third step in estimating sale value is relating each comparable sale to the appropriate value estimates or indexes as reported by federal and state agencies. Where the market has moved up or down since the sale it is necessary to use the reported values to adjust the sale price to bring it up to date.

A fourth and important step is a detailed comparison of each of the farms sold with the farm being appraised. Such features as soil productivity, property taxes, size, buildings, and location have to be evaluated in order to make valid and helpful comparisons. This process is in fact the heart of the procedure, leading as it does to the final selection of a market value for the farm appraised.

5

SALE PRICE RECORDS

IN THIS CHAPTER the discussion will center on the first step of the comparable sale approach—obtaining and verifying sales for use in estimating the market value of a farm.

COURTHOUSE RECORDS

Official records at the courthouse are the first and primary source of information on farm sales. The universal practice of recording land transfers, either by deed or by contract, makes this source a common one to use. But the appraiser must be skeptical and cautious in his use of these records because they are often misleading. The first rule is to get courthouse records, and the second rule is not to use them without further checking. It is not that the records are inaccurate or false but that they can be misleading, as for example the date of the deed which may indicate the approximate date of the sale agreement or may not.

Transfer records at the courthouse provide objective facts which the appraiser will include in his appraisal report. Such facts as the official notice of transfer, names and addresses of the buyer and seller, and legal description of the real estate transferred give the appraiser incontrovertible evidence that he can use in starting his farm sale inquiry.

Information on the exact sale price may or may not be provided by the courthouse records. Up until World War I most deeds were specific in giving the amount of consideration or sale price. Since that time the practice of stating a nominal figure plus an undisclosed amount has become more and more common, so much so in fact that in many areas it is surprising to find the exact amount stated in a deed. The usual deed now reads: "for one dollar and other valuable considerations."

Contracts, it happens, are one area of official records which do give the appraiser full information. In contrast to the deed the contract spells out the sale price in precise terms. Not only is the total amount given but in addition the exact schedule of payments, interest rate, and terms are given. Finding a contract in the courthouse records is a welcome event for the appraiser. Unfortunately, however, contracts are not always recorded. They are usually available in homestead exemption states which require recording if the owner under a contract wants to qualify for an exemption on his property taxes.

At this point an example of misleading information can be cited involving a case of a nonrecorded contract and a recorded deed both for the same sale. All that appears on the courthouse record is a deed made on a certain date. In this instance the actual sale date was five years earlier at which time a contract was agreed to and signed. In the contract was a provision that when a certain percentage of the amount was paid a deed would be issued. Five years after the contract was made the buyer fulfilled the contract provision entitling him to a deed to the property. The buyer in this example gave the seller a mortgage for the amount remaining due on the farm. Cases of this kind are a hazard the appraiser should always be alert to detect. The only reliable answer comes from information provided by the buyer, seller, or someone else familiar with the details of the transaction.

SALES PRICES FROM TRANSFER TAX STAMPS

An approximate sale price may be estimated from the tax stamps on the deed, if there are stamps on the deed. Even though the deed recites: "for one dollar and other valuable consideration," there may be tax stamps affixed to the recorded deed. How to estimate the sale price from the stamps depends on the tax rate in effect on the date the deed was recorded.

Congress in 1932 enacted a federal stamp tax on deed transfers which gave the public, including appraisers, tax assessors, lenders, and real estate brokers, a strong clue as to the actual cash consideration in the sale of a farm or other real estate. In 1932 the federal stamp tax on transfers of real estate, which applied to sales over $100, was 50 cents for each $500 or fraction of this amount. In 1941 the stamp tax was increased 10 percent to 55 cents for each $500 or fraction. Stamps indicating the amount paid had to be affixed to the deed or conveyance. The federal rate of 55 cents, which was in effect from 1941 to the end of 1967, applied as follows:

Tax Stamps	Consideration
$0.55	$ 101– 500
$1.10	$ 501–1,000
$1.65	$1,001–1,500
$2.20	$1,501–2,000

and so on, indefinitely.

No amount less than 55 cents and no amounts other than multiples of 55 cents were provided for; that is, provision was not made for any amount below 55 cents or for any amount between 55 cents and $1.10.

To find the indicated top price paid for a farm from tax stamps at the 55 cents for $500 rate, divide the total tax by 11 and multiply by 10,000. For example if the tax is $66, the indicated sale price is 6 times $10,000 or $60,000; if the tax is $37.95, the price is 3.45 times 10,000 or $34,500. The top price of $34,500 means that if the price was $34,501 the tax would be 55 cents higher or $38.50, indicating a price anywhere between $34,501 and $35,000. With the $37.95 tax the price could be anywhere between $34,001 and $34,500.

With the spread of $500 it is evident that rather wide variations in sale price would be possible on small amounts but relatively small variations on large amounts. For example a 10-acre tract selling for $110 an acre or for $150 an acre ($1,100 or $1,500 total) would result at the 55-cent rate in the same amount of stamps—$1.65. But a 300-acre tract selling for $110 or for $150 an acre ($33,000 or $45,000) would call for $36.30 in tax stamps in the first case and $49.50 in the second. In the latter case a difference of $2 an acre up or down would change the amount of the stamps by 55 cents.

Since sales are usually in round figures, either per acre or per farm, the stamps usually give a good indication of the amount paid unless a mortgage is included in the consideration or unless an error was made in the amount of stamps placed on the deed. If we assume tax stamps of $79.20, the consideration indicated is between $71,501 and $72,000. The farm consists of 160 acres so the most likely price is $450 an acre because that gives a total amount of $72,000. Any other assumption would not give a round figure like $450.

On December 31, 1967, the federal tax on deeds expired so that deeds after this date do not have federal stamps on them. However, many of the states have enacted transfer taxes to take the place of the federal tax. There is little uniformity in these state transfer tax laws, so the appraiser is required to familiarize himself with the specific details which apply to deeds in the state in which he is appraising. Where the state passed a law similar to the previous federal law (.55 per $500) and made it effective January 1, 1968, there was a smooth transition and little opportunity for mistakes. Unfortunately for appraisers there were only a few states which adopted a law similar to the federal law. A map showing the states which had a stamp tax on deeds in June 1968 is shown in Figure 5.1.

In estimating sale price from stamp tax, errors can creep in under various conditions. Since these conditions are important, a list of them with an explanation of each has been included.

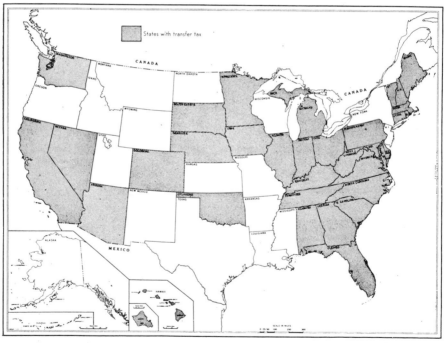

FIG. 5.1. *States with real estate transfer tax as of June 1968. From* Assessors News Letter, *International Association of Assessing Officers, Chicago, Ill., Vol. 34, No. 6, June 1968.*

1. Stamp tax not required on mortgages which ride through the sale. If John Jones sells his farm for $20,000 receiving $10,000 in cash and having the buyer assume an existing $10,000 mortgage, the tax may be required for only the $10,000 of cash. If the deed in this instance gives the consideration as "one dollar and other valuable considerations" and merely adds that there is an outstanding mortgage assumed without giving the amount, then the appraiser will be unable to estimate the sale amount from the stamps. The appraiser should always be on the alert for outstanding mortgages which figure as part of the sale consideration. A record of the mortgage can be traced in the county records, but the outstanding balance of the mortgage may have to be obtained from the lender or borrower.

2. Incorrect amount of stamps used. In some instances an error is made by the parties to a sale transaction. Sometimes the error is unintentional, other times extra stamps may be added to show a higher sale price than the actual one. Errors misled the author in two transactions; in one a lawyer acting for the buyer made a mistake in figuring the amount of the stamps required, and in the other the buyer put stamps on the deed to cover a mortgage assumed as part of the purchase price even though this was not required in this state. Another source of error is the addition of stamps to cover the value of personal property included with the real estate in the sale transaction. Stamps are usually required only for real estate; any personal property sold in connection with a real estate transfer is exempt from the tax.

3. Same transfer tax for a given range. Whenever the same tax covers a fairly wide range, such as $500 which was the federal law and is common in state laws, an error is possible in estimating the sale price. Although a rounded figure can be assumed, as described earlier, even the rounded figure can be an error. Some agencies estimating sale prices for a large number of transactions use the midpoint in the range as their estimate—a consideration for example of $22,750 for a tax indicating an amount anywhere between $22,501 and $23,000. However, this assumption also is likely to produce errors because many of the sales will not hit the midpoint. For individual appraisals where only a few sales are used the author favors the assumption of the rounded figure but favors even more, of course, verifying the sale price by discussing the sale with someone in a position to know the correct sale price.

A number of studies have been made to test the reliability

of the transfer tax as an indication of sale amounts.[1] In all of these studies the existence of the errors listed above has been noted. But these studies were designed principally to test the usefulness of the tax stamps in making sale price estimates when a large number of transactions were included. In such cases the tax was highly useful and reliable in the areas where it was checked (New York, Oklahoma, and Idaho). Errors which did exist were not large and were largely compensating.

A Montana study indicated considerable error in estimates based on transfer taxes.[2] The error would be mostly compensating if a large number of sales were included as is usually done in assessment-sale ratio studies, but of course the error would be important if only a few sales were used as in an ordinary appraisal. The two areas of error reported by Wicks and Simmons in their study were:

1. Affixing an incorrect amount of stamps on the deed.
2. Existence of a mortgage on the property which was part of the sale consideration.

Information on the extent of these two errors was obtained from 37 lawyers, 20 realtors, and 18 bankers in Montana.

On the first error of incorrect stamps, 17 percent of the group reported instances of excess stamps and 18 percent insufficient stamps. The authors estimated that excess stamps occurred about one percent of the time and insufficient stamps about the same percentage. In both cases the excess and deficiency averaged about 15 percent too much or too little which would give market values too high or too low by this percentage.

On the second point (existence of mortgages as part of the consideration) the Montana group estimated that such mortgages were present in about 30 percent of the sales. There is no trouble if the amount of the outstanding mortgage is stated in the deed and the stamps are figured without including this mortgage, but unfortunately there are cases where no mention is made of the outstanding mortgage on the deed. This situation

1. Floyd L. Corty, *A Comparison of Prices Derived from Federal Tax Stamps on Deeds with Purchase Prices of Farms and Rural Residences as Reported by Rural Property Owners in 15 New York Towns, 1954,* A.E. 997, Dept. of Agr. Econ., Cornell Univ., Ithaca, N.Y., 1955; Norman Nybroten, "Estimating Cash Considerations in Real Estate Transfers from Internal Revenue Stamps," *J. Farm Econ.,* Aug. 1948; Robert L. Tontz, Jeppe Kristensen, and C. Curtis Cable, Jr., "Reliability of Deed Samples as Indicators of Land Market Activity," *Land Econ.* 30(1):44–51, Feb. 1954.

2. John H. Wicks and Lee H. Simmons, "Error Sources from Tax Stamps as Estimates of Real Estate Value," *Nebr. J. of Econ. and Bus.,* Univ. of Nebr., Lincoln, 6(1):46–50, Spring 1967.

where the existing mortgage was not stated occurred in 9 percent of the deeds according to this study. In these cases the indicated sale price based on the stamps was below the actual price by the amount of the existing mortgage which was part of the sale price.

In addition there was another error where the existing mortgage was stated but not the total consideration, and stamps were placed on the deed for the total consideration even though the mortgage amount was exempt. This last error was estimated as occurring in 3 percent of all deeds.

A typical tax stamp situation was encountered in checking the farm comparison sales for the Farm *A* appraisal in Chapter 1. One farm sold on contract so there were no tax stamps in this case. The second farm was sold in two parts, one part on contract at $700 an acre and the other part, consisting of 29.28 acres, "for one dollar and other valuable considerations." There were tax stamps of $22.55 on the deed. Since the $1.10 tax stamp rate per thousand was in effect, the stamp total indicated a consideration of not more than $20,500. If $20,501 had been paid the stamps required would have been $23.10. Dividing $20,500 by 29.28 gives $700.13. The purchaser said the actual price paid was $700 an acre. In this instance, the tax stamps indicated almost exactly the sale price. The third farm was sold for one dollar with stamps totaling $77.00 which indicated $70,000 paid for the farm. The purchaser, however, said $106,000 was paid but part of the purchase price was the assumption of an existing mortgage for which no tax stamps had to be affixed to the deed. In this case the tax stamps did not give a reliable indication of the sale price.

OTHER SOURCES OF SALE PRICE INFORMATION

There are other sources of sale information besides the courthouse: real estate brokers, bankers, lawyers, farm mortgage lenders, assessors, farm managers, and farm owners and tenants. These individuals, who are often involved in farm sale negotiations in one way or another, are usually excellent sources concerning recent sales which have not yet been recorded in the courthouse. These unrecorded recent sales are of special importance because they provide the best evidence on the present land market, and it is this present or current market which the appraiser generally wants if he is valuing a farm at its current market price.

These persons outside the deed recording office may not know all the details regarding a recent farm sale. Actually their information may be sketchy or even unreliable if they were not in on the negotiations. But their knowledge that a certain farm or portion of a farm has been sold is the main point; with this lead the appraiser can check with the buyer, seller, or some other person who was in on the transaction. Since there is often a time lag of several months between the sale agreement and the recording of the deed, it is important for the appraiser to obtain from local people information about these current sales. It involves extra work but will often result in a better appraisal.

BONA FIDE SALES

Only sales representing agreements between buyer and seller at the current market price should be considered as usable by appraisers. If a sale is recognized and determined to be a certain percent above or below the market it can be used, but it is not as good evidence as the bona fide (good faith) sale which indicates the market.

To be classed as bona fide the sale must be what is referred to as an "arm's length" transaction; that is, it must be a voluntary agreement between a buyer and seller who are informed and essentially equal in their bargaining and negotiating power. Cases which are normally excluded because they do not meet this bona fide test are deeds and land contracts involving:

1. Family members—Transactions where the buyer and seller have the same name are usually of this class, but there are others between relatives with different names which are not apparent on the record; sales of this type have to be excluded later when verification reveals the family relationship between the buyer and seller.

2. Administrators and executors of estates, guardians, commissioners, etc. wherever family ties of some kind enter into the transaction.

3. Sheriffs (deeds in foreclosure), bailiffs, county clerks or treasurers (tax deeds), etc. where defaulted mortgages or failure to pay taxes have occurred.

4. Charitable and other nonprofit organizations.

5. Trades, deeds of convenience, and similar transactions made to correct or change the title.

The general practice to use in evaluating sales is to reject all sales where buyer and seller (grantee and grantor) have the same last name. However, there are family sales which are just as useful as indicators of market values as nonfamily sales. And there are frequently nonfamily sales that for good and sufficient reason should be rejected. In short there is no simple method of dividing the sales between those that are good indicators of market value and those that should be rejected. If you accept all nonfamily sales you will include some which are not good market indicators, and if you reject all family sales you will exclude some that are truly representative of the market.

Before using any family sales it is desirable to check the details carefully to make sure the transactions qualify as bona fide. Unless this is done the usual assumption that such sales are not indicative of the market is a valid one. There may be some sales between parties with the same last name where the parties are not related, but these cases are very infrequent. Much more common are the sales between parties with different last names but actually within the family. The sale of a farm by a father to his married daughter is a common example of this type. Checking the sale with someone involved in the transaction or familiar with it will make it easy for the appraiser to detect these family sales and discard them unless they can be established as bona fide indicators of market value.

A good rule to follow with family sales is not to use any of them unless there is a scarcity of comparable sales. If there have been no sales in the neighborhood of the appraised farm, then the use of a family sale which has been checked out as bona fide is justified. In situations like this it is even reasonable to use a family sale that is not bona fide if it is possible to determine from the family or other reliable sources what percentage the sale price represented of full market value at the time. As an illustration, the family may have agreed to a sale 10 percent below the going market. Of course use of such sales is justified only because other nonfamily sales representative of the market are not available.

VERIFICATION

The last phase of sale price determination, and in many respects the most important, is verification. This is simply checking the details of the transaction with the seller (grantor), or the buyer (grantee), or with the real estate broker or some other person who was in on the transaction. Unfortunately this is a phase which is often neglected. Time and again the importance of verification has been brought to the author's attention.

Verification not only gives a correct sale price figure but also reveals details about the transaction which assist the appraiser in using it as a comparable sale. For instance checking the sale price may indicate the reasons the seller had in deciding to sell and the reasons the buyer had in deciding to buy. It may indicate the amount of a mortgage used as part of the purchase price, special agreements made on payment of taxes, who is to get growing crops, and inclusion of certain movable property with the real estate. Finally, verification may indicate the influence of land contract terms, interest rates, or other modes of financing in the determination of the sale price. A contract at low interest may be offset by a sale price that is higher than would have been reached on a cash transaction.

Verification which gives qualitative background provides the essential information for an analysis of each sale. In the analysis we can be, as it were, an unseen spectator at the negotiations which result in the sale agreement.

QUESTIONS

1. What are the four major steps in the market value approach?
2. If a trip has been made to the courthouse to get sale price data, explain the procedure you followed in getting your data. Give examples of land contracts and deeds you saw at the courthouse.
3. Explain main provisions in transfer tax stamp law in your state, if your state has such a law. Indicate how the federal law which expired December 31, 1967, operated.
4. What are the usual difficulties in estimating sale price from transfer stamps?
5. What is meant by a bona fide sale?
6. What kind of policy do you recommend in the use or nonuse of family sales? Give your reasons.
7. Why is verification of sale price important? If you have verified any sales give information you obtained from your verification interviews.
8. Do you have enough sales in your home community to make reliable market value estimates? Discuss.

REFERENCES

Appraisal of Real Estate, 5th ed., American Institute of Real Estate Appraisers, Chicago, 1967.
Rural Appraisal Manual, 2nd ed., American Society of Farm Managers and Rural Appraisers, DeKalb, Ill., 1967.

6

ANALYSIS OF SALE
TRANSACTIONS

JOHN DOE agrees to sell his 200-acre farm to David Smith for
$100,000 or $500 an acre. The transaction is completed with
a handshake and a verbal agreement to draw up legal papers
in the lawyer's office the next day. Simple as this sale appears
at first glance, an analysis of the conditions that are possible
in this example indicates a wide variety of situations.

The $500-an-acre sale may be a straight cash sale with
no mortgage—this would be one extreme. Or it might be a
long-term contract with a small down payment—this would be
the other extreme. Between these extremes there is a multitude
of conditions not only as to amount paid down but as to interest
rate, principal payments, and special agreements on crops and
tax payments. It is also important to know something of the
ideas or reasons which motivated the buyer and seller in reaching their agreement on the price of $500 an acre. The price
itself may be above, below, or right at the current market for
land similar to the farm being sold. Our purpose in this chapter
will be to go behind the sale price for an understanding of the
conditions resulting in the price finally reached in the sale.
First we will consider deed transfers and then contracts.

DEED TRANSFERS

CASH SALE

The simplest sale is a cash transaction with no financing involved. In the case of Doe and Smith we will assume that Smith has $100,000 in government bonds which he cashes. Doe provides an up-to-date abstract of title which Smith's lawyer examines and finds satisfactory. Doe's lawyer has a warranty deed drawn up stating that for one dollar and other valuable considerations John Doe and his wife sell the 200 acres (legally described) to David Smith and his wife. The Does warrant the title and state that the property is free of mortgage and not subject to any outstanding leases. The Does sign the deed and Smith hands Doe a cashier's check for $100,000. The sale is completed, the Smiths own the 200-acre farm with no mortgage on it, and Doe has an additional $100,000 in his bank account.

When the deed in this Doe-Smith sale is recorded at the courthouse the appraiser notes the names and addresses of the buyer and seller, the date of the deed, the legal description, and the tax stamps which total $110 (a 55-cent rate being in effect in the state). From this information the appraiser makes a preliminary estimate of a cash sale at $500 an acre. But he does not stop here. Instead he calls on John Doe to get further details. Doe tells him that his estimate of the sale price is correct, that there was no financing, and that since the farmstead is now vacant the new owner has immediate possession. Doe also tells him that there is no family connection between him and Smith, and that he has had the farm on the market for about one month.

When asked why he sold the farm Doe replied that he was 65 years old and had decided to retire from farming. Smith on the other hand was a well-to-do farmer who lived several miles down the road and wanted a farm for his second son who had graduated from the University, had just finished his military service, was married, and wanted to farm. According to Doe there were no special conditions connected with the sale. He knew Smith wanted to buy a farm and in the month's time that the farm had been on the market he had not been able to find anyone who was willing to pay the $525 he was asking. Smith had offered him $500 an acre. There was no personal property involved in the sale, he had removed several portable grain bins, he had paid the 1968 taxes payable in 1969, and since this was March 1969 there was no growing crop to be considered and all of last year's crop had been sold and removed from the farm.

What has been described is the kind of clear-cut case that an appraiser seldom runs across. One variation would be for the seller to have a realtor sell the farm for him. In this case the broker might be able to find a buyer who would pay him more than $500 an acre. This type of case, with or without the broker or realtor, is almost too simple to be true, but of course it does occur occasionally and should be recognized as the extreme cash sale.

In this instance our appraiser obtained his first lead on the transaction when he went to the courthouse in April 1969 to obtain comparable sales for a farm he was appraising in the area. He saw the recorded deed but did not stop with the courthouse record; he called on John Doe from whom he got the information described above. The courthouse gave him only the skeleton of the sale, the interview gave him the flesh and blood of the transaction. If Doe had been unwilling to cooperate, the appraiser could have tried Smith the buyer, the lawyers involved, or some other individual like the local assessor who often knows the conditions connected with a farm sale.

CASH-MORTGAGE SALE

A second type of sale by deed is also for cash but the buyer has to borrow a sizable percentage of the purchase price. He may have only half the amount or $50,000 to pay down, the remaining $50,000 he has to borrow. If a lender such as a federal land bank or an insurance company agrees to provide the other $50,000 at the going rate of interest, the transaction could conceivably go through exactly the same way as the first one described where there was no financing. However, there are cases where the financing is not easily arranged. In one instance brought to the attention of the author, the seller sold to a cash buyer who needed no financing in preference to a buyer who had to have financing because the first buyer was ready to buy at once while the other buyer had to take time to arrange the financing.

CASH-EXISTING MORTGAGE SALE

An important variation of the cash-mortgage sale is the case of the buyer taking over an existing mortgage rather than borrowing on a new mortgage, including the existing mortgage as part of the sale consideration. If the outstanding mortgage at the time of the sale amounts to $47,635 and the total sale price is $100,000, then the buyer either assumes or buys subject to this $47,635 indebtedness and pays the remainder of $52,365 in cash. The difference between assuming this mortgage and buying subject to it is that in the first case the buyer agrees to pay the note and mortgage while in the second case he does not specifically agree to pay the note and mortgage but buys the property subject to this lien. In case of nonpayment the lender can sue the buyer if the mortgage is assumed but can take the property from the buyer only if the buyer bought it subject to the mortgage. The seller of course is interested in having the buyer assume the mortgage and thus become personally liable for payment, and of course the buyer would rather buy the farm subject to the mortgage so that in case of difficulties he would not be liable beyond losing the farm itself.

The interest rate on an existing mortgage which is part of the sale consideration can be a point of special importance when the mortgage rate of interest is below or above the going rate at the time of the sale. A common situation in the late sixties when interest rates were high was the existence of many old mortgages with much lower rates. When a low-rate mortgage figured as consideration in a sale, the question immediately arose as to how much of a premium was represented by this existing low-rate mortgage. The main factors determining the premium were the amount of the existing mortgage, the period it had to run, the schedule of required principal payments, and the difference between the existing mortgage rate of interest and the rate on current mortgage financing. After sizing up all the factors involved in such a situation the buyer might be willing to pay $510 an acre instead of $500 just because of the saving made possible by the low-rate mortgage which was part of the sale consideration.

On the other hand, if the existing mortgage carried a higher rate than the current rate and could not be paid off at the time of the sale, then the seller had a handicap because the buyer was forced to pay a higher rate than he would pay if he borrowed on a new mortgage. The same factors of mortgage amount, term, principal payments, and difference in rates would determine the amount of the handicap. It might be that the handicap would cause a reduction in the sale price of $10 an acre, reducing the price from $500 to $490. Since mortgages often allow payment in full when the farm is sold, this situation is not a likely one. However, the reverse situation where the mortgage rate is lower than the rate on new mortgages is a common one because the lender is usually not able to force payment of the loan at the time of sale. Of course the lender would like to have the loan paid off so he could relend at the higher current rate. In the late sixties for example many of the earlier 5% mortgages still had years to run but there was no way the lender could force payment when a farm was sold. In these cases it was to the advantage of both the seller and the buyer to continue the loan instead of taking out a new loan at a rate of 7% or more. Assuming that a new buyer has the advantage of a 2% saving in interest by taking over an existing mortgage, the value of this premium can be shown in precise terms.

The calculated saving from a 5% to a 7% mortgage can be set out with the use of the present value of future amounts. (See Table C in the Appendix showing the present value of 1.) In this instance a mortgage of $40,000 is assumed with a remaining term of 10 years with no principal payments. The interest savings would be the difference between $40,000 at 5% and at 7%, or a difference of 2% on $40,000 which is $800 a year. For ten years this amounts to $8,000. The present value of this $8,000 works out as follows:

Present Value at 7% of $800 Interest Saving

Year	Interest Saving	Year	Interest Saving
1st	$ 748	6th	$ 533
2nd	699	7th	498
3rd	653	8th	466
4th	610	9th	435
5th	570	10th	407
		Total	$5,619

The present value of the $8,000 in interest savings over the 10-year period amounts to $5,619 at 7%. The 7% rate was used because this was the going mortgage rate at the time the sale was made. For example, the savings of $800 at the end of the first year are not actual savings at the time of the sale but $800 after the lapse of a year's time. The savings for the first year are worth $748, which is the amount of money put at interest at 7% which will equal $800 in one year. And similarly $407 put at interest at 7% will equal $800 in ten years. From this it is evident that the buyer would have an advantage from the existing 5% mortgage of $40,000 of a present value of $5,619 or approximately $28 an acre.

A more likely situation than the one above would be principal payments which would reduce the premium for the existing low-rate mortgage. If everything were the same except that there were annual principal payments of $2,000 a year, the interest saving would have a present value of $4,512 instead of $5,619, the arithmetic being as follows:

Present Value at 7% of Interest Savings with Principal Payments

Year	Principal	Interest Saving	Present Value
1st	$ 40,000	$ 800	$ 748
2nd	38,000	760	664
3rd	36,000	720	588
4th	34,000	680	519
5th	32,000	640	456
6th	30,000	600	400
7th	28,000	560	349
8th	26,000	520	303
9th	24,000	480	261
10th	22,000	440	224
Total		$6,200	$4,512

With the principal payments the total interest savings are reduced from $8,000 to $6,200, and on a present-value basis from $5,619 to $4,512. On a per acre basis this means a reduction from $28 to $22.50.

The mathematical calculations in the examples just described should not be substituted for actual negotiations in a sale. If a sale of a 200-acre farm did occur with savings on an existing mortgage equal to the $4,512 shown above, there is a good question of whether the buyer would pay a premium of $22.50 an acre, say $522.50 instead of $500. And from the seller's point of view it would be a question of whether he could get the full $22.50 above the going $500 price for land of his type. What is paid by the buyer depends on his attitude, which will include not only the arithmetic of this 2% interest saving but also how much weight the buyer puts on this saving. In this case, for example, the buyer may have to get additional financing besides the $40,000 outstanding mortgage because he does not have sufficient cash to make up the difference. Another mortgage loan would of course be a second mortgage loan, subject to the prior claim of the $40,000 existing lien. However, whatever saving there is should be evaluated in terms of the weight given it by the buyer in agreeing to the price paid for the land.

CASH-FORMER OWNER MORTGAGE SALE

In this fourth deed sale we will assume that the buyer David Smith is a tenant who has been farming John Doe's farm of 200 acres plus other land for 15 years. He has accumulated a net worth of $30,000 and is able to raise $25,000 in cash for purchase of the farm. There is no family relationship between Doe and Smith, but Doe has been especially pleased with Smith as his tenant and is eager to help him purchase the farm. Doe, who is a physician in a nearby city, is willing to sell because the buildings, especially the house, are in need of repair and re-modeling. Besides Doe has almost completely depreciated the buildings on his income tax return over the last 30 years that he has owned the farm.

Doe agrees to sell the farm to Smith for $500 an acre with a down payment of only $25,000. Doe also agrees to finance the remainder of the sale price of $75,000 at a relatively low rate of 5½% with principal payments of $1,500 a year over a term of 20 years. At the end of the 20 years the remaining principal of $45,000 will come due. Doe also agrees to a provision in the mortgage allowing Smith to pay any amount on principal at any time. This is a former owner mortgage and in this case one that is especially favorable to the purchaser. The papers are drawn up and the sale completed with Doe giving Smith a deed to the farm and Smith giving Doe $25,000 in cash and a mortgage of $75,000 on the terms described above.

Our appraiser notes the deed at the courthouse with the same tax stamps, $110, as in the first cash sale. He estimates the sale price at $500 an acre but calls up Doe or Smith to get the details. What he finds is that in this transaction Doe is giving Smith a discount on the market price. This is a sale not to the highest bidder but to a buyer preferred by the seller. Although Doe is not getting the full market price when the favorable features of the mortgage are considered, he is getting the owner-operator that he wants to have on the farm, and he has a fixed contract to receive a definite sum each year of principal and interest in place of the variable crop share rental returns he has been receiving.

In using this sale as an indicator of farm market value, the appraiser should recognize that the $500 an acre is somewhat lower than Doe might have received if he had put the farm on the market for a period of time to attract the highest bidder under a variety of conditions that might be proposed by potential buyers. The upward adjustment made by the appraiser in this case may be as much as $25 an acre. It will be based on the facts he obtains from one or more of the parties involved in the transaction.

ESTATE SALES

One type of sale that usually deserves special attention is sale of a farm in an estate. Since there are many farms that are sold in estates it is important to consider the conditions that are likely to dominate these transactions. Frequently a member of the family is the buyer and, as has been pointed out in the previous chapter, this situation results in the transaction not being classed in most cases as a bona fide sale. But even when an outsider buys the farm the appraiser should be cautious about accepting the selling price as going market value. Before accepting such a sale as bona fide a careful check of the efforts made by the estate administrator to obtain competitive bidding is necessary. There is a tendency for an estate sale (to an outsider as well as to a member of the family) to be on the low side of market value because the seller may not have as much personal interest in getting the highest price as an individual owner would have. If there are several heirs it may be difficult to reach agreement on the method of selling, or it may be that the heirs are more interested in getting their money as soon as possible rather than in getting full market value for the property. Finally the terms of the sale may make it difficult for most buyers to arrange the necessary financing.

In one estate the buyer, not related to any of the estate family, obtained a farm at least 10 percent below the going market price. It was a case where the estate sold the farm as soon as it could to a cash buyer without making any extensive study of the market. The farm was sold as a whole unit—land and buildings together—although it would undoubtedly have brought considerably more if the farmstead had been sold separately. The farm was located close enough to town to attract a purchaser working in town who would have bought the farmstead, and neighboring farmers were interested in the land to add to their existing farms. Even a short inquiry of surrounding farm owners would have indicated a strong demand for unimproved cropland. If the estate had auctioned the farm in two parts or given it to an active real estate broker to sell, at least a 10 per cent higher price would probably have been realized.

LAND CONTRACTS

Usually contract sales are designed to take care of young buyers who can make only a small down payment, a payment so small that the seller prefers to keep title rather than to give a deed. In recent years, however, the contract has also been favored by sellers who have such a large capital gain out of their farm sale that they prefer a contract arrangement with payments coming in over a period of years. This enables them to spread their capital gain over a number of years and thus reduce the income tax they have to pay. If all the gain comes in one year as with a deed sale, the amount of gain puts them in a much higher income tax bracket. Illustrations of both types of contract, those designed for buyers with small down payments and those designed for sellers wanting to minimize capital gain, are described below.

BUYER-TYPE CONTRACTS

Smith is a young tenant farmer operating a farm some distance from Doe's 200-acre farm. Smith wants to buy Doe's farm but he has only $10,000 for a down payment. Doe in this instance has never met Smith before and is not particularly anxious to sell unless he gets a good price. Since Smith has only $10,000 to pay down, Doe offers him a contract calling for a down payment of $10,000 with interest on the outstanding balance at 7% and annual principal payments of $2,000. The contract price is $500 an acre or $100,000 and runs for 25 years or until the principal amount has been reduced to $40,000, at which time Smith or his successor on the contract is entitled to a deed to the property providing he can raise the remaining amount due on a mortgage or through other means.

Usually the price of a farm on a buyer-type contract is higher than it would be if the farm were sold for cash. The buyer is in a relatively weak position in the negotiations because of the small down payment which he has to offer. The seller is taking more risk in providing the high proportion of credit, so he is able to demand a somewhat higher price. This higher price is in a sense payment for the higher risk which the seller takes.

Instead of offering the farm for $500 an acre, Doe might raise the price to $525 or even $550 and lower the interest charge from 7% to 6% or even to 5%. In this way Doe would be getting more in capital gain and less in interest. The buyer on the other hand would be paying approximately the same total amount but paying more principal and less interest. Doe would be better off on his income tax to the extent that his capital gain is taxed at a lower rate than the money he receives as interest. But Smith loses on his income tax in that he can deduct interest as a cost but not payments on principal.

SELLER-TYPE CONTRACTS

Contracts designed to fit the wishes of sellers may and often do fit the wishes of buyers also. But there are certain contracts which suit the sellers but do not necessarily suit the buyers. The cases that come under this last grouping are those which call for much larger principal payments than most buyers can handle. For example a principal payment that is much larger than the farm is likely to produce in the way of income means that many buyers could not meet the terms.

The major reason a seller uses a contract is to spread the principal payments over a period of years, thus preventing all of the capital gain, if there is a gain, from being received in one year and increasing his income tax. While he is waiting for the gain over the period of years he is receiving interest to compensate for the extra time it takes to get the gain. In order to take advantage of this income tax saving, the percentage of the sale proceeds received at the time of sale must not exceed a certain amount, 30 percent as of this writing.

In the buyer contract described earlier the down payment was only 10 percent or less and the annual payments on principal were only $2,000 or roughly 2 percent of the purchase amount, so this contract fully qualified the seller to spread his capital gain over the first twenty years of the contract.

Sellers who are interested in spreading their capital gain generally want to complete the payments in a relatively short period, much shorter than the buyer with limited resources is able to handle. Insistence by the seller on a short contract can reduce the number of buyers and may decrease the price the seller can get for his farm.

If Doe wanted to sell on a short-term contract, his tenant or a neighboring farmer might have to pass up the opportunity to buy because he could not meet the terms. Doe, in order to spread his capital gain, might offer a ten-year contract calling for a down payment of $10,000 and annual payments of $9,000 for ten years plus interest on the unpaid balance. These terms would not only be out of reach of the tenant who might be interested in buying but would probably not be satisfactory or attractive to other buyers who preferred to pay cash either with or without mortgage financing. Instead of obtaining $500 an acre, Doe might have to settle under this contract for a lower price, say $480 an acre, a figure which might reduce or even eliminate any saving Doe might obtain by spreading his capital gain over the nine years. How much saving Doe would realize on his income tax would depend of course on his total taxable income; the higher his income the higher would be the income tax bracket into which the gain for any one year would push him and the more advantage he would have in receiving his capital gain income over a period of years.

The reason there is so much emphasis by sellers on spreading their capital gain over a period of years is that many of the farms being sold were purchased or inherited in years when farm real estate values were much lower. A look at the farm value record, as pictured in Figure 2.1, shows values in the late sixties higher than in any previous year in the history of the country, and much higher than in the years 1930–50 when many of the farms being sold were bought or obtained through inheritance or gift.

SPECIAL SALE TERMS

The price at which a farm sells is often affected by special terms such as agreements on payment of taxes, on who gets the growing crop, and on who is to have certain movable items that may or may not be sold with the farm. In many cases these items do not enter into the negotiations in any significant manner. For example a farm sold in December with possession given on March 1 will usually provide that the taxes as of March 1 are paid and that all crops in storage and other movable items are removed by the seller before March 1. Since there is no growing crop at this time of year in many areas of the country, this item will not be an issue. However, if the sale takes place during the summer growing season, the taxes and the growing crop are items on which agreement is important. For instance if the buyer gets the landlord's share of the crop, this could mean as much as $40 an acre to the buyer and be the reason he would be willing to pay a higher price for the farm than the going market price for farms where the growing crop does not go to the buyer. If a sale price appears out of line on the high side, some factor like this might explain it. Taxes and growing crops are items which the appraiser can clear up by obtaining details of the transaction directly from one of the parties involved in the sale. The only cases where this is not necessary are recorded contracts which usually spell out in minute detail all of the terms in the sale agreement. But even in the case of a contract an interview with one of the parties will usually provide additional and valuable information on the sale.

SUMMARY

Each farm sale needs to be analyzed carefully to bring out the particular conditions that were involved in the transaction. Sales with a deed differ from those with a contract. And of course there are different kinds of deed sales as well as different kinds of contract sales. The buyer in a deed sale is usually able to make a substantial down payment while the contract buyer usually is not. But this is no longer the whole story because frequently the seller is specifying a contract to spread his capital gain over a period of years to reduce his income tax.

Of paramount importance is the fact that price alone is not sufficient to describe the sale. Interest rate, size of principal payments, attitude of the seller, and agreements on growing crops and personal property are pertinent factors which have to be known before the appraiser can interpret the sale correctly. Sale prices of $475, $500, $525, and even $550 an acre may actually represent the same selling price when adjustment is made for different interest rates or other factors. All of the conditions which make up the sale must be considered—not just the selling price.

QUESTIONS

1. If a trip can be made to the courthouse, have each student examine one or two farm sales by deed and one or two land contract sales. Have reports on these made and then follow this by having students obtain verification and additional information on the conditions connected with the sales.
2. Explain the difference between a deed and a contract sale.
3. How would you handle a sale with an assumed mortgage at a rate below the current rate?
4. How can the "Present Value of 1" table be used in analyzing sales?
5. Describe a "cash-former owner mortgage sale." How may it differ from a straight cash sale?
6. What are the danger signals in connection with the use of estate sales?
7. Point out the similarities and differences between a buyer-type and a seller-type contract sale.
8. What are some of the special conditions that affect selling price? Explain.

7

FEDERAL CENSUS VALUES
AS BENCH MARKS

IN MOST FARMING AREAS there is a real estate market with a current or going market value for farms and farm tracts. If you ask a farm real estate broker, a farmer, or a lender what farms are selling for, you will usually get an estimate of this current market value. It is true that in certain areas and at certain times there may be few if any sales, but if the size of the area is expanded, sales can be found. Even more important, this estimate of what a farm would sell for if it were placed on the market does exist in the minds of those concerned.

Individual sales have two weaknesses that can be checked by market value data. First, a sale may not be in line with the market, and second, a sale may be one or more years old and will need to be brought up to date. Information which is available on the real estate market by counties, districts, and states provides bench marks that can be used to check the reliability of individual sales and to adjust those that have occurred some time ago. This information comes from three main sources: federal census figures, USDA market value indexes, and value data published by state agricultural experiment stations.[1]

---

1. Federal census reports are usually available in libraries, but an appraiser should have his own copy of the most recent agricultural census for his state. Information on obtaining these reports can be secured from the Bureau of the Census, U.S. Department of Commerce, Washington,

TABLE 7.1. Bench Mark Values for Medium Value Farms in Central Iowa, 1959–68

Year as of Nov. 1	Farm Value per Acre			
	State Average	District Average	County Average	Farm K
1959	$252	$305	$351	
1960	237	287		
1961	237	283		
1962	241	293		
1963	250	295		$300[a]
1964	265	324	385	
1965	293	354		
1966	331	403		
1967	362	444		450[b]
1968	375			

Source: State and district figures are from annual survey of Iowa farm values conducted by Agricultural Experiment Station at Iowa State University. County figures are from federal census.

[a] Actual sale price.
[b] Estimate.

BENCH MARK TABLES

One of the distinct advantages in appraising farms as compared to other real estate is the availability of periodic farm value estimates from which bench mark tables like Table 7.1 can be prepared. The peculiar position of farmland as a fixed nonreproducible factor makes it possible to make these valid comparisons from year to year.

Farmland in most cases does not change like buildings. Farmland is not produced in any appreciable amounts nor is it reduced by large amounts. Consequently it is possible to estimate the value of farmland in a state like Nebraska for 1959 and 1964, as the federal census did. The result in this instance was a value of farm real estate of $88 an acre in 1959 and $109 in 1964. In short we are valuing about the same property in both years.

What we have said about farmland is not so likely to be true of buildings because buildings depreciate and some are torn down and new ones are built. In an urban area the difficulty of comparing property from one year to another is apparent because new construction and depreciation are likely to change the

D.C. 20233. USDA market value indexes are published in *Farm Real Estate Market Developments*, a publication issued by the USDA, Washington, D.C. 20250. Every practicing appraiser should be on the mailing list to receive this publication. For state reports a request should be made to the bulletin department of the state agricultural experiment station or extension service for a list of available publications.

base so much that the comparison has little meaning. Since farm buildings are usually limited to the number of farms and since farm building value is a relatively small part of total farm value, it is possible and feasible to make the year-to-year comparisons on farm real estate including both land and buildings.

Bench mark tables indicate two things—the level of values in any area and what has happened to values in the area over the years. Table 7.1 provides such a table for use in evaluating Farm *K* located in Story County, Iowa. This table is designed for medium-grade farms in the central district of the state and represents below-average farms in Story County. A similar table for Story County was presented as part of the appraisal of Farm *A* in Chapter 1.

The owner of Farm *K*, who purchased the farm for $300 an acre in June 1963, asked the author what its approximate value was in June 1968. The answer was around $450 provided the owner had not made any large improvements since he bought it. Table 7.1 was used to get this answer. The first objective was to find a series of land values in the district which represented Farm *K*. A glance at a state report on farm values indicated that medium land in the central district was similar; in fact the price of $300 an acre for Farm *K* in June 1963 was only slightly above the farm values of November 1962 and November 1963. Since there was no reason for Farm *K* to vary from the upward trend of medium farm values in the district in which it was located, $450 was a logical answer for the estimated value in November 1967 and also for June 1968 since no appreciable change was evident in this period.

Actually Farm *K* was worth more than $450 in June 1968 because the owner had made a large investment in beef-feeding equipment during the five-year interval since he bought the farm, having added a silo and an automatic feeding unit. Consequently the value of Farm *K* was $450 plus what a buyer would be willing to pay extra for the beef-feeding equipment.

The value figures in Table 7.1 show the same upward trend depicted by the district medium values. Since there had been no federal census betwen 1964 and 1968 the county comparison is not very helpful. However, the state and district figures based on annual estimates show rising values with most of the increase occurring in the three years 1965–67. In brief the estimated increase in Farm *K*'s value between 1963 and 1968 was similar to that in the district and the state.

If there are no district figures available, the best procedure is to adjust the county figures each year by the state change re-

reported by the USDA in their publication *Farm Real Estate Market Developments,* discussed in the next chapter. This is not always a good alternative because there may be so much variation within the state that a state average change is not representative of what has happened in a given county.

FEDERAL CENSUS REPORTS

Every ten years the federal census obtains market value estimates for farmland and buildings and reports them by counties and states. These are the figures of the 1959, 1950, and previous decennial census years.[2] In addition a federal agricultural census has been taken at 5-year intervals between the regular 10-year census dates. These provide the figures of the 1964, 1954, and earlier agricultural census reports. In all cases the census farm values are the response obtained when the census taker asks the farm operator to state the market value of the farm he is operating. All farmers were asked this question in 1950 and before; since 1950 only a sample (20 percent) of the farm operators have been asked this question.

Farm values furnished by the federal census provide an important part of the bench marks or guides helpful to the appraiser. Federal census figures come out regularly every five years so that they offer a continuous series possible over a long period, and they are available for counties as well as for states. The county census figures are particularly useful since there is no other nationwide source of farm values by counties.

COUNTY COMPARISONS

Federal census figures by counties make it possible for the appraiser to tie his individual farm value to a county bench mark. State figures include so much area, much of it differing from the local area, that the appraiser often finds it difficult to make a meaningful comparison between a specific farm value and a census value for the state. But variation within the county is usually not large so a comparison can be made with a feeling of confidence. Where county values fail this test is in areas of wide variation, where for example irrigated and dry land acreage are both well represented, or where grazing and citrus groves are about equally present in the county.

An example of county farm comparison maps is presented in Figures 7.1 and 7.2. An appraiser operating in a state with

2. Beginning in 1954 the federal census started reporting values in the fall rather than the spring of the following year.

relatively uniform conditions will find it advantageous to prepare and have available a federal census map series similar to
these. A map similar to Figure 7.1 provides the value figures
which can be used in the appraisal report, and a map similar
to Fig. 7.2 gives the relative position of each county at a glance.

Counties have marked value changes over time. Examples
are provided by taking any two counties and comparing their
federal census values over a period of time. The comparison
can be done readily for counties anyhwhere in the country with
the use of the reference book edited by Thomas J. Pressly and
William H. Scofield entitled *Farm Real Estate Values in the
United States, by Counties, 1850–1959.* The book is a remarkable data collection and explanation of county and state farm
values per acre from every federal census during a period of 110
years. One of the advantages of this volume is the description of
the method followed in each of the 16 federal censuses recorded.
Since the dates and definitions as well as the material gathered
differed somewhat from one census to the next, this summary is
a welcome addition to the appraiser's reference list. With this
volume it is possible to trace farm values for any county in the
48 states from 1850 through every census up to and including

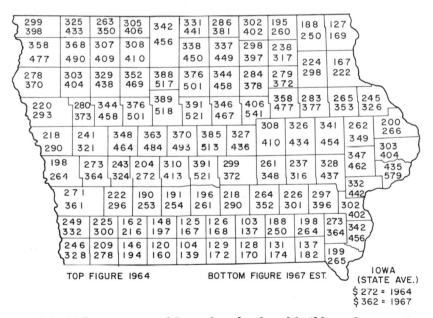

TOP FIGURE 1964   BOTTOM FIGURE 1967 EST.

IOWA
(STATE AVE.)
$ 272 = 1964
$ 362 = 1967

FIG. 7.1. *Value per acre of Iowa farmland and buildings by counties,
1964 and estimated 1967.*

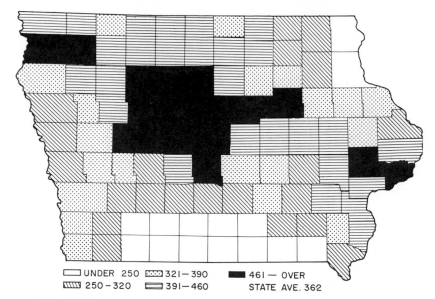

UNDER 250   321—390   461 — OVER
250 - 320   391—460   STATE AVE. 362

FIG. 7.2. *Value per acre of Iowa farmland and buildings by value groupings by counties, 1967. (Source: Federal Census.)*

1959. By obtaining the census reports for 1964 and subsequent censuses it is possible to have at hand the records of all counties from 1850 to the present.

An illustration of the county comparisons that can be made, using two counties in Florida and two in Iowa to show widely varying conditions, is shown in Table 7.2. In Florida, Flagler County showed a drop between 1920 and 1964 while Orange County, as a consequence of citrus development, experienced a remarkable increase. In Iowa, Wayne County, which suffered

TABLE 7.2.   Values per Acre for Selected Counties in Iowa and Florida, Federal Census, 1920, 1940, 1950, and 1964.

	Florida		Iowa	
Census Year	Flagler	Orange	Wayne	Hancock
	(dollars)			
1920	86	61	160	214
1940	22	133	35	91
1950	28	265	72	195
1964	66	1,132	129	338

Source: Thomas J. Pressly and William H. Scofield, eds., *Farm Real Estate Values in the United States, by Counties, 1850–1959,* Univ. of Wash. Press, Seattle, 1965, pp. 34, 35, 49, 50.

severe erosion damage, had lower values in 1964 than in 1920, while Hancock County, a level area, had values in 1964 more than 50 percent above 1920. Changes like these, which are going on all the time, give the appraiser an opportunity to study the forces which are constantly at work in the farm real estate market.

STATE COMPARISONS

State comparisons, like county comparisons, give the appraiser more understanding and confidence in his values. A collection of census figures by states such as that in Table 7.3 provides a ready source for state-by-state comparisons over a long period of time. An appraiser valuing lands in Illinois or Iowa or both would find the farm value record of these two states highly significant. In 1910 Illinois was more than 10 percent above Iowa in average value per acre of farmland and buildings. But in 1920 and 1930 Iowa's values were higher than those in Illinois. In 1940 Illinois took the lead again and held it

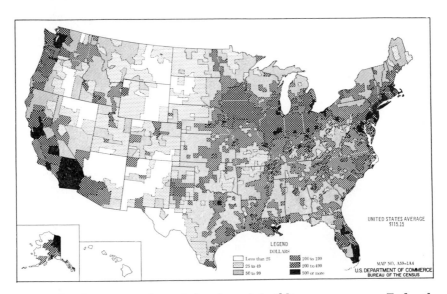

FIG. 7.3. *Average value of land and buildings per acre, Federal Census of 1959. Note the relatively high values near the heavy population centers on the east and west coasts, in the Corn Belt, and along the Mississippi River. Value per acre drops quickly west of the Corn Belt with the drop in rainfall. However, river valleys like the Platte, where irrigation is common, provide high values even in areas of low rainfall. (Courtesy Bureau of the Census.)*

TABLE 7.3. Farm Real Estate Value per Acre by States

Region	State	Average Value per Acre, Land and Buildings						
		1920	1930	1940	1950	1954	1959	1964
		(dollars)						
New England	Maine	38	42	29	54	61	83	100
	New Hampshire	35	39	34	73	86	106	132
	Vermont	38	37	30	56	61	81	109
	Massachusetts	99	130	109	190	224	305	386
	Rhode Island	80	124	119	232	343	418	485
	Connecticut	100	151	135	248	291	430	561
Middle	New York	69	73	55	92	111	145	177
Atlantic	New Jersey	110	170	122	293	404	539	662
	Pennsylvania	75	79	59	107	137	184	228
East North	Ohio	113	79	66	136	185	245	295
Central	Indiana	126	72	63	137	194	266	310
	Illinois	188	109	82	174	231	320	357
	Michigan	75	68	51	99	133	190	233
	Wisconsin	99	79	52	89	101	132	155
West North	Minnesota	109	69	44	84	108	154	166
Central	Iowa	227	124	79	161	199	252	272
	Missouri	88	53	32	64	81	111	150
	North Dakota	41	25	13	29	36	52	67
	South Dakota	71	35	13	31	39	52	62
	Nebraska	88	56	24	58	72	88	109
	Kansas	62	49	30	66	80	101	122
South	Delaware	69	74	61	114	157	243	322
Atlantic	Maryland	81	81	65	125	177	276	422
	Virginia	55	51	41	82	106	140	183
	West Virginia	43	39	30	59	68	76	91
	North Carolina	54	47	39	99	128	193	252
	South Carolina	65	36	30	69	87	139	171
	Georgia	45	26	20	43	60	100	135
	Florida	47	84	39	57	107	227	286
East South	Kentucky	60	44	38	81	95	133	181
Central	Tennessee	53	41	36	77	93	132	179
	Alabama	28	29	21	49	58	92	125
	Mississippi	43	33	25	55	74	109	150
West South	Arkansas	43	34	25	60	77	103	178
Central	Louisiana	47	45	35	82	112	171	233
	Oklahoma	43	37	24	51	63	85	121
	Texas	32	29	19	46	62	87	112
Mountain	Montana	22	12	8	17	24	36	42
	Idaho	69	45	33	70	90	126	132
	Wyoming	20	9	6	13	15	22	28
	Colorado	35	22	12	32	40	56	70
	New Mexico	9	7	5	15	21	29	35
	Arizona	30	18	6	15	26	78	53
	Utah	48	39	21	43	48	71	71
	Nevada	28	16	13	19	27	36	38
Pacific	Washington	69	57	39	85	115	149	154
	Oregon	50	38	27	60	78	93	115
	California	105	112	71	154	227	375	468
	Alaska							9
	Hawaii							205
United States Average		69	49	32	65	84	119	144

Source: Federal census.

through 1964, the latest lead being the most substantial of any in the 45-year period.

Reference is made later (Chapter 22) to a sharp rise in Illinois corn yields which outdistanced those in Iowa after 1940. This explains in part the larger gain in Illinois values. Another reason is the growing importance of metropolitan centers in Illinois and their effect on surrounding land values.

There is practically no end to the state comparisons that can be made. A comparison of the four leading (in value per acre) agricultural states, listed in alphabetical order—California, Illinois, Iowa, and Texas—and the United States average is presented in Table 7.4. Illinois was slightly ahead in 1940 and had extended its lead in 1950. But in the 1959 census and again in 1964, California took over the first-place position by a wide margin. In terms of percentage gain, Texas was second to California, with Iowa in last position among the four states. An explanation for the rise of values in California can be found in the heavy demand for irrigated lands devoted to specialty crops and for agricultural lands surrounding urban centers that have been expanding rapidly. Iowa and Illinois lands, on the other hand, were handicapped by a large surplus of feed grains.

QUESTIONS

1. Make for your state two county value maps like Figures 7.1 and 7.2, based on the two most recent federal census reports. Make a third map showing percentage changes in each county which occurred between the two census dates. What conclusions do you draw from this third map? If your state is a large one the assignment may be limited to a portion of the state.
2. What is the purpose of bench mark values? How are they used?
3. Why are bench mark values for farms easier to provide than for other types of real estate?

TABLE 7.4. Value of Farmland and Buildings in Four Leading Agricultural States and the United States, 1940–64

	Value per Acre of Land and Buildings in Census Years				Percentage Increase
State	1940	1950	1959	1964	1940 to 1964
			(dollars)		
California	71	154	375	468	559
Illinois	82	174	320	357	335
Iowa	79	161	252	272	244
Texas	19	46	87	112	489
United States	32	65	119	144	350

Source: Federal census. Alaska and Hawaii not included in United States average until 1964.

4. What specific federal census data are available on farm values? Look up the report on agricultural data from the most recent census for your state. Note the material on farm real estate values by counties and also note in the explanatory section how these data were obtained.

5. Discuss the extent to which the results of one census can be compared with another. Point out variations if any in the procedure and dates. In answering this question the reference book on federal census values edited by Pressly and Scofield will be helpful.

6. From the federal census data in Table 7.3 plus any additional reports since 1964, prepare a statement on what has happened in your state and in adjoining states since 1954.

7. With the same data you used in the preceding question indicate what regions in the United States have had the most change and what have had the least change since 1954? What reasons do you think account for these changes?

8. Comment on changes in your county and surrounding counties on the basis of federal census reports on farm values since 1954.

9. Explain the differences in county values as reported in the last census for your county and surrounding counties. Are these differences based on physical factors like tillable land, yields, and the like or on other factors such as urban influence?

REFERENCES

Federal Census Reports, Bureau of the Census, Washington, D.C. (See state reports of Agricultural Census.)

Pressly, Thomas J., and Scofield, William H., eds., *Farm Real Estate Values in the United States, by Counties, 1850–1959,* University of Washington Press, Seattle, 1965.

8

USDA ESTIMATES
AS BENCH MARKS

The USDA provides current reports as of March 1 and November 1 on the farm real estate market in a publication entitled *Farm Real Estate Market Developments*. Each issue presents state index figures and other pertinent information making it possible for readers to keep up to date on what is happening to farm values throughout the country. William H. Scofield of the USDA has been responsible for many years for this publication and also for the research that has provided the material contained in this helpful report for those working with farm values.

Widespread use is made of the USDA indexes of farm values which are reported for March 1 and November 1 of each year. The source of the data, as reported in the April 1968 issue, is as follows:

FARM REAL ESTATE INDEX

Index numbers of the average market value of farm real estate per acre express the value for a given date as a percentage of the average value in the 3-year period 1957–59 which is taken as a base of 100. For example, the November 1967 index of 166 for the 48 States means that the average value for that date was 66 percent higher than the 1957–59 average value. The chief use of the index numbers is for measuring the change in value between any two dates.

The basic data used for constructing the State index number are obtained from USDA crop reporters as of March 1 and

November 1. Nationally, about 15,000 estimates are available. Crop reporters are asked to provide estimates of the prevailing market value of farmland, including buildings, in their localities. They are instructed to exclude properties valued chiefly for nonagricultural uses. State offices of the Statistical Reporting Service review and edit the individual reports and prepare summaries for each crop reporting district in their States.

Average values for each district, typically 9 in each State, are then multiplied by the total acres of land in farms to obtain weighted average values for each State. Constant acreage weights, derived from the 1954 and 1959 Censuses of Agriculture, are used to minimize the possible effects of shifts in the acreage of land in farms on the land price index.

The various censuses of agriculture obtain estimates of market values of farm real estate at the county level. These data are useful for determining the actual dollar level of farm real estate values, but are less suitable than the index numbers for measuring changes in the value of comparable land over time.[1]

To supplement the index the USDA obtains added data on the land market from a semiannual questionnaire. This information which is provided by real estate brokers, bankers, and others gives the appraiser excellent background material to explain what is happening in the market. For example the extent of purchasing for farm enlargement, the size of tracts sold, classification of buyers and sellers, and reasons which motivate buyers and sellers are the kind of qualitative comment and statistical reporting which is based on the information furnished by the survey taken twice a year.

Use of the USDA indexes in bringing federal census values up to date is shown in Table 8.1 which gives a list of the 50 states ranked in order of their average per acre value as of March 1, 1967. Each state value in Table 8.1 represents the 1964 federal census value as shown in Table 7.3 in the preceding chapter adjusted with the USDA estimates for the change which occurred between November 1964 and March 1, 1967.

Variation in Table 8.1 ranged from a high of $756 an acre in New Jersey to a low of $10 in Alaska. There were eleven states centered around the national average of $167 an acre, ranging from $189 in Minnesota and Mississippi to $143 in Kansas and Oklahoma. The five highest states, with values of $500 or more an acre, were all located on the northern East Coast except for California on the West Coast. All low-value states (those below $60 an acre) were concentrated in the Rocky Mountain area stretching from Arizona to Alaska. Another interesting group

---

1. USDA, *Farm Real Estate Market Developments*, Apr. 1968, p. 31.

TABLE 8.1. States Ranked According to Value per Acre of Farmland and Buildings as of March 1, 1967

Rank	State	Value	Rank	State	Value	Rank	State	Value
1	New Jersey	$756	18	Virginia	$224	35	Vermont	$136
2	Connecticut	659	19	Arkansas	216	36	Nebraska	129
3	Rhode Island	570	20	Kentucky	207	37	Oregon	127
4	Maryland	526	21	Tennessee	207	38	Texas	124
5	California	500	22	New York	205	39	Maine	110
6	Illinois	446	23	South Carolina	204	40	West Virginia	96
7	Massachusetts	445	24	Missouri	190	41	Colorado	79
8	Indiana	395	25	Minnesota	189	42	Utah	77
9	Delaware	387	26	Mississippi	189	43	North Dakota	77
10	Iowa	350	27	Wisconsin	181	44	South Dakota	72
11	Ohio	347	28	Washington	176	45	Arizona	56
12	Louisiana	294	29	Georgia	176	46	Montana	48
13	Florida	291	30	Alabama	155	47	Nevada	42
14	North Carolina	289	31	New Hampshire	154	48	New Mexico	40
15	Michigan	275	32	Idaho	149	49	Wyoming	31
16	Pennsylvania	264	33	Kansas	143	50	Alaska	10[a]
17	Hawaii[a]	238	34	Oklahoma	143		U.S. Average	167[b]

Source: *Farm Real Estate Market Developments,* USDA, CD-70, Apr. 1968, pp. 42–43. Values are based on the 1964 federal census adjusted to March 1, 1967, by applying the index changes reported by the USDA.
[a] Hawaii and Alaska are estimates made by increasing the 1964 federal census values by the average percentage increase which took place in the 48 states.
[b] Average does not include Hawaii and Alaska.

was the four Corn Belt states of Illinois, Indiana, Iowa, and Ohio, all in the first eleven. One of the dominant characteristics of these four states is their relatively uniform soil with most of the land being tillable and adapted to corn and soybeans. All of the other states in the highest eleven were in either the north Atlantic area or the West Coast where heavy population or irrigation (as in California) was an important factor.

Between 1964 and 1967 there were several major changes in rank. Missouri and Vermont moved up three places while Iowa, Louisiana, Arkansas, Mississippi, and Alabama each jumped two places. Among the first ten states Maryland, Illinois, and Indiana each moved up a notch and Iowa moved up two. In general the largest percentage increases in value in this period were in the Delta states of Arkansas, Louisiana, and Mississippi and in the Corn Belt.

The USDA indexes can also be used to show what has happened in the four leading farm states for which federal census data were presented in Table 7.4. In Table 7.4 the latest year was 1964, in Table 8.2 the record is brought up to November 1967. The record between 1964 and 1967 indicates a decided change among the four leading states. California and Texas, which had been rising much more rapidly than Illinois and Iowa, slowed down. Both Iowa and Illinois experienced a substantial rise. This rise in the Corn Belt, which was noted in the ranking of the states, was largely the result of better prices stimulated by export demand for corn and soybeans, advances in technology, higher yields, and farm enlargement pressure.

Appraisers interested in the changes in values caused by changes in physical and economic factors will find a study of year-to-year reports of the USDA especially worthwhile. Developments like those in the Corn Belt and the South are good examples. The effect on values of irrigation, expanding urban centers, and rapidly increasing crops like soybeans and citrus are other examples.

SPECIALIZED FARM VALUES BY STATES

Bench marks for special types of farms provided by the USDA are presented in Tables 8.3, 8.4, and 8.5. In Table 8.3 value estimates for three types of land—irrigated, nonirrigated, and pasture—are presented for Nebraska and Texas. In Tables 8.4 and 8.5 market value estimates for various groves, orchards, and vineyards in California and Florida are given.

The importance of separating irrigated from nonirrigated lands is shown in Table 8.3. In both Nebraska and Texas irrigated land is valued about twice as high as nonirrigated cropland. It is easy to see that any market value which combined these two types would fail to give the kind of information that is needed. For example, Table 8.3 shows that in Texas nonirrigated land rose 19 percent between 1964 and 1967 while irrigated land rose only 5 percent. In Nebraska the situation was different with irrigated climbing 38 percent and nonirrigated rising 28 percent. A striking change in Texas was a 28 percent increase in pasture value, a larger percentage increase than for

TABLE 8.2.   USDA Indexes of Value per Acre of Farmland and Buildings in Four Leading Agricultural States and the United States, 1964–67
1957–59 = 100

State	1964	1965	1966	1967	Percent Increase 1964–67
California	159	167	176	181	14
Illinois	119	128	143	152	28
Iowa	114	124	140	151	32
Texas	156	158	168	180	15
United States	136	144	156	166	22

Source: Various issues of *Farm Real Estate Market Developments*, USDA.

irrigated or nonirrigated land. Just the opposite occurred in Nebraska with pasture registering the lowest increase.

If only the average values for Nebraska and Texas had been available between 1964 and 1967 no information would have been provided on the wide difference between the increase in irrigated land and pasture in the two states. Certainly the $28 average increase in value of all farm real estate in Nebraska between 1964 and 1967 would not have indicated the low increase on pasture or the high increase in irrigated land—$10 and $102 respectively.

Finally Table 8.3 reveals a large amount of change between 1964 and 1967 that would not have been available if only federal census figures were reported. Five years is too long to wait for reports on the land market. Fortunately the USDA service fills the gap not only with state estimated values on March 1 and November 1 of each year but also with specialized land figures like those in Table 8.3.

One other advantage in the USDA series is the relatively fast reporting compared to the federal census which understandably takes considerable time getting its results in shape for publication. Since county figures for the 1964 federal census were not available until 1966 and 1967, the results were not an indication of the current market when they were released. But the 1964 census figures did serve as bench marks to which the percentage increases of the years 1965–67 could be applied. In short, the estimates made twice a year by the USDA make it possible to keep the federal census figures up to date.

Fruit and nut crops come from groves, orchards, and vine-

TABLE 8.3. Market Values of Farm Real Estate for Three Types of Land in Nebraska and Texas, 1964–67

	Value per Acre, Nov. 1				Percent Increase
	1964	1965	1966	1967	1964–67
		(dollars)			
Nebraska					
Irrigated	268	288	317	370	38
Nonirrigated	142	151	161	182	28
Pasture	58	60	64	68	17
All land	105	111	119	133	27
Texas					
Irrigated	333	357	347	350	5
Nonirrigated	149	157	168	178	19
Pasture	84	91	94	107	28
All land	106	114	118	130	23

Source: *Farm Real Estate Market Developments*, USDA, CD-70, Apr. 1968, p. 10.

TABLE 8.4. Estimated Market Values of Orchards, Vineyards, and Groves for Certain Regions in California, 1964–67

| Crop | | Value per Acre, Mar. 1 | | | Percent Change |
	1964	1965	1966	1967	1964–67
		(dollars)			
English walnuts					
Central coast	3,175	3,400	3,500	2,950	—7
State average	2,800	3,075	2,975	3,075	10
Pears					
Central coast	4,300	4,550	4,350	3,550	—17
State average	3,650	3,800	3,800	3,150	—14
Grapes (Wine)					
San Joaquin	1,500	1,625	1,625	1,575	5
State average	1,800	2,025	1,975	1,900	6
Oranges (Navel)					
Southern	6,150	6,650	7,200	6,825	11
State average	4,800	5,075	5,375	4,900	2
Orange (Valencia)					
Southern	6,025	6,625	6,825	6,925	15
State average	5,675	6,175	6,350	6,325	11

Source: *Farm Real Estate Market Developments*, USDA, CD-70, Apr. 1968, p. 11.

yards which have unusually high values per acre as a result of large inputs of capital. The land is usually irrigated, which involves a large investment in such items as land leveling, irrigation equipment, and facilities to get the water to the land. Then there is the investment in the trees and after that the waiting period before the trees start bearing.

The high values shown in Table 8.4 of groves, orchards, and vineyards in certain areas of California indicate the size of investment behind these fruit and nut crops. The variety of

TABLE 8.5. Estimated Market Values of Orange and Grapefruit Groves in Florida, 1965–67

| | Value per Acre, May 1 | | | Percent Change |
	1965	1966	1967	1965–67
		(dollars)		
Oranges				
Valencia				
Nonbearing	1,750	1,150	950	—46
Bearing	2,500	2,375	1,850	—26
Grapefruit				
Seedless				
Nonbearing	1,000	1,150	950	— 5
Bearing	2,025	2,200	1,850	— 9
All groves[a]	1,875	1,875	1,475	—21

Source: *Farm Real Estate Market Developments*, USDA, CD-70, Apr. 1968, p. 13.
[a] Includes all citrus groves, not just those listed in this table.

changes shown in Table 8.4 provides ample evidence of the high priority of such information for use in valuation work. While English walnut orchards in the central coast area declined 7 percent between 1964 and 1967, those in the state as a whole rose 10 percent in the same period. In comparing different crops it is evident that pear orchards were down substantially in the three-year period while vineyards specializing in wine grapes and orange groves were both up. A look at the citrus groves shows all averages down from 1966 to 1967 except the Valencias in southern California which were up slightly.

The importance of figures like those for California is borne out by the following statement from the August 1965 issue of *Farm Real Estate Market Developments,* page 15:

> Because of the extremely wide range in the kinds and uses of land in California, an average value for the State has little meaning. Market values range from less than $100 per acre for grazing lands in the northern part of the State to $10,000 or more for citrus groves in the path of urban development in southern California. However, about 5.3 million acres of irrigated land devoted to bearing orchards, vineyards and groves, and to annual field crops account for nearly half of the total value of farmland and buildings in the State.

Reports on market values of citrus groves in Florida, like those in California, give localized information that neither county nor state figures can provide (see Table 8.5). It is apparent from the citrus area map of Florida, shown in Figure 8.1, that even though citrus is concentrated in Florida there are many small areas which represent only a small part of a county. An appraiser valuing citrus groves is obviously not going to get much if any help from the federal census figures for Florida counties in the citrus area since these county figures include grazing land and cropland along with the high-value citrus groves.

What happens to citrus land values in Florida is not necessarily going to be the same for citrus lands in California. Florida had a huge crop in 1966–67 with a big decline in values as a result. Information like that shown in Tables 8.4 and 8.5, localized by types and areas, offers the bench marks or guides which an appraiser can use in comparing his individual tract values with levels existing in an area.

QUESTIONS

1. Start with the last two federal census reports for your county and state, including in the assignment counties which surround your county and states which surround your state. With the use of federal census reports and recent issues of *Farm Real Estate*

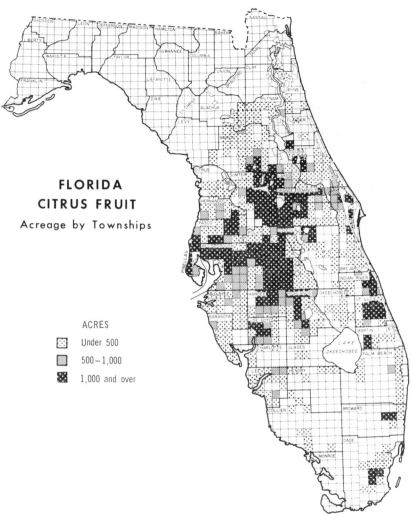

FLORIDA CITRUS FRUIT

Acreage by Townships

ACRES
- [ ] Under 500
- [ ] 500 – 1,000
- [ ] 1,000 and over

U. S. DEPARTMENT OF AGRICULTURE   NEG. ERS 3042–64(8) ECONOMIC RESEARCH SERVICE

FIG. 8.1. *Map of Florida showing citrus fruit acreage by townships. (Courtesy USDA, Econ. Res. Serv., Washington, D.C.)*

Market Developments prepare a table showing average value estimates for farmland and buildings for each county and state in your assignment for the last ten years. Discuss the conclusions indicated by your table.

2. Explain the source of the land value indexes and related information reported in *Farm Real Estate Market Developments.*

3. What information does the most recent issue of this publication have about your state?
4. What does the most recent issue of this publication tell you about the situation in your state compared to surrounding states and to the country as a whole?
5. Where does you state rank among the 50 states in average value per acre of farm real estate? Has this position changed in recent years? If so, give what you consider to be the reasons for the change.
6. Comment on the problem of using state indexes when there are wide differences in farm types within the state.
7. Explain how USDA indexes and federal census figures can be used to supplement each other.
8. Compare changes in farm values in Corn Belt states with those in the citrus states of California and Florida.
9. Can you see any connection between expanding metropolitan areas and rising farm values?

REFERENCES

USDA, *Farm Real Estate Market Developments*, Washington, D.C., various issues.

9

ANNUAL STATE REPORTS
AS BENCH MARKS

STATE UNIVERSITIES and experiment stations along with the
USDA have been increasingly concentrating on district land
values on an annual basis. This service has been welcomed not
only by appraisers but also by farm owners who are interested
in estimating the current value of their farm. Examples from
Minnesota and Iowa are presented here to show the type of bench
mark information available from these district state surveys.

The main contribution of the Minnesota survey, illustrated
by the material in Table 9.1 and Figures 9.1 and 9.2, is the series
of district averages which has been extended back to 1910, thus
providing a continuous annual value record from that year to the
present. Philip M. Raup, who started this district average series
in Minnesota, also established a series of actual sale value records
by districts. However, these do not give as accurate a measure
of the land market as the value estimates. On this point J. M.
Bambenek and P. M. Raup, in their 1967 report, comment as
follows:

> The estimates of farm land value are a more reliable basis
> for comparing year-to-year trends than are the reported sale
> prices received in actual sales. This is because of the erratic
> and occasionally wide variations in the qualities of land and
> buildings actually sold and the small number of sales that may
> occur in any given year and location.

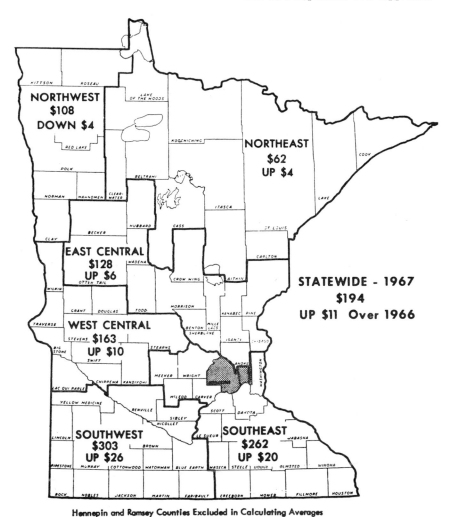

Hennepin and Ramsey Counties Excluded in Calculating Averages

FIG. 9.1. *Map of Minnesota showing districts in land value survey, and results of the 1967 survey by districts.* (Source: Econ. Study Rept. S 68-1, Feb. 1968.)

Typically, there are 20 to 50 voluntary farm sales per year in a representative Minnesota county. A reported change in average sale prices may primarily reflect a variation in quality of land or buildings on farms sold during the period studied, or a change in average sale prices may actually represent a change in local land prices.[1]

1. *Minnesota Rural Real Estate Market 1967*, Econ. Study Rept. S 68-1, Univ. of Minnesota, St. Paul, Feb. 1968, p. 3.

TABLE 9.1. Estimated Average Value per Acre of Farmland, by District, Minnesota, 1962–67

District	1962	1963	1964	1965	1966	1967
			(dollars)			
Southeast	192	194	206	219	242	262
Southwest	250	246	252	261	277	303
West central	138	142	145	146	153	163
East central	99	103	111	112	122	128
Northwest	104	114	115	113	112	108
Northeast	69	68	59	51	58	62
Minnesota	159	161	166	171	183	194

Source: Econ. Study Rept. S 68-1, Feb. 1968.

Table 9.1 indicates the information an appraiser can get from district averages which is not available from the state averages. Farm values between 1966 and 1967 increased 6 percent for the state as a whole, but in the southwest district they increased 9 percent and in the northwest district they declined 4 percent. In fact values in the northwest district fell for three consecutive years from 1964 to 1967. An appraiser working in this district would certainly have obtained little help from the state averages which went up every year between 1964 and 1967 while values in his district were declining. A picture of the land value movement by districts in Minnesota for the years 1945–67 is provided in Figure 9.2. The northeast and northwest districts

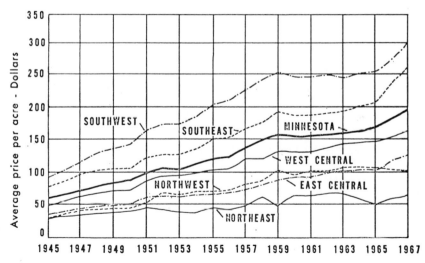

FIG. 9.2. *Estimated average value per acre of Minnesota farmland by districts, 1945–1967. (Source: Econ. Study Rept. S 68-1, Feb. 1968.)*

do not follow the trend which is relatively consistent in the other districts.

Reports on values for different qualities of farmland are also obtained on an estimate basis in the Minnesota survey. An example of how much difference there can be in the changes on the basis of quality is illustrated in Table 9.2 which shows what happened in the northwest district where high-quality lands rose 16 percent between 1959 and 1968 while low-quality lands declined 13 percent in the same period. This is the same district in which values on the average declined between 1964 and 1967. Information of this kind not only provides the appraiser with movement of the average in his district but also tells him what is happening to different qualities of land. If he is appraising a below-average farm in the northwest district he can be aware of the fact that values for this quality of farm are declining while values for high-quality farms are rising.

Annual surveys of farm values in Iowa were started by the author in 1942 to obtain values for three qualities of land for the five types of farming districts within the state. This series has been continued each year and in addition a classification has been made for the nine crop-reporting districts in the state. Values for the three qualities of land in the nine crop-reporting districts are shown in Figure 9.3 and Table 9.3.

A portion of the data for the bench mark table in the appraisal of Farm *A* in Chapter 1 and for Table 7.1 in Chapter 7 comes from Table 9.3. With the three qualities of land for the nine districts it is usually possible to find a reasonably close bench mark value for any given sale that occurs or has occurred in recent years. For Story County in the central district of Iowa

TABLE 9.2. Estimated Land Values per Acre and Changes by Quality of Land, by District, Minnesota, 1959–67

| | Quality of Land | | | | | |
| | Highest | | Medium | | Lowest | |
District	Estimated value 1967	Change since 1959	Estimated value 1967	Change since 1959	Estimated value 1967	Change since 1959
	(*dollars*)	(*percent*)	(*dollars*)	(*percent*)	(*dollars*)	(*percent*)
Southeast	354	24.2	259	31.5	182	44.4
Southwest	371	17.8	279	11.5	205	17.1
West central	219	21.7	165	23.1	118	32.5
East central	160	28.0	121	42.3	77	75.0
Northwest	162	16.5	104	−1.0	53	−13.1
Northeast	71	11.1	41	−2.4	25	31.6
Minnesota	250	18.5	188	19.7	129	25.2

Source: Econ. Study Rept. S 68-1, Feb. 1968, p. 7.

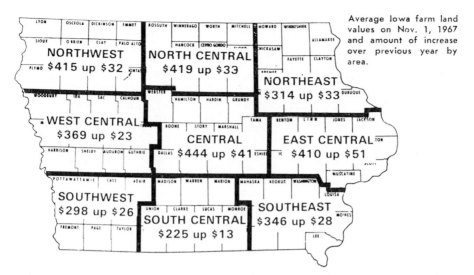

FIG. 9.3. *Map of Iowa showing nine crop-reporting districts, with results of the 1967 annual farm value survey.*

where Farm A is located, there are three values given for 1967—$602 for high-quality land, $441 for medium-grade land, and $290 for low-grade land. Although there will be sales between these grades (such as a sale at $525 or $360 an acre), it is not difficult to fit these sales into the table by interpolating, as for example figuring the change over a period of years as the average of two grades. These value estimates are not absolute figures; rather they are estimates representing the best judgment of farm brokers, who are in the business of buying and selling land, as to what different grades of land have sold for on November 1 of each year. The reason November 1 is used is that this is a high point of activity following harvest and a time when plans are being made for next year's operations.

Another use of Table 9.3 is comparing the movement of values for the different qualities of land over time. What for example has been happening to low-grade land compared to high-grade? In the southeast district in Iowa, low-grade land since 1957 has risen more percentagewise than the other two grades and more than low-grade land in any of the other districts in the state. In the high-grade group the central district which was second to the east central district in 1957 is on top in 1967 and has been every year since 1964.

One of the features of the annual survey in Iowa is the

TABLE 9.3. Farm Values for Three Grades of Land for Nine Crop-Reporting Districts in Iowa, 1957–67

Year	North-west	North Central	North-east	West Central	Central	East Central	South-west	South Central	South-east
				High-Grade Land					
1957	$370	$350	$272	$330	$393	$407	$270	$223	$359
1958	388	375	306	347	411	440	292	248	376
1959	392	348	326	368	427	445	292	256	373
1960	359	362	317	350	403	416	281	238	366
1961	363	361	324	355	397	409	272	247	369
1962	374	370	316	364	405	409	286	249	378
1963	384	378	334	378	413	419	299	245	391
1964	412	401	332	396	443	432	309	267	402
1965	454	452	358	422	481	478	339	297	436
1966	509	506	400	497	550	525	392	328	498
1967	554	546	445	518	602	593	417	341	538
				Medium-Grade Land					
1957	$275	$261	$193	$219	$272	$254	$178	$129	$191
1958	291	285	207	243	291	276	195	141	210
1959	291	291	219	251	301	310	192	150	210
1960	271	272	205	235	283	277	184	140	208
1961	274	266	216	235	281	274	189	142	214
1962	280	279	205	243	293	273	193	145	219
1963	287	288	220	254	292	278	210	143	220
1964	308	313	221	271	322	288	210	161	239
1965	337	348	241	299	349	319	236	185	263
1966	379	383	272	331	402	352	260	200	301
1967	412	418	307	362	441	401	290	215	330
				Low-Grade Land					
1957	$178	$178	$113	$129	$166	$126	$102	$ 61	$ 83
1958	189	196	126	143	177	143	119	71	93
1959	191	198	131	149	183	158	118	75	98
1960	178	180	200	131	175	148	107	70	97
1961	172	177	131	139	170	145	108	71	94
1962	176	189	121	144	180	148	112	71	98
1963	182	206	138	153	180	153	123	73	106
1964	202	218	135	163	206	156	120	83	114
1965	230	238	153	185	232	180	143	91	130
1966	261	268	169	212	256	201	162	108	155
1967	280	295	191	228	290	237	187	118	171
				All Grades					
1957	$274	$263	$193	$226	$277	$262	$184	$138	$211
1958	289	286	213	244	293	286	202	153	226
1959	292	279	225	256	305	305	200	160	227
1960	270	271	214	238	287	280	191	149	224
1961	270	268	224	243	283	276	192	153	226
1962	277	279	214	250	293	277	197	155	232
1963	284	291	231	262	295	283	211	154	239
1964	307	311	229	277	324	292	213	171	252
1965	340	346	250	302	354	325	239	191	276
1966	383	386	281	346	403	359	272	212	318
1967	415	419	314	369	444	410	298	225	346

blank space provided for comments. Many of the brokers fill this space with specific information on what is happening in the land market in their area. Frequently they indicate the major factors operating such as farm enlargement pressure, anticipation of higher values, outside investment buying, or higher interest rates and higher property taxes. It was from these comments in the 1955 annual survey that indications of farm enlargement as a value-boosting factor were first noted. In the report for 1955, which showed values up $10 an acre, the following statement appeared:

> The pressure of farm owners wanting additional land to enlarge their present farms was a strong factor in the rising values. In most cases these farmers have enough machinery and labor to operate larger farms than they now have. And they're often willing to pay more than the prevailing market price to get an adjoining tract of unimproved land.

A sample of the qualitative information provided by the annual survey is contained in the summary of the 1967 survey in Figure 9.4.

Highlights of the Survey

1. Farm land values up an average of $31 an acre or 9 percent.

2. Values up most in eastern and northeastern Iowa; up least in southern and western Iowa.

3. Below-average land values rose most percentagewise.

4. Farm enlargement major factor in increase.

5. Outside investors were an important factor in eastern and southern Iowa.

6. Use of contracts and scarcity of listings also contributed to rise in values.

7. Fewer sales reported than last year.

8. High interest rates, low grain prices and poor weather were depressing factors.

9. Most of value rise occurred early in year.

FIG. 9.4. *Summary of 1967 Iowa annual farm value survey.*

Timeliness has been stressed in the Iowa survey. Question-naires are sent out to around 800 real estate brokers asking for their estimates as of November 1. On the questionnaire, pictured in Figure 9.5, values reported for the previous year are in-serted for those who reported in the previous year so the broker will have last year's estimates to use in estimating values for the current year. Around three-fourths of the questionnaires go to those who reported the previous year and one-fourth to brokers who were not on the previous year's list. Response to the ques-tionnaire is prompt and relatively high; around 75 percent of the brokers send in their estimates within a four-week period. This makes it possible to issue a preliminary report at the end of the year on land values as of November 1. The final report appears in the February issue of *Iowa Farm Science* published by Iowa State University.

Among agencies providing district farm values are the federal reserve banks. An example is the Federal Reserve Bank of Chicago's report as shown in Figure 9.6. One of the features of this report is a forecast of the trend in values. In the spring of 1967, as an illustration, over 50 percent of the bankers report-ing forecasted rising values; a year later only 36 percent said the trend was upward.

Reports like the one issued quarterly by the Federal Re-serve Bank of Chicago are especially useful in measuring changes occurring in the land market. This type of report does not pro-vide any bench mark values. However, these surveys showing change are an excellent aid in adjusting the annual reports of the USDA and the state agricultural experiment stations. Often there is considerable change in values in the course of a year, and it is these changes which the quarterly reports can provide the appraiser. For example the USDA report for November 1, 1967, gave the estimated values for farms in Iowa, and the state survey also reported values for the same date, not only for the state as a whole but for districts as well. An appraiser in May and June of 1968 had these annual surveys and also the report of the Federal Reserve Bank of Chicago issued on May 3, 1968, giving the change in values from January 1 to April 1, 1968. Since this bank report indicated the value of "good" farms in all areas of the state, the appraiser had the basis for an adjustment to the November 1967 value in his bench mark evaluation.

Each year since 1964 Herbert B. Howell of the Extension staff at Iowa State University has conducted a survey on future land values at the Annual Soil Management and Land Valuation Conference attended by farm lenders and appraisers. The re-sults of these surveys covering average land values in Iowa over the five-year period are shown in Table 9.4. Each year the estimates have been below what turned out later to be the actual values, both for estimates later in the current year and for estimates made for the following year. It is likely that the estimates will be below the actual values that will exist in 1970. It should be pointed out that values have been rising at a fairly rapid pace

IOWA STATE UNIVERSITY
Department of Economics and Sociology
Ames, Iowa

October 30, 1968

Dear Sir:

Your cooperation in supplying information for the land value survey last year was appreciated. The report appeared in the Iowa Farm Science, and a copy was sent you. We would like to have you cooperate again this year. Please fill in the answers to the questions and return in the enclosed envelope to make this fact-finding survey possible.

Sincerely,

Professor, Agricultural
Economics

Farm Land Values in Your Territory as of November 1, 1968

1. Values for average-size farms:

	A Year Ago*	Present
High-grade farms	$_____per acre	$_____per acre
Medium-grade farms	$_____per acre	$_____per acre
Low-grade farms	$_____per acre	$_____per acre

2. Number of sales in 1968 compared to same period in 1967 is:
    More_____ Same_____ Fewer_____ (check one)

3. In your opinion, what were the most important factors operating in the land market in your territory this year?
    a. _____
    b. _____
    c. _____

4. Other comments on present land markets (continue comments on back if desired):

* This was your report to us last year

P.S. We will send you a copy of the results of this survey.

FIG. 9.5. *Copy of questionnaire used in Iowa farm value survey.*

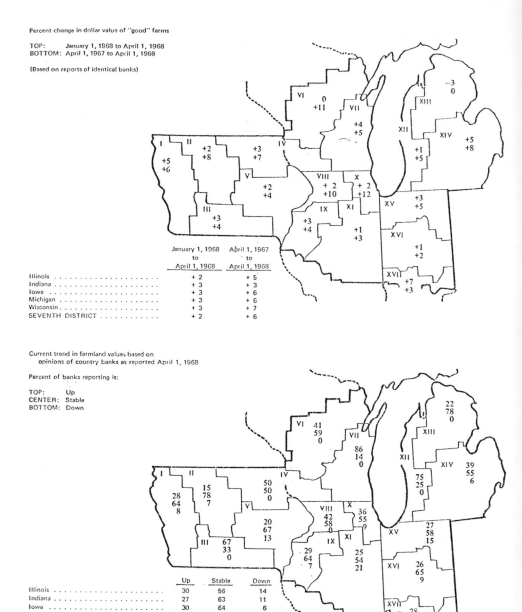

FIG. 9.6. *Farmland value report of the Federal Reserve Bank of Chicago covering areas within states in the Seventh Federal Reserve District for three-month period ending April 1, 1968. Top chart shows change in values of "good" farms, bottom chart, shows current trend in farmland values as reported by bankers.*

TABLE 9.4. Iowa Land Value Forecasts by Farm Lenders and Appraisers at Annual Soil Management and Land Valuation Conferences, 1964–68

Meeting Dates	Forecasts for Nov. 1							
May–June	1964	1965	1966	1967	1968	1969	1970	1980
				(dollars)				
1964	258	258	...	...	...	...	274	...
1965	...	273	279	...	...	...	296	...
1966	...	...	312	324	...	...	354	...
1967	...	...	...	332	344	...	374	156
1968	...	...	...	...	372	388	409	491
Actual value Nov. 1	265	293	331	362	375			

Source: Data furnished by Herbert B. Howell, Iowa State University, Ames.

in Iowa since 1964, faster than the regional or national averages have risen.

Appraisers of course are not in the business of forecasting farm values, but they are naturally interested in keeping up with value changes. They want to understand the factors which make value and cause values to change. In one sense this forecasting of values offers the appraiser an opportunity to do some logical thinking about the forces operating in the land market. There is no reason why appraisers should be perfect forecasters, however, because there are many unknowns in the farmland market which have their effect between the time of the forecast and the actual date when the forecast applies.

QUESTIONS

1. What are the reasons for collecting and publishing average values by districts within a state in addition to the state averages reported by the USDA? Use Minnesota and Iowa surveys as examples; also use any other state surveys that are available to you.
2. Would averages of actual sales be preferable to estimates? Discuss.
3. How reliable do you think the estimates of farm values by quality are? How would you explain the estimates on quality for the northwest district in Minnesota where the high-quality lands went up and the low-quality went down?
4. What other features do the annual state surveys have besides the value figures? Discuss.
5. What advantages if any does a survey like that of the Chicago Federal Reserve Bank have over the USDA and state surveys?
6. State in your own words the conclusions you draw from Table 9.4. Do these estimates of future values indicate that farm lenders and appraisers usually underestimate land value changes?

REFERENCES

State publications on farm real estate values; also other publications
carrying farm value information, such as Agricultural Letter of
the Federal Reserve Bank of Chicago.

10

SALE PRICE
COMPARISONS

In this chapter we come to the crux of the market value approach—selection of a market price for the farm appraised. In reaching this objective we consider (1) adjustments of sale prices of comparable farms to bring them up to date and (2) the tabular or quantitative comparisons of the farms sold and farm appraised. After the final judgment on market value is a discussion of the sale value tour as an aid in sharpening the appraiser's judgment skill.

ADJUSTING FOR TIME

Valid comparisons cannot be made unless the sales are all placed on a current basis. Since the objective in the appraisal is to obtain market value as of the day the appraisal is made, it is absolutely essential that the sale prices of comparable farms be adjusted so as to represent what these farms would sell for as of the day of the appraisal—not one month, six months, one year, or several years ago.

Values change and at certain times change rapidly, as we observed in the three preceding chapters on bench marks. Our problem, therefore, is to bring all sales up to the current moment of the appraisal. In doing this our best tools are the bench mark tables. The importance of having current values, at least annual figures, is readily apparent. Federal census values are useful for some purposes but not in providing directly the figures needed in adjusting sales to current sale values. The federal census, coming out only every five years and not reported for a year or more after the date of the census, gives the appraiser little direct assistance. Where the census does help is in providing county bench marks every five years. Then the annual USDA or state experiment station figures can be applied to these county census figures to bring them up to date. Since there are no other county values except those provided by the federal census, it is evident that the census plus the USDA and related annual estimates make an excellent combination for the use of the appraiser.

Use of the bench mark data to make sale price adjustments provides only an estimated correction. For instance there may be some peculiar local conditions not even countywide which affect a specific sale. Values in a district or a county may have risen 10 percent in a given year while values in a small local area where the farm is being appraised might not have risen more than 5 percent. The problem is particularly noticeable where an annual change for a state is applied to a local area or a county. In cases like this the appraiser is well aware that the statewide change is not necessarily the same as the change in the local area or county. But if there are no other more reliable estimates available, this procedure of adjustment can be used provided it is recognized for what it is. Rough approximations figured in this way can be given an additional adjustment by the appraiser after he has made inquiries to determine how the local or county land market has changed compared to the statewide change. It is important for the appraiser that he have first of all the state or district change; then on the basis of this bench mark he can make additional adjustments based on his judgment of the local situation.

The adjustment procedure is simply establishing the percentage or ratio of change between two dates and applying this percentage to the old value to bring it up to date. An example follows:

Date	District Value	Sale Price
November 1965	$230 an acre	$275 an acre
November 1968	$280 an acre	?

The adjusted sale price for November 1968 can be obtained using the following equation:

$$\frac{230}{280} = \frac{275}{X}$$

$$X = \$335 \text{ an acre.}$$

If the sale date were somewhere between the annual dates on which estimates are available, an interpolation between the reported estimates is recommended. If the sale in the example occurred in May 1965, the appraiser could use an estimate between the $202 an acre average for November 1964 and the $230 estimate for November 1965. The appraiser may know that most of the change took place between June and November of 1965, in which case he will use the $230 estimate; but if he has no knowledge of when the change occurred or if he knows that it happened evenly over the year, then he can interpolate by using an estimate halfway between or $216 which would give an estimated sale price of $356 an acre as compared to $335 an acre. The $356 estimate is based on a sale in May 1965 of $275 an acre, while the $335 estimate is based on a sale in November 1965 at $275 an acre. However, if the appraiser, after obtaining the $356 estimate for a May 1965 sale, knows that the farm and the local area in which it is located have not had as much of an increase as the district has had, he will reduce the price to a figure which in his judgment reflects fairly the change which has occurred in the local area since 1965. This appraisal judgment may reduce the price to $325 an acre.

Caution must be exercised to make allowances for any improvement changes or extensive depreciation that have occurred on the farm between the two dates. For example a field may have been tiled so that it can be cultivated and thus converted from pasture to a much higher yielding crop; or a new set of buildings or a new home may have been built in this interval. On the negative side serious erosion as the result of a severe storm may have reduced the yielding ability of a large area of the tillable land. In cases of this kind, either plus or minus, the appraiser will want to make adjustments based on his knowledge of the situation. If in the example above a new home were built in 1966, this would require an additional upward adjustment by the appraiser bringing the estimated sale price in November 1968 to perhaps $425 an acre.

This same general procedure has other uses besides bringing older sales up to date. Sometimes it is desirable, especially for tax purposes, to estimate a value at an earlier date. In this case the present sale price may be known, but in order to establish the amount of capital gain, it is necessary to estimate the price at the date it was inherited or acquired (other than by purchase). An example follows:

Date	District Value	Farm Value
November 1941	$ 75 an acre	?
November 1968	$297 an acre	$250 an acre

Using the same equation arrangement we have:

$$\frac{75}{297} = \frac{X}{250}$$

$$X = \$63 \text{ an acre.}$$

In this example the capital gain based on the district bench marks would be \$187 an acre. It would be necessary of course for the appraiser to make added adjustments for any major changes in the farm covering improvements or depreciation. These major changes would be those which were above or below the average of those taking place in the district average over the years covered. If the changes on the farm in question had been about average for the district, then the district change could be assumed to be the correct one to apply to the farm. If the farm had been allowed to run down with serious erosion and extensive building depreciation occurring, then the \$250 sale price in 1968 would represent a poorer farm physically compared to the district average than had existed in 1941, and consequently the estimated price in 1941 would have been somewhat higher than \$63 an acre with a smaller capital gain.

The compendium of county census values edited by T. J. Pressly and W. H. Scofield[1] can be an excellent aid to the appraiser estimating farm values in an early year. Since this collection of census data goes back to 1850, estimates can be made back to this year. Annual estimates by the USDA did not start until 1912, so estimates before that year have to be made by interpolating between federal census years. An estimate for 1895 for a farm in a given county would lie halfway between the census values for 1890 and 1900. In Tompkins County, New York, for example, the 1890 average value was \$47 an acre and for 1900 was \$36 which gives an average of \$41.50 for 1895.

Pressly and Scofield explain the percentage method of obtaining early estimated values using county census values as follows:

> Land appraisers frequently are confronted with establishing fair market values of properties at a specific historical date. In such retrospective appraisals, a value may be known for a current date, and a measure is needed of the relationship between the value for this current date and an earlier date. Reference to the average values for the county in which the property is located provides a first approximation of the probable change in the value of the subject property. This initial indication should then be modified to take into account special features or charac-

1. *Farm Real Estate Values in the United States by Counties, 1850–1959*, Univ. of Washington Press, Seattle, 1965.

teristics of the property that would have caused its value to have changed more, or less, than property values generally in that county. Where long periods are involved between the two dates required for the appraisal, county values often provide the only basis for measuring the probable change in values.

To illustrate, assume that a farm in Adair County, Kentucky, was sold in 1959 for $110 per acre. It had been acquired by inheritance in 1930, and a value is needed as of that date for determining capital gains for federal tax purposes. The census shows an average value of $22 per acre for this county in 1930, and $85 per acre in 1959. Dividing the 1930 value by the 1959 value indicates that values in 1930 were only about one-fourth (26 per cent) of the 1959 level. If this percentage is applied to the actual sales price of $110 in 1959, an indicated value of $28 per acre is obtained for 1930. Additional steps are required, of course, to arrive at the adjusted cost basis required by federal tax regulations, but the chief concern in such cases is usually the comparative level of market values of the real estate at the date acquired, and at the date of sale.[2]

In the example from Adair County, Kentucky, cited above, it will be observed that the two dates are in federal census years. If the years happened to be 1933 and 1958 the same general procedure and the same principle would hold but additional adjustments or interpolation would be necessary because the years do not coincide with census years. Interpolation between census years would be possible. The Adair County figures for 1930 and 1935 are $22 and $17 which would give a proportionate figure of $19 for 1933. The figures for 1954 and 1959 are $68 and $85 which gives an interpolated figure of $80 for 1958.

Another method of estimating between census years is to use the USDA state estimates to adjust for the years between the censuses.[3] The USDA index for Kentucky as a state was 33 in 1930 and 22 in 1935, or a decline of one-third. This decline was slightly more than the decline in the county, but it is reasonable to assume that the annual changes in the state and county were approximately the same. But the interesting point is that values for the state according to the USDA estimates went down from 33 in 1930, on an index basis, to 20 in 1933 and then back up to 22 in 1935. This would indicate that the 1933 estimate should be lower than the 1935 census figure. In short, the appraiser may want to modify the interpolated census figures by use of the USDA annual estimates or by other district averages available on an annual basis within the state. On the basis of the an-

---

2. Ibid., pp. 8–9.
3. For USDA annual estimates from 1912 to 1964 on an index basis, see *Farm Real Estate Market Developments,* CD–66, USDA, Washington, D.C., Oct. 1964, pp. 48–53.

nual USDA estimates for Kentucky it would appear that the Adair County estimated sale price for 1933 was approximately $15 or slightly below the average for 1935. What had happened between 1930 and 1935, as borne out by the annual USDA state estimates, was a rapid decline from 1930 to 1933 and then a slight recovery from 1933 to 1935.

A similar method of making adjustments on a percentage basis is recommended by William H. Scofield of the USDA. He uses the USDA state indexes which are reported usually twice a year for November 1 and March 1. The procedure Scofield suggests was presented in the April 1968 issue of *Farm Real Estate Market Developments,* pp. 31–32.

USING LAND VALUE INDEXES

State indexes of average value per acre are useful in many valuation problems where appraisals of specific properties are not made. Several applications of the use of price indexes are illustrated in the following examples:

1. Deriving a current value for a property having a known value at some earlier date.

$$\frac{\text{Index of current years}}{\text{Index of earlier date}} \times \text{known value} = \text{estimated current value}$$

Example: A property in Iowa purchased in March 1951 for $225 per acre

$$\frac{\text{Index for 1967}}{\text{Index for 1951}} \quad \frac{147}{84} \times \$225 = \$394$$

This implies that improvements to the property were average for the State, as was also the effect of general market factors. Adjustment of estimated value is necessary if market factors for this specific farm differ from property throughout the State.

2. Deriving the value of a property at an earlier date when the current value is known.

$$\text{Current value} \div \frac{\text{Index for current date}}{\text{Index for earlier date}} = \text{earlier value}$$

Example: Same situation as before: $394 \div \dfrac{147}{84} = \$225$

This result does not include factors usually involved in establishing an adjusted cost basis needed for determining capital gains. Depreciation on farm buildings and possible capital investments in the property must also be considered in arriving at the adjusted cost.

3. Adjusting the cash rental rate of a long-term lease in order to yield a constant ratio of gross rent to market value.

Example: A farm valued at $200 per acre in 1964 is rented for 10 years at the initial rate of $15 per acre (a gross rental of 7.5 percent of the value). Both tenant and owner agree that the rent for subsequent years will be adjusted to yield this rate of return on market value.

Rent for 1967:

$$\frac{\text{State index, 1967}}{\text{State index, 1964}} = \frac{150}{125} = 1.20 \times \$200 = \$240$$

Rent $= \$240 \times 0.075 = \$18.00$ per acre in 1967.

COMPARISON TABLES

Farm sale comparisons can be made either in a qualitative or quantitative manner. In Chapter 6 the qualitative or descriptive method was discussed with examples. In this section the quantitative or numerical rating method will be considered.

An appropriate form for presenting the quantitative measures of the comparison farms and the subject farm is a table which lists the factors which can be rated by the appraiser. Such a table was included in the appraisal of Farm *A* in Chapter 1. A similar table for Farm *X*, including Farm *A* and its appraisal in 1959, is presented in Table 10.1.

The factors listed in these tables vary with the area of the country and the type of farming practiced where the appraisal is located. For the Corn Belt a list of factors similar to those in Table 10.1 is suggested as logical and useful in sale price comparisons. A few pertinent comments regarding these quantitative factors are offered to show how important a table like this can be in summarizing significant aspects of the farms sold and the farm appraised.

FARM SIZE AND VALUE

Size, usually expressed in acres, is often a vital point in a sale price agreement. Since the mid-fifties when farm enlarge-

TABLE 10.1 Comparability Test for Farms near Farm X

	1	2	3	X
Acres	40	80	160	120
Percent tillable	all	all	93%	all
Soil rating	1.5	1.5	2.5	2.0
Corn yield estimate (bushels)	75	75	60	70
Dwelling	poor	excellent	good	very good
Service buildings	poor	good	good	good
Tax assessment value				
buildings	$900	$3,135	$2,040	$2,835
land per acre	$ 69	$ 78	$ 58	$ 68
Location	excellent	excellent	excellent	excellent
Road type	asphalt	gravel	paved	paved
Year sold	1959	1960	1959[a]	1961[a]
Sale price per acre	$462	$ 525	$ 410	$ 500
Adjusted 1961 price	$475	$ 525	$ 425	$ 500

[a] Appraised, not sold. Farm 3 is the same as Farm *A*.

ment became a potent factor in the land market, it is frequently true that a small tract will sell higher than a larger one of the same quality, and this in the face of the fact that a large operation is more profitable than a small one. What happens is that several farm owners, usually located nearby and eager to get additional acres to enlarge their existing farms, will bid high to get this extra acreage. These "farm enlargers" will bid higher than the would-be buyer who wants to farm just this acreage as a unit.

It is often true in an area of active farm enlargement that an average-sized farm will sell for more if sold in parcels than if sold as a whole. Neighboring farmers will bid high for portions of the farm, a worker from town may bid for the buildings, and the total of these bids is likely to exceed what any one buyer will pay for the farm as a whole. Experience of autioneers who have sold a given farm both ways supports this general tendency.

If a farm large enough for profitable operation comes on the market, chances are that it will not sell as high as a small nearby tract of equal productivity because the bidder cannot afford to pay a premium for the large farm. But buyers will pay a premium for the small tract because this premium is offset by the cost savings made by operating a larger farm. The purchaser of the small tract figures he can operate the additional land with no increase in machinery and with little increase in labor. In one sense there is a two-price system for land in those areas where farm expansion is taking place: one price for large economic units which do not need to be enlarged and the other price, a higher one, for small tracts being bought by farm owners who need to expand their existing farms in order to have an economic-sized operation.

The premium often paid for the small tract must be checked against the prevailing situation in the local land market. The author attended an auction sale of 40 acres which three owners of adjoining land wanted. The going market price for land of this type was around $600 an acre. The three adjoining owners carried the bidding to $930 an acre, at which point one bidder dropped out. The other two went to $990; then the second bidder quit, letting the third bidder have the farm for $1,000 an acre. A similar 40 acres equally productive and almost as well located sold at auction a week later at $560 an acre. This second 40-acre tract, which was assessed at approximately the same amount as the first one, did not have adjoining owners bidding against each other to get it. From this it is evident that the appraiser in evaluating size as a sale price factor must take into consideration the situation among neighboring farm owners.

OTHER VALUE FACTORS

The percentage of tillable land is a more significant factor in sale price than is generally recognized. Throughout the Corn Belt there existed years ago a tradition that a small permanent pasture with a stream running through it was a distinct asset to a farm, making it worth as much or even more than a farm without such a pasture. Along with tradition was the commonly expressed opinion that a good permanent pasture could be worth as much as good cropland because there was no cost in harvesting the pasture crop; the livestock did the harvesting without the farmer ever taking a machine into the pasture. However, with the coming of the professional rural appraisers who figured income value of pasture and of cropland separately, the premium for cropland was definitely established in most cases. And with the use of fertilizers, improved seed, and other new crop practices, the advantage for the tillable acres over the nontillable acres has increased. This means that it is highly important to compare the percentage of tillable land in the final tabular evaluation of the farms sold and the subject farm.

The importance of a numerical soil rating in farm comparisons is obvious. In later chapters devoted to the income approach, special treatment is given to the various aspects of the soil which affect value. Unfortunately, as the soil experts will agree, there is no easy way to measure the productivity of a soil. It is not possible, at least not yet, to take a few samples of soil from a farm, analyze them, and then come up with a numerical rating of productivity for valuation purposes. But it is possible, as we will see in later chapters, to make some meaningful estimates that are definitely helpful in rating the soil on a farm.

One of the most helpful procedures for soil rating in the Corn Belt is the rating in terms of the principal crop which generally has been corn but which now includes soybeans because of their importance in total crop value.

Along with the soil and crop yield ratings should be included an estimate of the erosion hazard and a pasture rating. The importance of this information in farm comparisons becomes especially clear if two farms are included with about the same crop yields but with one farm having a severe erosion hazard and one farm having a very good permanent pasture and the other a poor pasture. In the case of these erosion and pasture ratings the information, where it is applicable, is not of a numerical character but rather a breakdown into qualitative groupings.

The information on the house and service buildings also tends more to ratings in quality than to numerical ratings. But it is important to have these rating judgments because the appraiser needs to pin these factors down as definitely as possible in evaluating them for his final judgment on market value. A good example is a description of the house and service buildings including information on size, age, physical features, and attractiveness followed by a rating in which the house and service buildings are each placed in a definite grouping. Such a grouping system might include five letters from A through E with a plus and minus possible in the case of each letter, giving a total of 15 ratings.

The use of property tax assessment data will vary. Generally such use is on the increase in view of the increased effort assessors are making to provide systematic qualitative and quantitive data for their farm assessments. Soil ratings and building information are becoming more common. Since the assessment has a direct effect on the property tax paid, it is a good practice to use assessment comparisons for what they are worth; that is, the assessment comparisons will be included in the appraisal but will be discounted by the appraiser if he considers them not helpful or reliable. In making this judgment the appraiser naturally will have to be familiar with the assessor's work in the area of the farm being appraised.

The rating of location should be based not only on the distance to town but on the existence of other favorable or unfavorable factors connected with the specific location of the farm. Nearness to town does not mean much unless some information is included as to how fast the town is likely to grow in the direction of the farm being considered. In the case of Farm *A* the location includes proximity to a new flood control dam which has been approved but is still in the planning stage.

The condition of the farm road may be a highly critical factor, but where nearly all the farms are on hard-surfaced roads there may be little distinction based on road type. This factor will be given special attention in a later chapter on nonincome factors.

The final items in the farm comparison table are date of sale, sale price, and adjusted sale price. These may be included in the sale price record or in connection with the comparisons as in Table 10.1. There is good reason for including them in the comparison table because the adjusted sale price at the bottom of the table is the summarizing figure giving the current sale price estimate of each of the comparable farms and provides a spot where the appraiser can put the market price estimate arrived at for the farm appraised. Thus the appraiser and those who read the appraisal have one table which provides the major factor ratings and current sale price estimates for the comparable farms and the subject farm.

A method of comparison favored by some appraisers is to make plus and minus adjustments for each of the factors for each of the farms sold. The plus and minus adjustments represent for each factor the amount that the appraised or subject farm is better or poorer than the farm sold. If the appraised farm is poorer than the farm sold by $20 an acre on all of five major factors, and if the farm sold for $400 an acre, then the indicated value of the subject farm is $300 an acre. If the appraised farm is better by $20 an acre on three factors and poorer by $30 an acre on the other two factors, then the indicated value of the subject farm is the same as the farm sold—$400 an acre. This method has merit in being specific but presents difficulties in the making of such precise dollar measurements. As it becomes possible to make and to support reliable judgments in dollar figures for individual factors, this procedure will become common and useful.

FINAL JUDGMENT ON MARKET VALUE

The appraiser in making his judgment on market value can rely on a logical step-by-step procedure which leads to this final result. (1) The bona fide sales are reviewed. (2) Each of the sales selected for comparison is analyzed. (3) Bench mark tables are consulted and studied to check existing values in the county, district, and state for various years. (4) Ratings for the comparable farms and the farm appraised are set side by side in a table with adjustments made for time so that all sales are brought to a current basis. (5) One or two of the comparable farm sales are selected as being most nearly like the farm appraised. This fifth step is not always possible, but where it is the appraiser will find it helpful in giving him confidence in his final conclusion on his estimated market value of the subject farm.

Unfortunately there is no mathematical formula or easy rating system to give the market value estimate that the appraiser wants. It has to be a judgment, despite the emphasis that is being placed on mathematics, statistics, and computers. The reason for the necessity of relying on judgment is not hard to find. The factors which determine value are not subject to precise measurement so that the basic ingredients for a mathematical solution are not present. This lack of preciseness in factor measurement can be seen in evaluating erosion hazard, location, and quality of buildings. All of these can be rated but the ratings involve judgment rather than precise calculations.

Another feature of the land market which makes it difficult to handle by formula is the reaction of individuals in the market. These individuals, buyers and sellers, do not use precise calculations in forming their market value estimates; instead they measure the major factors in a rather crude way and then make up their minds on a buying or selling price. And as conditions change their reactions change, resulting in a new market situation which the appraiser endeavors to unravel and to understand.

One of the features of the land market which the appraiser must understand is that some sales are at prices above the market, some at the market, and some below the market. But how is it possible to say that a certain farm sold above or below the market, especially when sales are what make the market? There is general agreement among farm operators, real estate brokers, farm lenders, and appraisers on a going price for land of a certain quality in a certain area, even though it is based on sales below the market, at the market, and above the market. There is usually a reason for the sale that is below or above the market. In one instance that the author noticed, the seller was in a hurry to sell and did not wait for a better price he might have received by putting the land in the hands of a broker and giving him some time to find a buyer. Another case sold far above the market because the buyer in the deal had to have a certain farm, and he just kept bidding until he finally bought it at a price a full $100 above the going market for land of this type. It is this process of analyzing sales which makes it possible to establish the existence of a market price for a certain quality of farmland. With this in mind it is then not difficult to say that a given sale is above, below, or right at the market.

The appraiser's task, however, is not an easy one even though there is an existing market price level in the minds of those dealing with the market for land. Often there are not enough sales to make a good market price in a local area. Or there may be several sales but they do not represent the farm being appraised; the sales may be of relatively good land on good roads, while the farm appraised is of below-average productivity and poorly located. In fact there may be few farms like it so there has been no market established recently for this type of land.

Then there is the perplexing problem of estimating a market price for land in a changing situation. It is known that land has gone up recently in the surrounding area but not in the area where the farm is being appraised. There have been several sales, but either they have not been representative of the subject farm or there have been peculiar characteristics connected with the sales which make them unreliable as good market estimates. The appraiser has from nearby comparable sales no support for an increase in market value, but he knows that values have gone up in the surrounding areas and that there is no valid reason for values not going up in the area of his appraisal. The appraiser in this situation is forced to visualize what a normal sale of land like that being appraised would sell for under present conditions. It is not an easy task but one the appraiser may be called on to handle. In making his final estimate of market value in this case, he will use all of the factual evidence at his command and rely on his best judgment for the value.

The small size of the local real estate market is one of the hard realities that the appraiser has to face. The prices of crops like grain, fruit, or vegetables may be unstable but they are usually quoted for a wide area with variations depending on transportation, surplus and deficit conditions, and the like. If the price of wheat, for example, gets out of line in a local area compared with surrounding areas, the local farmers soon find out about it and the local price is brought in line by farmers selling in the surrounding higher-price areas. But with land the market is much more isolated and local. Sellers are limited to those owning farms in the area and, to some extent although not entirely, buyers are often limited to local farmers and others living nearby.

Philip M. Raup in his studies of the land market in Minnesota has noted this tendency for the land market to be local and somewhat isolated. Bambenek and Raup in a report on the 1967 Minnesota Rural Real Estate Market had this to say about the local land market:

> . . . in a geographic sense, there is no statewide market in which land is bought and sold. There are instead hundreds of local markets, in which buyers and sellers may be handicapped by lack of information about relative values, trends in sale prices, or available alternatives. It is often observed that the land market is an imperfect market. The data underline the restricted nature of the market in the areas where farming predominates. Perhaps the most important consequence is the small number of potential buyers for any given tract.[4]

The appraiser in his market value estimate has to reflect the local situation which Raup and Bambenek emphasize. If

---

4. J. V. Bambenek and P. M. Raup, *Minnesota Rural Real Estate Market 1967*, Univ. of Minn. Econ. Study Rept. S 68-1, Feb. 1968, p. 22.

there is a strong demand for rural residences by commuters, he has to include this in his evaluation. If there is an unusually small demand for land to expand existing farms, this must become part of the appraiser's factual inventory.

In the end the appraiser weighs the evidence, giving each factor its proper consideration in its effect on the market value. The decision is in the last analysis a judgment of the market for the farm being appraised. After the appraiser has reached a tentative conclusion he may find it helpful to compare this value with values he has reached on other farms he or his colleagues have appraised, and with other farms he knows that have sold outside the immediate area of the subject farm. Then he goes over the comparison table once again, concentrating on the farm sale or sales that most nearly coincide with the farm being appraised. If still in doubt he goes back over the steps he took in reaching his tentative decision. After checking and rechecking he makes his final market value estimate.

SALE VALUE TOURS

Practice in judging market values is a desirable objective. There are so many unusual and constantly changing situations in the land market that appraisers are well advised to make it a habit to study individual farm sales over and above the comparable sales required by their appraisals.

One enjoyable type of sale price study is the sale value tour. These tours are frequently organized for groups of appraisers, real estate brokers, lenders, or students studying appraisals. The enjoyable feature is the competitive spirit provided by allowing each member of the tour to estimate the market price of the farm at the time it sold, before the leader gives out the actual selling price. For this part of the tour to be successful it is necessary of course that the selling prices of the farms not be known except by the leader.

The leader has the responsibility of providing the group with information about each farm visited on the tour, including the type of information listed in Table 10.1. If there is time it is also desirable for the group to walk over at least a part of the farm and see any factors which are especially important in their bearing on the farm's selling price. The discussion among the members of the group and the estimation of selling prices can be useful as well as stimulating for the appraiser and others who are interested in improving their skill in estimating the market value of individual farms.

QUESTIONS

1. Select a list of comparable farms that have sold recently near a farm you are appraising or have appraised as to income value. Make a map showing the location of the farms similar to Figure 10.1. Summarize courthouse information and information from verification of the sales. Determine a sale value estimate for farm being appraised and give reasons in support of your figure.

2. Sale value tour. Instructor may find one of the best means of illustrating comparable sale approach and prevailing market values is a tour of an area in which a farm is being or has been appraised and in which several farms have been sold recently. Before setting out on the tour each of the students can be given a sheet similar to Figure 10.1 showing a map of the tour, a list of the farms to be viewed, and the dates when they were sold. Care

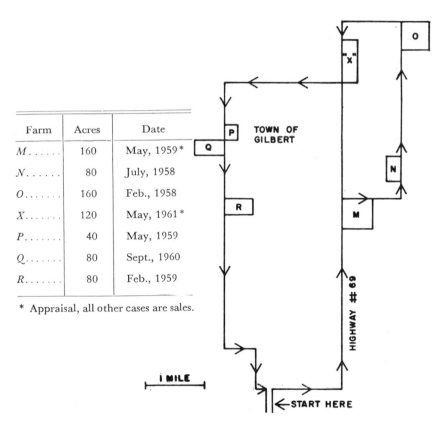

Farm	Acres	Date
M......	160	May, 1959*
N......	80	July, 1958
O.......	160	Feb., 1958
X.......	120	May, 1961*
P.......	40	May, 1959
Q.......	80	Sept., 1960
R.......	80	Feb., 1959

* Appraisal, all other cases are sales.

FIG. 10.1. *Map locating the farms visited on tour.*

should be exercised to make certain that no advance information is given out on sale prices for comparable farms, because one of the high points of the tour is having each of the students make his own estimate of the sale price of the farms.

A plan which has worked well on these tours is to divide the class into groups of 5 or 6, each group operating as a separate committee. After a comparable farm has been viewed and certain factual data given out by the instructor, each committee member makes his sale value estimate and hands it in to the committee secretary who reports it to the instructor. The instructor lists the committee estimates, reports the class average, and then, *and not until then,* announces the actual sale price.

The student on the tour keeps a card on which he records four estimated sale values for each comparable farm; first, his own sale price estimate; second, the average for his committee; third, the average for all the committees in the class; and fourth, the actual sale price. The advantage of this type of tour is that the student gets a chance to test his own ability to estimate sales prices in competition with the other members in his committee and with the other committees in the class. The result can be a spirited discussion in which all participate.

A tour of this kind for a group of bankers attending an Agricultural Credit School in the summer of 1961 indicated that the bankers did not realize that values were as high as they were in the Farm X area. However, one of the bankers came up with an overall average of $440 for the seven farms included compared to the average of $439 for the actual sale price and appraisal figures. The record of this top-scoring student follows:

TOUR RECORD CARD

Farm No.	Student Estimate	Committee Estimate	Class Estimate	Actual Price
M[a]	$375	$374	$349	$410
N	425	427	379	425
O	310	301	293	300
X	495	476	476	500
P	480	451	419	462
Q	525	498	483	525
R	470	464	486	450
Average	$440	$427	$412	$439

[a] Farm M is the same as Farm A in Chapter 1.

REFERENCES

Appraisal of Real Estate, 5th ed., American Institute of Real Estate Appraisers, Chicago, 1967.
Encyclopedia of Real Estate Appraising, Edith J. Friedmann, ed., Prentice-Hall, Englewood Cliffs, N.J., 1959.
Fischer, Loyd K., Burkholder, Richard, and Muehlbeier, John, *The*

Farm Real Estate Market in Nebraka, Nebr. Agr. Exp. Sta. Bull. 456, 1960.

Journal of American Society of Farm Managers and Rural Appraisers, Vol. XXIV, No. 1, Apr. 1960—especially articles by Bert P. Campbell and Charles H. Everett.

Rural Appraisal Manual, 2nd ed., American Society of Farm Managers and Rural Appraisers, DeKalb, Ill., 1967.

Stewart, Charles L., and Voelker, Stanley W., "The Mechanics of Land Transfer," *Land: The 1958 Yearbook of Agriculture,* USDA.

Wendt, Paul F., *Real Estate Appraisal,* Holt, Rinehart, & Winston, New York, 1956.

3

INCOME APPROACH

THE INCOME APPROACH to valuation requires estimates of income and expense which are used to obtain net income. Net income can then be capitalized to give an income value. In estimating income and expense either the landlord or the owner-operator method may be used. In both cases the major factor involved in obtaining gross income, assuming productivity has been estimated, is the choice of prices for the products sold.

Problems connected with the estimation of gross and net income are tackled first. These are followed by consideration of the capitalization process by which both an income value and an estimated sale value are obtained. The income value is based solely on the farm's estimated income and on an investment capitalization rate or rate of return. Nonincome features are not included in this value. The capitalized sale value is based on the rate of return received on farms similar to the farm being appraised; since the sale value of the similar farms is used in obtaining the rate of capitalization, the resulting value does include nonincome features.

11

INCOME AND PRICE
ESTIMATES

IN THIS CHAPTER the two methods of obtaining net income—landlord and owner-operator—are considered, followed by a discussion of price estimates used in obtaining gross income. In the next two chapters expense estimates and the capitalization process are considered.

LANDLORD AND OWNER-OPERATOR
METHODS COMPARED

The differences between the landlord and owner-operator income methods are significant. Landlord income estimates represent the returns received by the landlord for the annual use of land and buildings. Owner-operator estimates represent the amount which the owner has left as a return on land and buildings after all expenses have been deducted. The expenses to be deducted include not only operating expenses but an allowance for unpaid family labor and interest on the owner's investment in equipment. Either method can be used by the appraiser regardless of the plan being followed on the farm; the landlord method can be applied to a farm actually operated by the owner and vice versa. The landlord or rental-income method is usually preferred because the expense deductions on the owner-operator basis are so difficult to estimate. Furthermore, a decided advantage of the rental method is the existence of a competitive market in which cash rents and rental contracts are established by renters and landlords. Owner-operator income is often estimated, however, not so much to determine land value as to determine the likelihood of the owner's having sufficient income to pay the interest on a mortgage loan.

Comparisons of landlord and owner-operator income methods are available in farm bookkeeping records published by some of the state agricultural experiment stations and extension services. Unusually detailed and complete records for Illinois have been used for the comparisons presented in Table 11.1.

Landlord expenses, it will be noted in Table 11.1, are less than one-fourth of tenant expenses in this particular example. Purchases of livestock and feed by the tenant amounting to $7,026 accounted for a large part of the difference, but even if this livestock item plus the livestock expense of $524 is omitted, the remaining tenant expenses are still more than twice the amount spent by the landlord. Besides, it is not always easy to omit the livestock operations on a farm because they are mixed in with other operations including the use of pasture land and the use of the buildings.

Owner-operator expenses for a group of farms similar to those in Table 11.1 would be approximately equal to the total

TABLE 11.1.   Landlord and Tenant Expenses and Returns for a Group of Medium-Sized Crop-Share Rented Farms in Northern Illinois
(Number of farms in group—26; average size—231 acres; soil rating—average)

	Landlord	Tenant	Total
Expenses			
Taxes (property)	$ 1,414	$ 208	$ 1,622
Seed	235	389	624
Fertilizer	1,213	1,373	2,586
Other crop expense	222	525	747
Hired labor	. . .	466	466
Gas, oil, etc.	. . .	1,352	1,352
Machinery repair, etc.	. . .	2,243	2,243
Livestock purchase & feed	. . .	7,026	7,026
Livestock expense	. . .	524	524
Depreciation	952	2,427	3,379
Miscellaneous	425	563	988
Total	$ 4,461	$17,096	$21,557
Gross income	$10,185	$24,651	$34,836
Net income	$ 5,724	$ 7,555	$13,279
Investment per acre	$ 514		
Return per acre	4.8%		

Source: Franklin J. Reiss, *Landlord and Tenant Shares 1966,* aerr No. 88, Dept. of Agr. Econ., Univ. of Ill., Urbana, Sept. 1967, pp. 20–22.

of the landlord and the tenant, or $21,557 as shown in the last column in the table. Consequently a comparison of landlord and owner-operator expense estimates using Table 11.1 would be a comparison of the first and last columns.

Landlord operation has three main advantages over owner operation as indicated by Table 11.1. First, there are fewer types of expense: the landlord has only six major types of expense while the owner-operator has 11 expense groupings plus several others that will be mentioned later. Second, the landlord items are much easier to estimate than are the tenant or owner-operator items. Crop expenses which are usually divided about evenly between landlord and tenant are generally large expenditures made at one time. Property taxes also are a large item paid in a lump sum either once or twice a year. But when an effort is made to estimate all of the fuel bills for the tractor and the percentage of auto use charged to farm business, the task is much more difficult. The same can be said of machinery repair items which may vary widely from year to year.

Third and finally, we come to the expenses most difficult of all to estimate for owner operation—deductions for the operator's investment, his unpaid labor along with that of the family, and his return on management. Let us assume that all we have are the figures in the last column of Table 11.1. The difference between the total income of $34,836 and the expenses of $21,557 gives a net income of $13,279. From this $13,279 we must deduct interest on the investment of the operator in feed, livestock, and machinery and also deduct for unpaid labor and management.

The average investment in feed, livestock, and machinery was $23,760. A 7 percent return on this total means a deduction of $1,663 to cover the cost or the amount that the operator could have received by having this investment elsewhere bringing in a 7 percent return. How much should be estimated to allow for unpaid labor and management? If we estimate $500 a month or $6,000 for the year, this gives a total deduction of $1,663 plus $6,000 or $7,663 from the net income of $13,279.

The net income which can be allocated to the land and buildings under the assumptions made is $5,616 as compared to the $5,724 as shown in the landlord column in Table 11.1. But the important point should be clear, namely, that the estimates for return on investment and for unpaid labor and management are subject to a great deal of variation. If we had used an 8 percent return on investment and a labor and management cost of $600 a month the net remaining to the real estate would have been only $4,178 instead of $5,616, a reduction of approximately 25 percent. In the light of this situation the appraiser will find it more satisfactory in many cases to use the landlord method, applying it to owner-operator as well as to rented properties.

Tenancy, even though it may be infrequent, is useful to the appraiser in judging net return on other farms. Any rental agreement between a landlord and a tenant, whether it be for a whole farm or a part, is grist for the appraiser's mill. Every lease represents an agreement between owner and tenant as to division of the crops between them or the cash sum the tenant shall pay for use of the land. Whether the rent is a share of the crops or a sum of money, both parties have agreed upon an amount which represents the use value of the farm for an annual period. There are of course stock-share leases where the landlord contributes livestock and equipment in addition to land and buildings, but in the main the rental agreement establishes a figure which may be assumed to represent the yearly worth of a farm. This annual rent presents an excellent shortcut to farm value which is, after all, the right to receive the annual rental returns not for one year only but perpetually.

LANDLORD METHOD

The rental approach to net income places major emphasis on the crop pattern, avoiding to a large extent livestock operations which are generally carried on independent of the land. The effect of livestock operations is felt chiefly through buildings and pasture. An example of a crop pattern was provided in the appraisal of Farm A in Chapter 1. Another one, presented in Table 11.2, covers estimated gross income, expenses, net income, and a capitalized value for a typical farm that might be found in some area in Oklahoma. This Oklahoma example was presented in a bulletin as a guide to farm purchasers with the suggestion that they prepare a table of this type for any farm they are thinking of buying.

The first major decision in the income valuation is the crop pattern. In Table 11.2 the problem is the allocation of the 240 acres in the farm to various uses. If the pasture and hay land are permanent (that is, not tillable), there is little flexibility or choice in the use to be made of this part of the farm, as with roads and waste. With tillable areas, however, there may be considerable flexibility, the decision of use finally being determined by what would be the typical crop pattern for this farm. One of the difficulties appraisers are having in deciding what is typical

TABLE 11.2. Example of Income Appraisal from Oklahoma

Landlord Income:

Crop	Acres	Yield[a]	Rental Share	Landlord Share	Price[a]	Gross Income
Corn	25	21.0 bu.	⅓	175 bu.	$ 1.25	$ 219.00
Cotton	10	210 lbs.	¼	525 lbs.	.27	141.75
Alfalfa	10	2.5 tons	⅓	8⅓ tons	22.00	183.00
Hay	20	1.0 tons	⅓	6⅔ tons	12.00	80.00
Wheat	60	14.5 bu.	⅓	290 bu.	1.80	522.00
Pasture	110	20 AU[b] for 8 months = AU months at 1.50[c]				240.00
		plus 30 AU months for small grain at $6.00				180.00
Roads & waste	5	None				
TOTALS	240					$1,565.75

Landlord Expenses:

Taxes 240 acres @ $1.00	$240.00	(from county records)
Upkeep & depreciation	120.00	(based on improvements)
Insurance	22.00	(at $0.80 per $100 of value)
		(check local rates)
Crop costs paid by landlord	30.00	412.00
TOTAL	$412.00	

Gross income — Expenses = Net income $1,153.75

$1,153.75 capitalized at 3.3 percent $= \dfrac{973.75}{.033} = \$34{,}962$ which is the income

value of the farm. In other words, if a person paid $34,962 for the farm he could expect a 3.3 percent return on his investment. Five percent is the rate sometimes used by appraisers, but it is well known that many farms purchased at current market prices yield no more than 3 or 4 percent. Such buyers probably expect land prices to continue to rise so that in the long run they will realize a more adequate return on their investment.

Source: Jeffrey, Maynard, and Parcher, *Suggestions for Buying or Selling a Farm*, Okla. Circ. E-786, Okla. St. Univ., Stillwater, 1966, pp. 11–12.

[a] What a difference would occur if you estimated yields and/or prices 20 percent higher or 20 percent lower! Land value is greatly affected.

[b] AU = animal units.

[c] In some areas so much per acre per year is charged for the pasture. Use such a figure when it is available. In addition, the landlord typically gets ⅓ of the rent for small grain pasture. Also in some areas good buildings, especially good housing, will command additional cash rent.

is that crop patterns often are undergoing change, frequently to a more intensive use. Heavier cropping, however, is not all gain because expenses, especially fertilizer, rise at the same time. From this it is evident that the appraiser in choosing a crop pattern has to be constantly on the alert to keep in line with prevailing trends and changing technology.

Choice of crops is followed by estimated yields and the setting aside of the landlord's share as indicated in Table 11.2. Problems connected with yield estimation as well as with crop patterns are covered in later chapters devoted to the appraisal of farm productivity.

DETERMINATION OF RENTAL SHARE

Difficulty may be experienced in deciding on the proper rental figures to use even where tenancy is common. A first requirement is a general knowledge of rental terms within an area, including not only the average rental share of the crop or price per acre but also the extent of the variation above and below the average for farms of different quality. Uniformity in the crop-share division and only slight differences in cash rent per acre are customary in some communities. Competition in other communities is keen enough to bring about a close adjustment between the quality of the land and the rental charges. The difficulty is to decide on the difference, if any, between what is common in the community and what is appropriate in calculating value. As an extreme example, landlords in one community were able to ask unusually high rental terms because farmers from drouth areas were coming into the neighborhood to rent farms, causing active bidding for the farms. The rent was bid so high in some cases that it was difficult for the tenant to pay it. The reverse situation occurred in drouth areas where land was rented for an amount not even equal to the taxes. Unusual situations like these must be corrected by the appraiser before the rental terms used will reflect the long-period earnings of land and buildings.

The rental share to use, then, is that which is most likely to represent the long-period earnings of land and buildings. The appraiser desires a long-period division of earnings between landlord and tenant, not a temporary crop division which may be heavily weighted in favor of either the landlord or the tenant. It would be much simpler of course if the rental terms common at the time could be used invariably. Some may argue that this should be done because it keeps the appraiser close to reality. A long-range view is essential, however, since value represents the future prospects for farm earnings, not present earnings alone.

Once the rental share has been determined, the next step is applying it to the cropping system. If the rental terms are one-third of all grain to the landlord, then one-third of the crop production can be set aside as the landlord's share. Where the share varies with different crops, the appropriate division can be made and, as shown in Table 11.2, the annual estimated amount of each crop going to the landlord can be set out in a column designed for the purpose.

ESTIMATING PRICES FOR FARM PRODUCTS SOLD

What prices should an appraiser use in estimating gross income? Prices at the time of the appraisal would be easy to use but of doubtful value because they may be low due to bumper crops or a business recession or they may be high due to a drouth or a business boom. Nor is it possible to avoid an answer if a net income value is desired, because income value is based on gross income and gross income is based on the prices received for products sold.

The price problem will be simplified if it is divided into two parts:

1. Variations in price from year to year.
2. Variations in price from one region or area to another.

The first will be called price fluctuations over time and the second geographical or area price variations. Both types of variation are involved in an appraisal, but of the two, the time variable is more likely to give trouble.

PRICE FLUCTUATIONS OVER TIME

The set of prices used in valuation must represent the appraiser's best estimate of the prices a farmer is likely to receive for products he raises. The question immediately arises as to how far into the future it is necessary to figure. The most important year is this year, and after that the most important is next year, and so on, each succeeding future year being less important than the preceding, a fact illustrated by the computation of interest. The right to receive $10 today is worth more than the right to receive $10 a year from today, and much more than the right to receive $10 ten years from now. The appraiser is far more concerned with the probable prices received for farm produce in the next 20 years than with the prices in effect from 40 to 60 years from now.

Various procedures are followed in selecting farm product prices used in appraisal. A fixed average for a given period of years in the immediate past—5, 10, 20, or more years—might be used. Instead of a fixed average, a simple or weighted moving average might be used. Finally, an estimated or "normal" price might be used. An estimated price is merely the appraiser's best forecast of the price likely to prevail. A "normal" price is one that a lending agency or some group selects as being the one which will approximate the midpoint between the high and the low prices of the future. The appraiser, therefore, has a wide choice in methods to follow in arriving at the prices he uses in estimating gross income for appraisal purposes. Before making the choice, a study of the different methods will prove worthwhile. The first method to be studied will be fixed and moving averages гABLE 11.3.   Corn Price Averages

A. United States Seasonal Price Averages by Years					
Crop Year	Price per Bushel	Crop Year	Price per Bushel	Crop Year	Price per Bushel
1931	$ .32	1946	$1.56	1961	$1.08
1932	.32	1947	2.16	1962	1.10
1933	.52	1948	1.30	1963	1.09
1934	.81	1949	1.25	1964	1.15
1935	.65	1950	1.53	1965	1.16
1936	1.04	1951	1.66	1966	1.24
1937	.52	1952	1.53	1967	1.04
1938	.50	1953	1.49	1968	
1939	.57	1954	1.43	1969	
1940	.62	1955	1.34	1970	
1941	.75	1956	1.29		
1942	.92	1957	1.12		
1943	1.12	1958	1.12		
1944	1.09	1959	1.04		
1945	1.27	1960	1.00		

B. Fixed Period Averages					
Number of Years	Period	Price per Bushel	Number of Years	Period	Price per Bushel
5	1941–45	$1.03	15	1941–55	$1.36
5	1946–50	1.56	15	1946–60	1.39
5	1951–55	1.49	15	1951–65	1.24
5	1956–60	1.11	15	1953–67	1.18
5	1961–65	1.12			
5	1963–67	1.14	20	1946–65	1.32
			20	1948–67	1.25
10	1941–50	1.29			
10	1946–55	1.53	25	1941–65	1.26
10	1951–60	1.30	25	1943–67	1.29
10	1956–65	1.12			
10	1958–67	1.10	40	1928–67	1.05

Source: *Agricultural Statistics,* published annually by USDA. Prices for individual years represent the seasonal average price per bushel received by farmers, the 1967 price of $1.04 being the price received by farmers for the 1967 crop between Oct. 1967 and Sept. 1968.

using seasonal average prices for corn, shown in Table 11.3, as examples.

Fixed Period Averages

Fixed period averages take out the wide swings in year-to-year fluctuations. In the 37-year period from 1931 through 1967 the extremes in corn prices were a low of 32 cents a bushel in 1931 and 1932 and a high of $2.16 a bushel in 1947 (Table 11.3). The number of years to include in an average, whether 5, 10, 15, 20, 25, or more years, presents a puzzling problem. The difficulty is that there is no right or wrong number of years, and there is a difference in the price depending on the number of years included in the average as indicated in Part B in Table 11.3.

Fixed averages of 5 years, as shown for the years 1941–67, are too short to take out the major price cycle swings, the variation in this period going from a low of $1.03 a bushel for the 1941–45 average to $1.56 a bushel for the 1946–50 average. The 10-year average reduces some of the variation but not all by any means. Note for example the increase from $1.29 to $1.53 and then back down to $1.30. The 15-year average takes out more of the variation, and the longer averages still more. The 37-year average, composed as it is of low prices in the depression years of the thirties and high prices in the war and postwar years of the forties, provides a good average for the late sixties. But is this reasonable or purely a coincidence? Certainly the situation in the late sixties had no direct relation either to the situation in the thirties or that in the forties or to an average of the two.

Choice of a long-period average leaves the question of how long a period to include. If the period were lengthened to include World War I and the boom that followed, a higher price would be obtained. If the period were extended still farther back to 1901 this would lower the average price to 92 cents a bushel. A preference for long-period averages gives increasing weight to the distant past and less to recent years, a preference that obviously can be carried too far. There is no easy way, unfortunately, of telling just how long a period the average should cover. It would be difficult to prove that the appraiser using a 40-year average of $1.05 is right and that another using a 25-year average of $1.29 a bushel is wrong. One suggestion is that if a fixed average is used it should cover both high and low swings in price fluctuations. It is not always easy to tell what is high and what is low, but at least this is a reasonable approach to an answer.

Moving Averages

Whatever average is selected may become a moving average as the appraiser uses it over time. Each year a new figure is added to the average and an old figure at the other extreme dropped off. An example of 10- and 20-year simple moving averages during the 1962–67 period is shown in Table 11.4. The 10-year average for 1953–62 was $1.20 a bushel; moving to 1954–63, the price for the new year 1963 of $1.09 is added, and the price for the old year 1953 of $1.49 is dropped. The net change is a reduction of 40 cents for the 10-year period or an average annual decline of 4 cents, bringing the average down from $1.20 to $1.16 a bushel.

TABLE 11.4. Simple Moving Averages of Corn Prices (Based on data in Table 11.3)

10-Year Average		20-Year Average	
Period	Price per bushel	Period	Price per bushel
1953–62	$1.20	1943–62	$1.32
1954–63	1.16	1944–63	1.32
1955–64	1.13	1945–64	1.33
1956–65	1.11	1946–65	1.32
1957–66	1.11	1947–66	1.30
1958–67	1.10	1948–67	1.25

Similarly with the 20-year average in moving from 1962 to 1963, the price for 1963 is added and the price for 1943 is dropped.

An embarrassing situation may occur in using moving averages when changes may go in opposite directions. For example the 10-year moving average for 1955–64 was down 3 cents a bushel from the previous year, while the 20-year moving average for 1945–64 rose 1 cent a bushel from the previous year. If an appraiser had been using a 10-year moving average during the years 1963 and 1964 he would have been faced with rising corn prices and a declining moving average. Situations of this kind happen partly because an equal weight is given to all years—to the distant year as well as to the new year that is added. This kind of situation can be largely avoided by giving more weight to the current year. There is a good reason behind this type of weighting in that the current year is usually recognized as being more significant than a distant year in explaining current farm real estate values.

Weighted Moving Averages

The weighted moving average is supported by the fact that in forecasting prices the present price is more important than the price 10 years ago or even one year ago. According to this reasoning, the price this year, although most likely to indicate the future trend, may in some years be misleading either because of surplus or drouth or because of external events such as depression, prosperity, or war. Consequently, in addition to the most recent year other years in the past are included but with less weight.

Different systems of weights may be used, but in this discussion a uniform decrease in weight is given each preceding year. An example showing how the weights are used in calculating the average is presented in Table 11.5. Moving averages calculated on this 10-year basis are shown for the years 1962–67 in Table 11.6.

TABLE 11.5. Calculation of a 10-Year Weighted Moving Average of Corn Prices (Based on data in Table 11.3)

Year	Price	Weight	Total
1953	$1.49	1	$ 1.49
1954	1.43	2	2.86
1955	1.34	3	4.02
1956	1.29	4	5.16
1957	1.12	5	5.60
1958	1.12	6	6.72
1959	1.04	7	7.28
1960	1.00	8	8.00
1961	1.08	9	9.72
1962	1.10	10	11.00
Total		55	$61.85
Average			$ 1.12

A comparison of the weighted with the simple moving average on a 10-year basis for the 1962–67 period shows the tendency of the weighted to move closer to current prices (compare Tables 11.4 and 11.6). During the 1962–67 period the weighted average varied only between $1.10 and $1.13 while the simple average went from $1.10 to $1.20.

All of the moving averages—simple and weighted—are mechanical. They are calculated by formula without any judgment being used once the decision is made on the type of average, number of years included, and weights if any. This mechanical feature is generally regarded as a weakness of moving averages for use in income valuation. Even though a formula is devised that would have worked beautifully during the last 20 years, it may work poorly if prices start behaving in an entirely different pattern from the previous 20 years. This has led many appraisers and appraisal organizations to use estimated or "normal" prices in figuring gross income in their valuations.

TABLE 11.6. Weighted 10-Year Moving Averages of Corn Prices, 1962–67[a] (Based on data in Table 11.3)

Period	Price per Bushel
1953–62	$1.12
1954–63	1.11
1955–64	1.10
1956–65	1.11
1957–66	1.13
1958–67	1.12

[a] Weights: 10 for most recent year to 1 for most distant year, same procedure as in Table 11.5.

Estimated Prices

Estimated or "normal" prices as previously defined are nothing more than an intelligent judgment of what prices are going to average in the future. By the future is meant the next 10 to 15 years. This judgment is based on the one hand on an analysis of prices in the past in relation to such factors as farm technology, population growth, food habits, government programs, and the like; and on the other hand on these and other forces as they relate to the future (see Figure 11.1). In short, future price prospects are based on what has happened and on what is likely to happen.

Averages of past prices can be helpful in formulating a forecast of the future. Consequently, even though price averages similar to those in Tables 11.3, 11.4, and 11.6 are not used directly in figuring gross income, they serve a useful purpose in providing factual background for the reasonable estimate of what prices are likely to be in the future.

This forecast or estimate of prices has been dressed up with different names. During the depression of the thirties when the Farm Credit Administration was lending funds to prevent farm mortgage foreclosures, the "normal" price concept was widely accepted as a way to obtain income values above the current de-

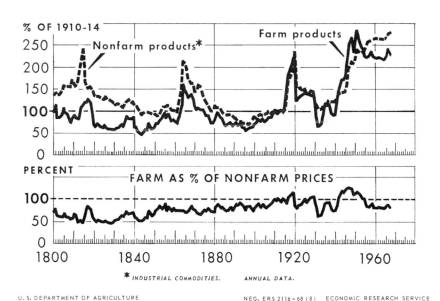

FIG. 11.1. *Wholesale prices of farm products. (Courtesy USDA.)*

pressed level. A "normal" corn price of 65 cents a bushel when corn was selling for only 35 cents made possible relatively high values and loans which helped financially distressed farm owners. Similarly, when prices get abnormally high, a "normal" price should be estimated that is well below the current high level so as to give conservative land values and loans which will prevent farm owners from getting caught with debts out of line with the long-term earning power of their land.

During the early fifties the American Society of Farm Managers and Rural Appraisers in cooperation with the Doane Agricultural Service adopted a "standard" price procedure which was pioneered by the Appraisal Institute of Canada. This procedure included a committee which recommended prices for appraisers to use in figuring gross income. The committee's task was to make an intensive study of all factors bearing on likely prices before making their recommendations. In both Canada and the United States the committee recommendations were a reasonably good forecast of the price levels which later prevailed.

In summary, selection of prices is a difficult task. Those who give scant consideration to this item are avoiding a critical issue. Of all the possible selections, the choice of an estimated average appears to have an edge over the others. A weighted moving average comes closest to giving a set of prices that correlates with current price trends. On the other hand, an estimated average answers the requirements of those who want a set of prices which they think will be the average in the future. The appraiser should recognize that in any selection, personal judgment has an important bearing on the decision. Unfortunately there is no escaping this conclusion because a set formula or moving average by itself will not give satisfactory results over a period of years.

GEOGRAPHICAL PRICE VARIATIONS

Geographical price variations apply to all products but more to some commodities like corn and wheat than to others like cotton. In general, products with a high value per pound— such as cotton and eggs—do not show much variation because the cost of transportation is only a small percentage of the value. Wool, however, does show considerable variation, the reason being variations in grade and quality in different parts of the country. On the other hand, products with a low value per pound—such as hay and straw—will vary widely in price depending upon location. Transportation costs from surplus to deficit areas, the main cause of the price differences, may equal or even exceed the price in the surplus area. The appraiser should give special attention to the local price factor when he is converting the landlord's share of the crop into value. It is not enough to decide on an average price; whatever price is decided upon must be adjusted to the local area in which the farm is located.

Local price quotations and price supports on a county basis provide the information an appraiser needs in selecting the farm product prices he uses in his income appraisals. Price supports for corn and wheat illustrate the local variation that exists and that should be reflected in income values. Corn price supports for a group of corn-growing states are shown in Table 11.7, with the high and low county prices in each state.

The highest county price supports for the 1968 crop of corn, among the selected corn-growing states shown in Table 11.7, were in Colorado; the lowest were in Minnesota, South Dakota, and Iowa. Actually in states not listed the corn price support was even higher—$1.57 a bushel in Hawaii and $1.31 in the New England states. The corn loan rates for support purposes are based on corn grading No. 2, with premiums for moisture below 15 percent and for 2 percent or less of broken kernels and foreign material. Discounts are also provided for damaged corn.

Wheat price supports vary widely even in the same general area, as shown in Table 11.8. For example, in South Dakota support prices for wheat vary from $1.34–$1.38 a bushel in the northeast corner of the state down to $1.11–$1.17 a bushel in the southwest corner. Similarly in Colorado the $1.13–$1.14 counties were along the eastern boundary while the $.92–$.94

TABLE 11.7. Corn Price Supports on 1968 Crop in Selected Corn-Growing States Showing High and Low Price Supports by Counties

State	County Price Supports	
	High	Low
Colorado	$1.25	$1.11
Tennessee	1.24	1.17
Kentucky	1.22	1.14
Ohio	1.20	1.08
Missouri	1.15	1.06
Kansas	1.14	1.05
Michigan	1.12	1.08
Indiana	1.12	1.07
Wisconsin	1.12	1.07
Illinois	1.12	1.06
Nebraska	1.10	1.01
South Dakota	1.08	.97
Iowa	1.06	.98
Minnesota	1.04	.97

Source: Commodity Credit Corporation, USDA, Washington, D.C.

TABLE 11.8. Wheat Price Supports on 1968 Crop in Selected Wheat-Growing States Showing High and Low Price Supports by Counties

State	County Price Supports	
	High	Low
South Dakota	$1.38	$1.11
North Dakota	1.36	1.13
Indiana	1.32	1.18
Washington	1.32	1.13
Kansas	1.31	1.11
Nebraska	1.31	1.11
Oklahoma	1.27	1.21
Montana	1.15	1.05
Colorado	1.14	.92

Source: Commodity Credit Corporation, USDA, Washington, D.C.

counties were on the western slope. The price supports for wheat are all based on wheat grading No. 1.

In addition to the variation in support prices quoted in Table 11.8, there were premiums and discounts to be considered. The premium for protein content on the 1968 crop started at 12 percent. Wheat with 15 percent protein was entitled to a premium of 10.5 cents a bushel added to the support price. On the other hand undesirable varieties were discounted 20 cents a bushel.

Appraisers valuing cotton lands will also want to use price support schedules. However, because of the high value per pound of cotton compared to corn and wheat, there is only a slight geographical variation in loan price rates. The big difference in cotton is in quality differentials. Premiums and discounts are especially important for staple length and quality grades. Cotton loan rates are based on cotton grading middling white one inch long.

If an appraiser is valuing farms over an area as large as a state or even over a smaller area in which prices vary, he will find a price map a useful aid. On this map he can insert county prices or price supports for the major crops grown in the area. With such a map he will not have to look up price data each time he appraises a farm in a county in which he has not worked before. In addition he will have the satisfaction of being able in all of his appraisals to reflect the local price situation in the income values he prepares.

QUESTIONS

1. With the material prepared in appraisal of a farm previously studied, set up an estimated production and income statement on a landlord basis. (If in an area where renting is relatively in-

frequent an owner-operator basis can be used.) Give reasons for your choice of crop share and cash rental for pasture. Do you consider this a fair division of the total crop produced? In approximate terms, how much do you think the tenant has as income after paying farm operating expenses?
2. For the farm used in Problem 1 above, figure the gross income on three levels of assumed prices for farm products.
3. Show how a 10-year simple moving average of prices would have worked in figuring gross income on farms in your locality.
4. Construct a weighted moving average for several years for a crop important in your community.
5. From data on prices by counties, or from county price supports for a major crop in your area, construct a price map that could be used in appraising farms.

REFERENCES

Hurlburt, Virgil L., "On the Theory of Evaluating Farmland by the Income Approach," USDA, Agr. Res. Serv., Apr. 1959.
Rural Appraisal Manual, 2nd ed., American Society of Farm Managers and Rural Appraisers, DeKalb, Ill., 1967.

12

LANDLORD EXPENSE ESTIMATES

NET AND GROSS INCOME to the landlord should not be confused. Expenses, which vary with every farm, are inescapable and always have to be deducted from the gross income of the landlord. Taxes on land and buildings are practically universal. Repairs and depreciation on buildings and other improvements, whether maintained or not, have to be figured. The expense of fertilizer, seed, and similar items is often a landlord obligation. Finally, a charge to cover the cost of landlord management is becoming common.

RISE IN EXPENSES

A comparison of landlord expenses in 1937, 1958, and 1965, as shown in Table 12.1, indicates the character and changes in expense outlays over a period of 28 years. The most spectacular increase has been crop expense, rising from only 21 cents an acre in 1937 to $1.52 in 1958 and $3.20 in 1965. Crop expense was the largest item in 1965 while it was next to the lowest in 1937. Increased use of fertilizer has of course been the main reason for the big rise in crop expense, a rise largely responsible for increased yields which in turn helped greatly in boosting gross income. Property taxes have been the other big factor causing increased landlord expenditures. The two items of taxes and crop expense accounted for 86 percent of the landlord expense listed in Table 12.1. Expenses for buildings and other improvements are underestimated in Table 12.1, however, because no allowance was made for depreciation; only cash expenses are included.

The rise in fertilizer expense in the sixties is shown by the record of landlord expenses in Illinois as indicated by the figures in Table 12.2. For the group of farms with a high soil rating, fertilizer expense in the 1960–64 period amounted to 25 percent of total cash expenditures by the landlords, while in 1966 the fertilizer expense amounted to 30 percent. Another important point in Table 12.2 is the rapid rise in total expense between the 1960–64 average and 1966, a rise of 48 percent. In this same period fertilizer expense was up 83 percent.

TAX ESTIMATES

Taxes are relatively easy to estimate in some areas but largely a guess in others. The records in some communities are fairly stable although in almost all areas the trend has been upward. In other communities tax rates may be highly variable, a period of heavy taxes followed by one of limited spending. In any event the tax estimate used in appraisal should be not just an average of the past but the best possible estimate of the future.

For an estimate of taxes it is essential to obtain actual tax figures for several years to give an idea of the fluctuation as well as the absolute amount. When getting these tax figures at the courthouse or from other sources, it is better to scatter the years over a long period than to get them for a consecutive series of

TABLE 12.1.   Landlord Income and Expense per Acre on Groups of Western Corn-Belt Farms, 1937, 1958, 1965 (dollars per acre)

	1937[a]		1958[b]		1965[b]	
Gross Income		5.39		18.09		24.95
Expenses						
Taxes	.96		1.95		2.95	
Insurance	.08		.35		.57	
Repairs	.31		.60		.40	
Fertilizer, seed	.21		1.52		3.20	
Total		1.56		4.42		7.12
Net income		3.83		13.67		17.83

Source: Farmers National Co., Omaha, Nebr. Expenses do not include management fee.

[a] Group of 25 farms chiefly in Iowa and Nebraska.

[b] Group of 50 farms chiefly in Iowa and Nebraska.

TABLE 12.2. Landlord Expenditures per Tillable Acre on Illinois Grain Farms of
260 to 339 Acres in Highest Soil-Rating Group, 1960

Expenditure	1960–64	1963	1964	1965	1966
Property taxes	$ 6.48	$ 6.58	$ 7.14	$ 7.19	$ 7.49
Fertilizer	3.57	3.86	4.39	5.21	6.53
Other	4.45	4.78	4.50	5.36	7.42
Total	$14.50	$15.22	$16.03	$17.76	$21.44

Source: Franklin J. Reiss, *Landlord and Tenant Shares 1966,* AERR No. 88, Dept.
of Agr. Econ., Univ. of Ill., Urbana, Sept. 1967, p. 11.

recent years. For example, it is preferable to have taxes for 1960, 1963, 1965, 1967, and 1968, than to have them for the years 1965–68. Although it is not always easy to get records for earlier years, the data for the longer period give a much clearer picture of local tax fluctuations. One way of using a long tax record in estimating is to get the tax rate applicable to all farms in a tax district for the number of years desired. This tax rate may then be used as an index for all farms appraised in the area in which the tax rate applies. In using this procedure, however, allowance must be made for any changes in assessments during the period.

An illustration of the erratic but generally upward tendency of property taxes is evident in the following millage rates or levies during an 18-year period in which there was no reassessment of property in the district and no change in the specific property being checked between 1951 and 1961:

Millage Levies

Year	Levy	Year	Levy	Year	Levy
1951	67.6	1957	97.2	1963	116.2
1952	83.9	1958	100.5	1964	116.4
1953	85.4	1959	115.0	1965	114.7
1954	88.3	1960	120.8	1966	104.4
1955	88.7	1961	123.0	1967	108.8
1956	94.4	1962	112.2	1968	112.8

Although the 4-year period 1952–55 in the above series showed little change in property taxes, the other years registered such heavy increases that taxes in 1961 were almost double those in 1951.

In 1962 the taxing district was enlarged, and since then a reassessment and another change in the district boundary has made the millage levy a poor indicator of total property taxes. For example the drop in millage in 1966 was caused by a reassessment which raised values substantially. Actually property taxes in 1966 were higher in the district than in 1965, and they continued to increase in 1967 and 1968.

Information should be gathered on any special taxes such as drainage assessments. Here again a search of county records and other sources will disclose actual payments over a relatively long period. It may be necessary to make an investigation of drainage district finances to ascertain the amount of outstanding bonded indebtedness, because this bonded debt represents a lien prior to any mortgage which may be placed on the land. Another type of expense that should not be overlooked is the charge for maintenance of a drainage system; heavy expenses on this score might soon be assessed on a farm served by a ditch which needs to be redredged.

An appraiser will find it profitable to add to his background knowledge all the information he can get on property taxation. The study can start in the home county with an examination of (1) the total taxes assessed on land and the use made of the funds collected, and (2) the trends from year to year for the county as a whole, for minor civil divisions, and for individual farms. As he continues his study, he will piece together a large portion of the county's financial history from available records and at the same time provide himself with the background necessary for estimating future taxes. He will have an understanding of the way in which bond issues are floated for schools, the rate at

TABLE 12.3. Farm Real Estate Taxes per Acre, Average for United States, 1890, 1900, 1910, and 1920–67

Year	Tax	Year	Tax	Year	Tax
	(dollars)		*(dollars)*		*(dollars)*
1890	0.13	1932	0.45	1950	0.69
		1933	.49	1951	.72
1900	.13	1934	.37	1952	.76
		1935	.37	1953	.79
1910	.19	1936	.38	1954	.82
		1937	.39	1955	.88
1920	.51	1938	.38	1956	.91
1921	.54	1939	.39	1957	.97
1922	.54	1940	.39	1958	1.03
1923	.55	1941	.39	1959	1.11
1924	.55	1942	.38	1960	1.22
1925	.56	1943	.38	1961	1.29
1926	.56	1944	.40	1962	1.36
1927	.57	1945	.44	1963	1.43
1928	.58	1946	.49	1964	1.51
1929	.58	1947	.57	1965	1.61
1930	.57	1948	.62	1966	1.74
1931	.53	1949	.66	1967	1.89

Source: "Farm Real Estate Taxes," *Agricultural Statistics*, published annually by USDA. After 1959 all 50 states are included.

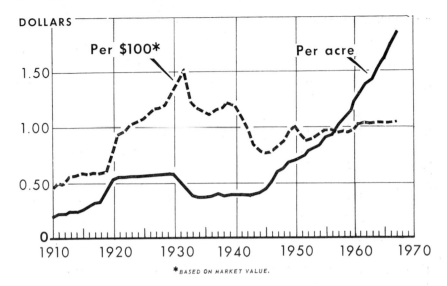

FIG. 12.1. *Farm real estate taxes for United States not including Alaska and Hawaii for the years 1910–68, using index numbers with 1946–48 equal to 100. (Courtesy USDA.)*

which they are paid off through sinking funds, and the resulting effect on property taxes. Further, he will see by comparison how differences in spending policy have materially affected the taxes in different school districts throughout the county.

A perspective on the trend of farmland taxation is given in Table 12.3 which shows the average taxes levied per acre in the United States at different dates from 1890 through 1967.[1] Taxes levied in 1966 and payable largely in 1967 were roughly double those in 1955, much higher than in any other year on record. Taxes, it will be observed in Figures 11.1 and 12.1, fluctuate less than farm prices and tend to lag behind farm price movements. The price decline, for example, struck a low point in 1932, but taxes continued to decline until 1934 and the rise did not start until 1944 as indicated in Figure 12.1. This tendency of taxes to lag behind price prosperity and depression is accounted for in the main by the time elapsing between decisions on expenditures and the levying of taxes.

An example of the increase in property taxes between 1959 and 1967 is provided by the record for Farm *A* as shown in

1. Taxes by individual state are presented in Appendix B by periods from 1910 to 1967.

TABLE 12.4. Landlord Expenses on Farm *A* over Nine-Year Period, 1959–67

Year	Property Taxes	Build-ings	Insur-ance	Seed	Fertil-izer	Insect. Herb.	Mg't & Misc.	Total
				(dollars)				
1959	511	671	11	232	307	..	464	2,196
1960	593	116	60	188	517	8	380	1,862
1961	664	190	60	138	417	9	509	1,987
1962	660	111	68	164	653	24	410	2,090
1963	708	677	107	147	502	13	408	2,562
1964	739	513	79	110	533	12	411	2,397
1965	890	466	95	234	658	32	402	2,777
1966	839	1,104	90	211	875	170	481	3,770
1967	1,016	143	90	147	738	64	514	2,712
Annual average	735	443	73	175	578	36	442	2,484
Per acre average	4.60	2.77	.45	1.10	3.61	.22	2.77	15.52

Source: Doane Agricultural Service.

Table 12.4. Property taxes on this farm doubled between 1959 and 1967.

BUILDING DEPRECIATION AND REPAIR EXPENSES

The landlord, after he has paid the taxes, has next to consider building and other improvement expense. This expense is the amount which the landlord pays out for repairs and maintenance of buildings and other improvements such as fences, water system, tile drainage, and terracing—plus the amount, if any, that needs to be set aside to replace present improvements when they wear out or become obsolete. If all buildings on a farm were new at the same time, and after a certain period of years all had to be replaced at the same time, the necessary sum to set aside to build new buildings could be figured with little difficulty. As it is, however, buildings on a farm are usually in different stages of their probable life. Even though they all may have been built at the same time, some will last longer than others. Moreover, some buildings will be repaired, painted, and possibly improved, while others will be allowed to depreciate. In fact a building with a good foundation may last indefinitely, provided parts like the roof are repaired when they give out.

Depreciation and repairs should be considered at the same time in figuring building expense. If the depreciation is high, the repairs may be low because the buildings are being allowed to deteriorate. There will be cases, on the other hand, where the depreciation will be very low or nonexistent because a large amount is being spent in maintaining the buildings. The total figure of both depreciation and repairs will vary with the upkeep program, since a policy of painting and repairing buildings to keep them in good condition may result in less annual cost in the long run than a policy of allowing the buildings to depreciate. It is obvious that the allowance for depreciation will vary within wide limits, not only with the amount spent on repairs but also with climate, quality of construction, and use made of the buildings. For example the life expectancy of a new, well-built barn could be 67 years. A new $6,000 barn would be depreciated at $90 (1½ percent of full valuation); at 33½ years it would be evaluated at half its original cost, or $3,000. The depreciation rate applied to the $3,000 value would have to be doubled (3 percent) to give the same $90 depreciation. As deterioration increases, the value goes down and the depreciation rate on the lower value goes up. Near the end of the building's useful life the valuation could be $150 and the depreciation rate 60 percent.

How much expense is it reasonable to figure for repairs and depreciation? Actual cash expenditures on repairs to buildings and other improvements have been reported in various record-keeping studies. The figures vary widely because the building investment as well as the policy followed by farm owners in building maintenance varies so much.

Depreciation not covered by repairs is a cost which is often overlooked because it is not an annual out-of-pocket expense. However, when the farm owner is suddenly confronted with the necessity of building a new house or service building or of remodeling a barn, he is facing the accumulated depreciation which has been going on for years. If he has been making an allowance annually for depreciation in addition to minor repair expense, he will be prepared for the heavy building outlay which comes all in one year.

An indication of the life span of farm buildings is provided by an Indiana study summarized in Table 12.5. The service life of farm buildings in this study runs mostly from 50 to 100 years, which means an average annual loss in value of between 1 and 2 percent of the original cost. If the buildings are not new the depreciation rate on the present value has to be adjusted upward to cover the amount.

The actual money spent by the landlord of Farm *A* for the nine-year period 1959–67, shown in Table 12.4, indicates a wide variation from year to year. Only $111 was spent in 1962 while $1,104 was spent in 1966. Variations like this are common on many farms. When a major repair and painting program was carried out on Farm *A* in 1966 the expense turned out to be the

TABLE 12.5. Average Length of Life and Depreciation Rates for Buildings of Different Types on 166 Farms in Indiana

Buildings	Years Already Used	Estimated Remaining Years Service	Total Years Service	Approximate Rate of Depreciation on Initial Cost[a]	Approximate Rate of Depreciation on Present Value[a]
				(*percent*)	(*percent*)
Barns	36	58	94	1	2
Double cribs	29	41	70	1.5	2.5
Single cribs	22	24	46	2	4
Implement houses	33	14	47	2	7
Poultry houses	20	30	50	2	3
Stationary hog houses	20	30	50	2	3
Granaries	25	37	62	1.5	3
Milk houses	10	40	50	2	2.5
Brooder houses	11	20	31	3	5
Movable hog houses	8	14	22	5	7

Source: Lynn Robertson, Purdue Univ. Agr. Exp. Sta., Bull. 435, p. 26.
[a] Based on remaining years of life.

highest building total for the nine-year period. The average yearly expense of around $450 or $2.80 an acre, which included some capital outlays, did maintain the buildings. In fact the buildings in 1968 appeared to be in as good condition if not better than they were in 1959. An annual charge of $2.80 an acre for buildings on Farm *A* figures roughly 2.5 percent of their 1968 value of $17,500. However, this is physical condition and does not cover obsolescence, an aspect of depreciation which will be taken up in Chapter 14.

In general, an expense allowance of between 2 and 4 percent of present value is a reasonable charge against landlord gross income. This is an estimate of a required amount which if spent annually would maintain the buildings in their present condition. In the usual situation with some buildings relatively new, some half worn out, and a few really old, the appraiser's problem is to estimate the annual amount which will preserve this average situation, replacing the worn-out buildings or remodeling them at the appropriate time. In some years more will be spent than is allowed; it may be a new roof, a remodeling job, a new building, a large fencing project, or painting. But in other years the amount spent may be small.

INSURANCE

Insurance against fire and windstorm for the buildings, both dwelling and service buildings, must be included as a landlord expense. The amount to charge can be estimated by noting current and past rates in the territory. The annual charge for fire insurance on all farm buildings may average around $3.00 per thousand, and for windstorm insurance around $1.50 per thousand, or a total of $4.50 per thousand. These are average figures, however, and do not indicate the wide variations that exist between states and between districts within states. The appraiser can check the situation easily in his own district by obtaining rates from local insurance companies and agents. A recent development that should be noted is the use of package policies which include other types of insurance such as liability along with fire and windstorm.

OTHER IMPROVEMENT ITEMS

Fences are often forgotten in an appraisal. Farm owners, however, know that fences are a real item of expense. The average expense will depend on the size of fields, type of fences (woven wire, barbed wire, or electric), amount of wear and tear on the fences, and how well they are built. Variation in fence expense, as with other improvement expenses, will be extremely wide within regions and among farms.

Another item likely to be overlooked is maintenance of the drainage system. If the farm is in a drainage district, previous assessments for maintenance may help in estimating future expense, which is often for redredging. When the district is comparatively new, the experience of neighboring districts may be used as a basis for estimate. Many farms, however, will have private branches or their own tile drainage system, both of which require repairs by the owner. Tiles get clogged, break, and wash out at the outlets, and unless repairs are made when needed, crop yields may be cut drastically in the area drained by the tile. Maintenance of an adequate water supply represents still another expense, one especially important in livestock areas and where water is difficult to get.

FERTILIZER, SEED, AND LIMESTONE

Several references have already been made in this chapter to the rapid increase in the use of fertilizer. Tables 12.1, 12.2, and 12.4 all indicate this amazing expansion in fertilizer use. Equally amazing has been the absence of any rise in the prices of fertilizer. As Table 12.6 indicates, not only has there been no rise, there has actually been a large decline in the price of anhydrous ammonia which supplies nitrogen.

TABLE 12.6.   Average Prices per Ton Paid by Farmers for Selected Fertilizers, United
States, Apr. 15 Prices, 1957–59 Average and 1964–69

Year	Anhydrous Ammonia	Superphosphate 46% P_2O_5	20% P_2O_5	Ammonium Phosphate 16–20–0	Potash 55% K_2O and over[a]	Mixed Fertilizer 6–24–24
			(dollars)			
Average:						
1957–59	149.00	82.20	37.00	89.60	52.10	91.10
1964	126.00	78.90	40.30	82.30	53.90	85.20
1965	122.00	79.10	40.70	80.70	53.60	85.60
1966	119.00	80.90	41.40	71.10	54.90	85.10
1967	113.00	84.10	42.10	80.70	53.60	85.70
1968	91.40	78.40	43.20	78.40	49.10	81.80
1969	75.60	74.00	43.80	77.70	47.80	73.20

Source: *Agricultural Prices,* Statistical Reporting Service, USDA, Washington,
D.C.
    [a] 60% K_2O after 1967.

In estimating landlord expenses for fertilizer, seed, and limestone the appraiser has to take into consideration the typical cropping system and fertilizer practice in the area in which he is appraising, and the apparent needs of the particular farm being appraised. In addition it is necessary to know the customary rental agreements between landlords and tenants on sharing fertilizer, seed, and limestone expenses.

MANAGEMENT EXPENSE

Another expense sometimes included is for management. An absentee landlord usually has some expense in connection with supervising his farm, or he pays a fee or a percentage of the gross rental income to have a farm management service take over this supervision. A charge ranging from 7 to 12½ percent of landlord gross income or its equivalent in a per acre charge is sometimes made for management, the range reflecting the size of the income and the intensity of the management service provided. On a highly productive cash-grain farm a 7 percent charge may return as much as a 12½ percent charge on a low-producing farm. A higher charge is usually made on a livestock-share arrangement than on a straight grain-share lease because so much more supervision is required.

Either totals or per acre figures may be used in setting down expense deductions from landlord income. Some appraisers prefer totals for the whole farm because they are easily compared with totals actually expended. Per acre figures tend to be unrealistic when applied to a combination of cropland and pasture or to small and large farms, but are preferred when making farm-to-farm comparisons. Taxes, for instance, are usually spoken of in per acre figures. Since both methods of giving expenses have advantages and disadvantages, it is desirable to include both total and per acre figures in the appraisal report.

NET INCOME TO LANDLORD

With the completion of landlord expenses, net rental income is obtained by deducting the expense total from the gross income. Care should be taken not to exclude any items which should be included in either the income or the expense. A list may be made for the appraiser's territory of all possible income or revenue items, and likewise a list of all common expense items. These lists may be checked, item by item, against each appraisal report to make sure that all incomes and expenses are shown.

Landlord net income, total and per acre, is an excellent index for comparative purposes. Comparisons of course will have more significance in a community where renting is common than where renting is almost unknown. The more renting there is, the more reliable and useful is this method of estimating landlord gross income, expense, and net income.

QUESTIONS

1. Estimate and substantiate with reasons the landlord expense items on the farm used in the problem at the end of Chapter 11. If time permits, find out the tax situation in the district where the farm is located. Show in detail your computations of building depreciation, repairs, and maintenance; these items should check with the building inventory.
2. Obtain information on costs of fire and windstorm insurance in your locality.
3. Determine from a study of typical cropping and rental practice in your area the average annual landlord expense for fertilizer, seed, and limestone.
4. Are there farms in your community managed for landlords by professional farm managers? If so, what are the charges?

13

VALUE BY
CAPITALIZATION

ONCE an annual net income for a farm has been estimated it is relatively easy to compute either an independent income value or a sale value estimate by capitalization. The main problems are an understanding of the capitalization concept and the selection of a capitalization rate.

TWO CAPITALIZATION VALUES

It is important to see clearly at the very beginning that there are two entirely different values that can be obtained by capitalizing net income. One is an income value which has no connection with sale or market value. This value is based entirely on income and on a rate that is arbitrarily selected. The other is an estimated sale value obtained by using a capitalization rate which is determined by the rate of return on similar farms that have been selling.

An illustration may help to clarify the difference between the two capitalized values. A farm being appraised has an estimated net income of $15 an acre. The farm has several attractive features that have little if any bearing on its income: it is located on a paved highway, has a new well-built house, is near town, and has a recreational lake nearby. This farm on a strict income basis and using an arbitrary interest rate of 6%

183

has an income value of $250 an acre; if 5% is selected the income value is $300 an acre. These income values of $250 and $300 an acre are based solely on estimated net income and on interest rates which though selected arbitrarily do have some relation to the rates being received on investments. In short, the independent income values are determined by net income and by interest rates related to the investment market, with no weight at all being given to the attractive nonincome features of the farm.

This same farm has a capitalized sale or market value that is much higher than its independent income value. Farms with the same attractive features as this one have been selling at prices which return only 3% on the investment. Using this 3% as our capitalization rate we get an estimated sale or market value for this farm of $500 an acre.

Farms farther from town than this one and without the other attractive features have been selling at prices which return 4%. Those farms with a net income of $15 an acre are selling at a capitalized value of $375 an acre. If you ask why these farms are selling at $375 an acre instead of $300 to yield 5%, the answer may be that buyers are anticipating future inflation with rising farm values and consequently are willing to take a 4% return on a farm rather than a 5% return on the investment market. Another reason may be that buyers just prefer owning a farm at 4% to investing their money elsewhere at 5%.

Therefore, the market capitalization rate represents what farms are actually selling for, and the investment capitalization rate indicates what farms would sell for if they had no features other than net income. This gives us a market or sale value estimate based on a capitalization rate which is derived from the sale of comparable farms, and an income value estimate based on a capitalization rate from the money market. The two values, it should be clearly recognized, are independent of each other because the two capitalization rates are entirely separate and distinct.

Why not stick to the capitalized market value and forget the independent income value? Why should we be concerned with farms purely as a money investment when we know they have other features? The answer is that in the purchase of a farm both the buyer and the mortgage lender are interested in what the farm will produce in income separate from the other features. Interest and principal payments on the loan are not going to come out of the attractive features. In the example cited above the loan will be based on the income value of $250 or

$300 an acre even though the farm is actually worth $500 an acre.

Before taking up the various questions involved in selecting the capitalization rates for the two values, it is important that the reader have an understanding of what capitalization means, of how value can be obtained by using a rate.

THE CAPITALIZATION PROCESS

Capitalization, the process by which value is figured from annual income, requires two factors: an estimate of annual income and a capitalization rate. Capitalization is commonly expressed by the following equation:

$$\text{Value} = \frac{\text{Annual Net Income}}{\text{Capitalization Rate}}$$

Example: Income Value of Farm $A = \dfrac{\$17}{.05} = \$340$ an acre.

The independent income value per acre of Farm A was obtained by capitalizing (dividing) the estimated net annual income per acre by the capitalization rate, which in this instance was assumed to be 5%.

Another expression of this same capitalization process is that commonly used in Great Britain where real estate is quoted as so many years' purchase. The number of years to use is established by dividing 100% by the rate; thus the equivalent of a 4% rate would be 25 years, 5% would be 20 years, and 6%, 16⅔ years. In applying this as a formula:

Annual rent to landlord × number of years = value per acre. Using Farm A again: $27 × 20 (years) = $540 per acre.

Capitalization, according to a somewhat different point of view, is the present worth of all future incomes. The capital value of property expected to yield $100 annually is the sum total of all future incomes of $100 a year discounted to the present. This year's income of $100 is worth more than next year's, and still more than that of the following year, and much more than the expectation of $100 in 20 years. It is necessary, however, to have a rate to use in determining how much more $100 is worth this year than next year. At 5%, $95.24 today is equal to a payment in one year of $100; the amount which, if put at interest at 5%, would total $100 in one year's time. Similarly, $90.70 today is equal to the right to receive $100 two years hence. A curve of discounted values like those in Figure 13.1 is the result of following this procedure. The present worth of an income in a future year may be computed by using the general formula:

$$\text{Present value} = \frac{1}{(1 + \text{interest rate})^n} \times \text{future income.}$$

In this formula "n" equals the number of years between the present and the time when the future income is received.

An advantage of the present worth analysis, as graphically shown in Figure 13.1, is the numerical weight which it gives each future year in the determination of value. The contribution of each future year is a definite figure. The result of totaling the present worth of all future incomes is the same as dividing the annual income by the interest rate, which is the formula given previously. It will be observed in Figure 13.1 that the major portion of land value is contributed by the years relatively close to the present; with a 5% interest rate, over 60 percent of the value is accounted for by the income in the first 20 years.

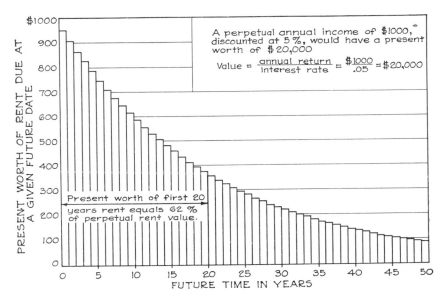

FIG. 13.1. *Present worth of future annual returns. Illustration of $1,000 net annual rent discounted to the present 5%. Although only fifty years are shown in this chart, the present worth continues to decline indefinitely. (Courtesy Herbert Pike.)*

VALUE OF APPRECIATING OR DEPRECIATING LAND

The analysis of present worth of future income provides a useful method of valuing farms that are appreciating (improving) or depreciating. In the case of appreciation this method is especially appropriate where a farm will be improved in the future by such measures as drainage, erosion control practices, a soil-building rotation, or the planting of fruit trees. If a farm is likely to yield net returns of $600 the first year, $700 the second year, $800 the third year, $900 the fourth year, and $1,000 a year from the fifth year on, the value may be computed by assuming a constant income of $1,000 a year and deducting for the smaller amounts in the first four years. Thus the $1,000 income at 5% equals $20,000 minus the present worth of $400 in one year, $300 in two years, and so on, or a total deduction of $908 which gives a net value of $19,092.[1] The details follow:

Years Hence	Dollar Amount		Present Value of 1		Present Worth in Dollars
1	400	×	.9524	=	380.96
2	300	×	.9070	=	272.10
3	200	×	.8638	=	172.76
4	100	×	.8227	=	82.27
Total	1,000				908.09

$$\text{Value} = \frac{\$1,000}{.05} - \$908 = \$19,092.$$

Depreciating farms are those with declining net incomes. Examples include losses from sheet and gully erosion, from the washing away of land along a river bank, and from deposition of gravel or other undesirable material on good soil. Another type of loss is an increase in taxes or other expenses over a period of years. The method of valuation based on present worth of future incomes may be used advantageously in all cases of declining net returns, including coal mines, stone quarries, and the like, as well as farms. The value of a farm with an income expectation of $500 the first year, $300 the second year, $100 the third, and no net income at all after that, would be the present worth of $500 in one year, $300 in two years, and $100 in three years, or at 5% a total of $835. If an income of $500 is estimated for two years, $400 for the following two years, and $300 continuously after that, the value of the farm would be $300 capitalized at 5% plus the present worth of the extra amounts

---

1. See Appendix C, a table of present values of 1, which may be used in working problems of this type.

during the first four years at 5%, or $6,000 plus $541. The use of 6% instead of 5% would capitalize this same farm at $5,000 plus $530. The difference between the two values, $6,541 and $5,530, is due entirely to the rate of capitalization.

VALUATION OF ORCHARDS AND TIMBERLANDS

Capitalization of future income through the present worth process has a special application to orchards and timberlands. In these two cases there may be no income for a period while the trees are reaching the fruit-bearing or merchantable timber stage. Since the estimated income will not be received for a period of time, it is necessary in determining present value to obtain the present worth of estimated future income and to discount estimated future expense. A simple example like the following may be helpful: valuation of a young orchard which will begin bearing in two years and reach a sustained average in four years. We will assume expenses of $100 at the end of the first year and $200 each year afterward. Also we will assume the first gross income which comes at the end of the second year is estimated at $300, in the third year, $400, and in each succeeding year, $500. The present value of the property can be obtained by capitalizing the sustained average net income of $300 a year and subtracting the present worth of the deficiencies in the first three years. Following this procedure and assuming an interest and capitalization rate of 6%, we obtain a present worth of $5,000 less the deficiencies. The deficiencies are $400 at the end of the first year, $200 the second year, and $100 the third year. The present worth of these deficiencies at 6% is $377.36, $178.00, and $83.96, or a total of $639.32. Therefore, the present capitalized value of the orchard would be $4,361. A year later the value would be $4,722, two years later, $4,906, and three years later, $5,000, assuming in all these cases that the estimated income and expense followed the same pattern as was estimated in the beginning.

The same procedure of determining value by present worth can be applied to woodlots and timberlands. If woodlots yield an extra income not figured in the ordinary rental income, the present worth of this net income should be included in the value of the farm. If, for example, the stumpage value of a woodlot or timber tract is estimated at a net of $1,000 five years hence, then the present value of this $1,000, which is $783.50 assuming a 5% rate, can be added to the value of the farm.

CAPITALIZATION RATE FOR INCOME VALUE

How does one proceed in selecting a capitalization rate to use in obtaining an independent income value? Before selecting a rate it is desirable to understand what this rate is and what it is expected to do. Then we can consider the actual procedure followed in selecting a rate.

The capitalization rate used in obtaining an income value which is independent of sale value has a close connection with interest rates. In fact we can use interest rates as an aid in explaining the rate we are after. The estimated annual income from the farm is in one sense similar to an interest return on a bond investment. One man buys $50,000 worth of bonds while another buys a farm; the first man receives interest on his bonds as an annual return or income while the second receives an annual income from his farm. It is true that the interest rate on the bonds is fixed while the annual return from the farm is not, but for our purposes at this point the two investments have a similarity in that they both yield an annual return.

A simple example to show the relationship between bond interest and farm income is a perpetual bond (one which runs indefinitely) bearing a 5% interest rate, paying $50 annually on a $1,000 bond. This bond will be priced in the market at $1,000 as long as the going interest rate on investments of this kind is 5%. However, if this interest rate declines to 4%, the bond will sell not for $1,000 but for $1,250 ($50 divided by .04). At this price of $1,250 the return of $50 will be exactly 4%. If the going rate rises to 6% the bond will sell for only $833.33 ($50 divided by .06), which will yield a 6% return. In this last example the net income of $50 was capitalized at (divided by) 6% to give an income value of $833.33.

From the perpetual bond it is only a step to the value of common stock in a corporation with relatively stable earnings. This step, however, does involve a shift from interest on the bond to earnings on the common stock, but the concept of earnings on an investment is the same. Investment in a common stock is much the same in general principle as investment in a farm; both are investments in a going business. The price of the common stock in the market will generally run from 15 to 20 times the annual earning rate per share, or at a rate of 5 to 6⅔%. In short the value of common stock per share is usually the annual return capitalized at between 5 and 7%. Hence, whether it be a bond, a share of common stock, a farm, or some other income-bearing property, its value can be determined by means of an annual income estimate and an appropriate capitalization rate. If value and annual income are known, then the capitalization rate or rate of return can be obtained by dividing the value into the annual income.

The capitalization rate we are after is set by borrowers and lenders. It is also set by buyers and sellers of common stocks where the income is relatively stable with no anticipation of increased returns. In the case of borrowers and lenders the interest rate is set in a money market which balances the forces of money supply and money demand. At times, as in the sixties, the rate may be high—6 or 7% interest rate on farm mortgages and comparable loans. At other times, as in the forties, the rate may be low—4 or 5%. A similar situation is present in the market for stable common stocks. Their annual return over a long period of years varies usually between 4 and 7%.

In our capitalization rate for income value we are seeking an alternative to the long-term rate of return received by investors in the money market, particularly investments in bonds, mortgages, and stable common stocks. There is no provision in this rate for growth, for inflation, or for appreciation due to other factors. We are selecting a rate which will give a conservative value, and we do this by concentrating on the estimated income that can be *counted on* as we look ahead—and only on income, not on other nonincome features or anticipated future developments.

In selecting the rate to use for income value the choice is actually an arbitrary one. What do you consider a reasonable return on a farm as an investment? The farm mortgage lender is likely to select a rate which is not far from the rate he charges the farm owner on a mortgage loan. The buyer of a farm may choose a similar rate, especially if he has interest and principal payments to meet annually.

The choice of a specific rate is not as important as giving emphasis to the rate which is selected. If the annual income estimate is $15 an acre and the rate selected is 6%, the point to emphasize is that at 6% the income value is $250 an acre; if the rate selected is 5%, the income value is $300 an acre.

As an appraiser the author has been using 5% as a rate to obtain income value and usually a lower rate to obtain sale value. One of the advantages of this double-value procedure is the emphasis it places on the difference between a value based wholly on income and a sale value which includes nonincome features and anticipated future increases in value. In appraisals for mortgage lending and also in assessments for tax purposes, it is often desirable to distinguish between the value based on income and the sale value based on income plus a number of other factors, some of which may be highly speculative.

The 5% income capitalization rate used by the author in his appraisal of Farm *A* and other examples is an arbitrary choice, but it is supported by the long-term movement of interest rates which was under 5% in the forties and over 5% in the sixties. If in the years ahead interest rates and the return on stable common stocks should shift upward for a long period, an increase in the capitalization rate above the 5% level would certainly be justified.

CAPITALIZATION RATE FOR SALE VALUE

To obtain a capitalization rate which gives an estimated sale or market value, it is necessary to compare the selling price of a group of farms with the estimated net income from these farms. Thus we are establishing a ratio between selling price and income from a group of farms and applying this rate or ratio to the farm being appraised. It is essential of course that the group of farms used in arriving at this rate be similar to the farm being appraised.

This method of determining a capitalization rate to estimate sale value does not provide an independent sale value estimate. We have used sale values to obtain a sale value estimate, which is somewhat similar to the comparable sale approach discussed earlier. In the comparable sale approach, an estimated sale value for the subject farm is obtained from sale values of similar farms. In the capitalization approach we are using income and sale values. Instead of making direct sale comparisons we are using the relationship of income to sale value as the comparison tool. In terms of an equation the sale value capitalization looks like this:

$$\frac{\text{Av. income of group of farms}}{\text{Av. selling value of same farms}} = \frac{\text{Income of appraised farm}}{\text{X (estimated sale value of appraised farm)}}$$

Using in this equation information from a group of rented farms in the north central cash-grain area in Iowa and Farm *A*, we get:

$$\frac{\$20.40}{\$495} = \frac{\$27.00}{X}$$
$$X = \$655 \text{ an acre}$$

In this illustration the $495 represents the estimated selling price of 146 tenant-operated farms in the cash-grain area of Iowa which were included in a survey sponsored by the Iowa Society of Farm Managers and Rural Appraisers and Iowa State University. For these 146 farms the managers reported an average net income per acre of $20.40. For Farm A $27 is the estimated net income per acre to the landlord as indicated in the appraisal of Farm A in Chapter 1. The $655 an acre is the estimated selling price of Farm A based on the estimated rate of return on current sale value received by landlords in the area where Farm A is located. The sale value capitalization rate for the group of farms was 4.1% ($20.40 divided by $495). This rate of 4.1% applied to the $27 an acre estimated net income for Farm A gave the $655 an acre estimated selling price of Farm A.

But $655 an acre is not the estimated selling price of Farm A because this farm has something that the comparison farms do not have, something that is not represented in its $27 net income—closeness to the city of Ames and to a flood control dam and recreation area in the planning stage. These positive non-income factors reduce the sale value capitalization rate. An estimate of the net income and selling price of comparable farms near Farm A indicated the following:

$$\frac{\$26}{\$700} = \frac{\$27}{X}$$
$$X = \$730 \text{ an acre}$$

The ratio of $26 to $700 which is 3.7% applies specifically to the situation in which Farm A is located. From this it is evident that in arriving at the appropriate sale value capitalization rate it is necessary to select a group of farms that have the same major characteristics as the farm being appraised.

Results of the Iowa farm survey on net income, estimated selling value, and rate of return on crop-rented farms are shown in more detail for Iowa and for the cash-grain area in Table 13.1. An important point regarding the survey data needs explanation. Since methods of accounting vary among managers, all reports on income are on a cash basis, which means that inventories and depreciation are not considered. This does not give precise accounting for a specific year because inventories may be heavier or lighter than usual, but over the years the results will average out correctly. As for depreciation which is difficult to estimate on farms, the reports include all repairs, maintenance, and capital expenditures which in the long run take care of depreciation and are likely to come out closer to the real situation than do depreciation estimates. Since most of the farms were not sold during the year, the managers were asked to estimate current sale value unless a farm was sold, in which case the actual selling price was reported.

Judgment is required in the choice of a sale value capitalization rate, because net income in a given year may not be representative. The net income and sale value for the cash-grain area in Iowa for 1967 as shown in Table 13.1 were more nearly representative of average conditions than was the report for the state as a whole. The state average return for 1967 went down to 3.7% from 4.4% in 1966. For Iowa as a whole 1967 was a relatively poor year whereas 1966 was a relatively good year. However, in the cash-grain area the two years were more nearly the same in net income. Under conditions like these the natural tendency would be to use the average of several years, but the appraiser will find this difficult in a period like 1965–67 because sale values were changing rapidly in these years. Thus the appraiser is forced to obtain what data are available, like those in Table 13.1, and rely on his own judgment in coming up with a sale value capitalization rate.

SUMMARY

Capitalization can be used to obtain two entirely different farm value estimates: an income capitalization rate gives an income value and a sale value capitalization rate gives an estimated sale value.

The income capitalization rate is an interest rate selected from the investment market. Since there are many interest rates existing at any time in the investment market the selection is an arbitrary one. The important point is not so much the rate

TABLE 13.1. Net Income, Estimated Sale Value, and Rate of Return per Acre of Crop-Rented Farms in Iowa and Cash-Grain Area, 1965–67

Factor	Iowa			North Central Cash-Grain Area		
	1965	1966	1967	1965	1966	1967
Number of farms	503	421	433	180	138	146
Av. gross income	$ 30.55	$ 32.22	$ 32.78	$ 34.45	$ 37.50	$ 41.45
Operating expense	$ 13.20	$ 14.40	$ 16.50	$ 14.10	$ 16.00	$ 19.75
Capital expendit.	$ 1.49	$ 1.52	$ 1.81	$ .90	$ 1.00	$ 1.40
Net income	$ 15.87	$ 16.30	$ 14.52	$ 19.45	$ 20.50	$ 20.30
Est. sale value	$348	$378	$392	$425	$472	$495
Rate of return	4.6%	4.4%	3.7%	4.6%	4.3%	4.1%

Source: Survey sponsored by Iowa Society of Farm Managers and Rural Appraisers and Agricultural Experiment Station of Iowa State University, Ames.

selected as the emphasis on what the farm would be valued at using this rate. This is the income value which, assuming the estimated net income, will give the owner a return equivalent to the rate selected. The important point is that income is stressed; in no part of this process are the nonincome features included. The income value arrived at by this method is likely to be of vital interest to the lender on farm mortgages and to the buyer who is a borrower, because it indicates the debt-paying ability of the farm. It is true that net income alone can show this debt-paying ability, but using the net income and a capitalization rate based on the investment market makes it possible to emphasize more clearly the farm's value strictly on an income basis compared to a sale value which includes nonincome features.

The sale value capitalization rate is obtained by using a group of farms similar in major respects to the farm being appraised. The main task is to get an estimated income and an estimated selling value for this group of farms and from these two figures compute the rate of return. This rate of return for the group of farms is then applied to the estimated net income of the farm being appraised. Whatever nonincome features are present in the group of farms will be represented in the rate of return obtained. If the farm appraised has these same non-income features the value obtained through the use of the capi-talized rate from the group will provide an estimated market value for the subject farm which includes these nonincome fea-tures. The nonincome features are included by virtue of taking sale values for the group of farms which have these features. In short, the rate of return on sale value (including nonincome features) received on a group of farms is applied to a given farm to obtain an estimated capitalized sale value for it.

QUESTIONS

1. Explain in your own words what is meant by the capitalization process.
2. What connection if any is there between interest rates and capi-talization rates?
3. By means of a simple example explain the idea of present worth of future income.
4. Show the difference in capitalized income value for a given farm caused by variation in capitalization rates from 4 to 6%, with other things such as income remaining the same. Change the situation by assuming (1) that income will increase $100 a year for three years and (2) that income will decrease $100 a year for three years.
5. Explain the difference between capitalized income values and capitalized sale value estimates.

REFERENCES

Appraisal of Real Estate, 5th ed., American Institute of Real Estate Appraisers, Chicago, 1967.

Encyclopedia of Real Estate Appraising, Edith J. Friedman, ed., Prentice-Hall, Englewood Cliffs, N.J., 1959.

Heady, Earl O., *Economics of Agricultural Production and Resource Use,* Prentice-Hall, Englewood Cliffs, N.J., 1952.

Nelson, A. G., and Murray, W. G., *Agricultural Finance,* Iowa State University Press, Ames, 1967.

Wendt, Paul F., *Real Estate Appraisal,* Holt, Rinehart, & Winston, New York, 1956.

4

FARM BUILDINGS IN COST, SALE, AND INCOME APPROACHES

THE ONLY significant phase of farm appraisal which involves the cost approach is the valuation of farm buildings. Consequently the first part of the discussion covers building inventory and cost estimates. The second part covers buildings as they are considered in the sale and income approaches with special emphasis on farmsteads sold separately or with the farm as a whole.

Buildings are a troublesome aspect of farm appraisal because in many cases they are surplus items—victims of the farm enlargement movement which has resulted in a marked reduction in the number of farms and of farmsteads since 1955. However, valuation of farm buildings independent of the land is important. In the first place, there is an increasing number of farmsteads with small acreages being sold as rural homes separate from the farmland. Second, it is necessary to value the buildings for fire and windstorm insurance. Third, it is common in many states to place a separate assessment on farm buildings for tax purposes. Fourth, it is often desirable to value buildings for bookkeeping purposes in order to figure depreciation charges for income tax reports and profit and loss statements.

To obtain as clear an understanding as possible of farm building values the subject is divided into three parts: (1) inventory, (2) value based on cost less depreciation, and (3) value of buildings by themselves and as part of total farm value. In Chapter 14 the first two parts are covered; in Chapter 15 the third part is considered in connection with income and sale value of the whole farm. Another and slightly different treatment of building values—tax assessment—is presented in Chapter 25.

14

BUILDINGS–INVENTORY AND COST VALUE

A DETAILED DESCRIPTION of farm buildings and other improvements is necessary before any values can be assigned to them. Such a description, if it presents a complete picture of the buildings, will include measurements, type of construction, condition, and age. Information will also be wanted on use capacity of buildings and the extent to which this use capacity can be profitably utilized on the farm. An inventory of buildings, therefore, may be divided quite naturally into two parts: the first a detailed description and the second, often more difficult, an analysis of use capacity.

BUILDING INVENTORY

DESCRIPTION

Measurements should be taken of each important building in order to provide reliable data on building size. Horizontal measurements may be made with a tape, by pacing, or by counting studdings. Vertical measurements from the top of the foundation to the eaves may be estimated. These measurements, plus a description of the roof, provide the information necessary to compute the size of each building in cubic feet. The more valuable a building, the more time usually spent in obtaining accurate measurements, although practice in this regard varies from region to region. One important exception to this rule is that buildings in bad repair should be measured and described carefully so repairs can be figured easily.

The foundation should receive special attention. One often hears the saying, "A building is no better than its foundation." This applies to many farm buildings, particularly to larger ones like barns. It is far too common to see barns with swaybacked roofs, sagging corners, or bulging sides—all caused by poor foundations. Since the foundation has such an important bearing on the life and maintenance cost of a building, it is essential that information on foundations be as complete as possible. Completeness means a description of material used and the depth of the foundation below ground if this can be ascertained. Depth is especially important where a building is located on a sidehill, because a slight washing away of soil may undermine the foundation. Furthermore, the depth figure will indicate whether or not a foundation extends below the frost line, a fact which has a decided bearing on building value in areas where freezing and thawing may cause trouble.

A second item under construction is type of material used— wood, concrete block, hollow tile, aluminum, or steel. Information on material is needed in order to compute the annual repair and upkeep allowance because this allowance varies with the type of material. A barn constructed of wood means a painting bill periodically, an item that can be avoided if the building is constructed of hollow tile, concrete block, or steel.

A third item under construction is bracing. Inadequate bracing, like a poor foundation, may lead to rapid depreciation of a structure. Windstorm and tornado insurance companies recognize the importance of bracing because their losses vary with the soundness of the construction in the buildings they insure.

In a frame structure the size of timbers used as compared with the size of the building and the load it will be expected to carry should be considered along with bracing. It should be noted whether joists, posts, and other framework in the barn are large enough and spaced often enough to carry the weight of the hay; and whether the sides of the crib and granary are braced sufficiently to prevent grain from spreading the building. A purchaser of a farm recently found the crib and overhead granary required repairs equal to half the expense of a new building the same size, all because in the original construction 20 years earlier, bracing was skimped and the timbers used were too small.

A final item under construction is the roof. Many types of roof construction are used, but the principal ones are shed, gable, gambrel, and gothic (see Figure 14.1). Since the cost and upkeep of these different roofs vary, information on the kind of roof construction and shingles should be included in the appraisal report.

Age and condition are two other aspects of building description that need to be noted. Age gives a first approximation of building condition and indicates within limits the manner in which the structure was built. It is difficult, however, to obtain the accurate age of buildings on many farms because the present owner or tenant may not know when they were built. A knowledge of the manner and types of construction in different periods will be helpful in estimating age in these cases. A common distinction, for instance, is that between old barns built of hand-hewn timbers and newer ones constructed of sawed lumber. The inside of a building should be examined before estimating age, because a new coat of paint on the outside may be an effective disguise.

Age is not always a reliable indication of condition. Many of the older barns in the eastern part of the country are in far better condition today than barns half their age built of inferior materials and not maintained as well. A building in poor condition should be so indicated in the appraisal report by remarks on inferior materials, inadequate foundation, and poor bracing, and also by a special notation on the need for repairs. Often shingles or siding will be loose or missing, paint will be needed, or the foundation will be cracked or broken away at some point. These and similar evidence of depreciation give a ready clue to building condition.

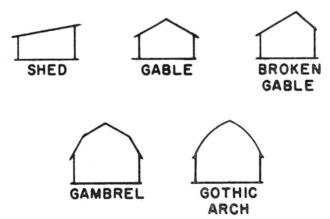

FIG. 14.1. *Roof types commonly found on farm buildings.*

TABLE 14.1. Appraisal Form for Building Information

Building	Measurements			Square Feet	Cubic Feet	Material	Foundation	Roof	Age	Condition	Adapted
	Length	Width	Height								
House, 1½ story	26	32	16	1,248[a]	19,970[a]	wood	concrete tile	gable	24	good	yes
Barn	32	40	24	1,280	30,720	wood	concrete	Gambrel	24	fair	no
Crib-granary	31	32	20	992	19,840	wood	concrete	Gambrel	24	good	yes
Silo	..	14	32	....	1,568	tile	concrete	none	40	poor	no
Machine shed	20	40	11	800	8,800	wood	pole-type	shed	20	good	yes
Feed tank	..	9	17	....	555	steel	concrete	none	3	excellent	yes

[a] Square footage for 1½-story house is 1½ times base area; cubic footage is figured by multiplying length times height to point on roof which makes an even cube.

Notes:

Water system. Electric pump with 100-foot well, equipment 10 years old. Adequate supply of water available in all seasons.

Fences. Boundary is three-wire barb, in good condition. Inside fences are mostly one-wire electric.

Shelter belt, landscaping, etc. Present shelter consists of a few old trees of little value. No landscaping around dwelling, not even a fence separating house from other buildings.

Building layout. Buildings are proper distance apart and from highway, except granary which is located too far from barn. More space is included in farmstead than is necessary.

In the sample building form presented in Table 14.1, a place has been provided for the appraiser to record his building measurements and other facts pertinent to the building description. It is not intended that each column be filled out for every building regardless of size or value. Some buildings are not worth much time or thought. An old hog house worth only a few dollars as salvage does not warrant much attention.

A set of pictures is a valuable addition to the building section of an appraisal report. A group of building pictures, such as those in Figure 14.2, not only conveys a better description of the general appearance and setting of buildings than any number of words but also stands as a permanent record of the buildings at the time of appraisal. Moreover, photographs often bring out details that might escape the appraiser's notice. Pictures are so inexpensive and easy to take that they should be considered as a regular part of the appraisal report.

A set of pictures, however, cannot take the place of a written description like that given in Table 14.1. Pictures do not show inside construction such as bracing, nor do they show the depth of foundation. Neither pictures nor tabulated description alone provides a complete inventory; both are needed to make a thorough appraisal report on buildings.

USE CAPACITY AND ADAPTATION

Physical measurements and description are only part of the building appraisal; a second and more difficult phase is a judgment or estimate of building adaptation. A barn may be large enough but because of poor arrangement may provide for a smaller number of livestock than might be expected. Moreover, much depends on the use that can be made of the barn. A well-equipped dairy barn on a wheat farm is far less important than the same barn on a dairy farm near a big city.

The information on adaptation, which takes only a small amount of space in the appraisal report, assists the appraiser or anyone reading the report in deciding how useful a given building is to the farm. Study of these use evaluations will indicate how adequately a farm is supplied with buildings—whether the granary will hold the grain that would normally be stored, and whether the barn will provide space for the livestock that would normally be kept on the farm.

It is not easy to say with exactness how well a building is adapted to a given farm; at least it is not possible to give a yes or no answer in every case. A large central hog house accommodating 12 sows may be larger than the farm warrants, because a farmer on this farm will keep on the average only 6 sows. The

FIG. 14.2. *Photographs of Farm* X *of the kind used in an appraisal report.*

present operator, however, may be keeping more than 12 sows and using the hog house to the limit. The next operator coming on the farm may have little use for the hog house because he uses portable hog houses. A similar problem arises with silos for cattle feeding; the present operator may use a silo, whereas the community practice may be evenly divided between those who do and those who do not use silos in their cattle feeding operations. Cases of this kind are likely to bewilder the beginning appraiser if he attempts to reach a yes or no answer for the adaptability of every building. To do the job well the appraiser needs to be familiar with the use made of buildings in the community, and he needs to know also the kinds of buildings that go with different types of farming. Difficult cases may be handled by simply identifying them as such or by indicating that the building in question has too much or too little capacity for the needs of the farm.

A desirable method of judging adaptation is to compare building capacity with cropping system and farm type. Poor adaptation exists if the farm contains a large amount of permanent pasture but does not have buildings designed to house the livestock that normally would be carried on the pasture. A crib and granary that will hold 2,000 bushels of corn and 4,000 bushels of small grain is poorly adapted if the farm will normally produce 6,000 bushels of corn and 2,000 bushels of small grain. A note indicating the lack of buildings or poor adaptation may be made in the appraisal report, as in the column labeled "Adapted" in Table 14.1.

CHANGING TECHNOLOGY

New developments in building design and use have brought about rapid obsolescence in many farm service buildings. Hay balers, new types of silos, grain drying equipment, and automatic feeding are examples of modern technology which are changing the types of buildings being built. In describing and evaluating the building situation on a given farm the appraiser may find himself faced with problems such as the following: Will the dairy barn and facilities meet Grade A milk requirements? If not, how much will it cost to meet these standards? How much would it cost to convert the present stall-type dairy barn into a loafing shed and milking parlor? How much expense would be required to convert present slat-type crib storage for ear corn into facilities for drying and storing shelled corn? Are lots and grain storage arranged so automatic feeding facilities can be installed economically? Is the cattle shed built so that a

tractor with a manure loader can clean it out? Has provision
been made for storage of baled or chopped hay convenient to
area where hay is fed? Questions which have a direct bearing on
the operator's use of buildings affect the value of buildings and
hence the value of the farm as a whole. To handle questions of
this kind the appraiser must familiarize himself with material
and equipment costs and with the current trends in building construction and use.

THE HOUSE

Farmers will probably attach more significance to their
houses in the future than they do now. This trend is suggested
by the added meaning which the farm home has to the owner in
the East where sufficient time has elapsed to develop attractive,
well-landscaped farmsteads. In the West and Midwest, on the
other hand, the early stage of development is just passing, and
with it some of the hastily built houses that will be and are being
replaced with more carefully planned and better constructed
dwellings. Another reason for the lack of attention given to the
house in some sections is a high percentage of tenancy. But
even on tenanted farms housing standards are being raised. In
the thirties many rural houses had no electricity and no plumbing. Now electric service is taken for granted, and so are running water, plumbing, and central heating in most areas. A
prospective farm operator considering rental or purchase of a
farm will in all probability make careful inquiry, at the insistence
of his wife who may accompany him, as to the facilities in the
house—including information on the reliability of electric service
and whether the house has a basement, modern plumbing, central heating, and insulation. The very fact that an increasing
number of farm homes do have these features means that our
dwelling standards are on a higher plane. In view of this added
emphasis on the dwelling, the appraiser should include in the
appraisal report a detailed description of the house.

OTHER IMPROVEMENTS

The water system, fences, windbreak or shelter belt, and
similar items which are a part of the farm inventory should
receive attention in proportion to their importance to the farm.
The appraiser will be justified in spending a large amount of
time analyzing the water system on a livestock farm or ranch in
a region where water is difficult to get and wells often go dry.
He will want to determine the flow of water, whether or not it is
adequate, how likely it is to dry up, and the condition of the

pumping equipment. Fences are given more time and thought where livestock has to be considered. A farm with 500 rods of hog-tight fence in good condition may be worth several dollars an acre more than a farm similar in every respect except that it has poor fence. Notes also should be made of the shelter belt or windbreak because of its value to the farm. An outline map of the farmstead, as in Figure 14.3, may be made by the appraiser where a detailed appraisal is desired by a prospective buyer. Such a map, by showing the distances between the various build-

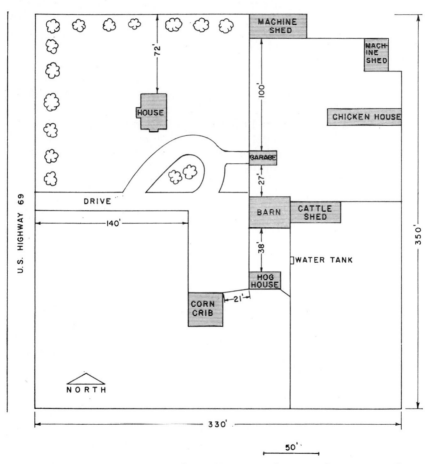

FIG. 14.3. *Map of farmstead on Farm* X *showing location and arrangement of buildings. A map of this kind is especially desirable in a detailed appraisal of a farm with an extensive outlay of buildings.*

ings, indicates whether or not they are economically arranged for work around the farmstead.

VALUE BASED ON REPLACEMENT COST
LESS DEPRECIATION

Since farm buildings do not lend themselves readily to either sale value comparison or income capitalization, replacement cost less depreciation is the principal method for figuring farm building values. For city buildings sale value and income methods can usually be used, because in the city rental income on buildings is available in some cases and can be estimated for others. Furthermore, rough sale value estimates can be made for land and for buildings. But in strictly farm areas neither building income nor building sale values are very reliable. In this chapter the method used will be replacement cost less depreciation, and the extent to which income and sale values of farm buildings can be used in farm appraisal will be considered in Chapter 15.

Two different farm building situations will be used to explain the replacement cost less depreciation method of valuation. The first is simpler and easier to grasp: a set of new buildings. The second is more complex and difficult: a set of old buildings which exhibit both physical deterioration and obsolescence. These two combined factors, physical deterioration and obsolescence, represent what is generally recognized as depreciation. With an old set of buildings, cost is based on replacement—the kind of building that would be built to replace the present one—not on reproduction of the old building.

COST-NEW SITUATION

The value of a new building is generally regarded as not more than its cost. We can look at this cost figure as the upper limit or ceiling to value. No one would be willing to pay more than cost because of the opportunity to build at the cost figure. If for a short period it is not possible to build, such as a wartime restriction, values could exceed cost because this alternative of building would not exist.

Cost, however, is not the lower limit, or floor, to value since an unwisely planned building may not be worth what it costs. For example, the value of a poorly arranged building may be less than the cost to everyone except the owner who may want this particular type of structure. But the owner's view would be subjective value and not objective value supported by public opinion. Cost value in this instance would be reduced by a dollar amount of obsolescence to cover the unwise planning.

Cost in this discussion is broadly defined. It covers all the items entering into the construction of the building including overhead, profit of the builder, insurance, and interest on the investment during construction, as well as labor and materials.

The cost method is actually not as simple as it appears at first sight. If five competing contractors are bidding on a new barn, it is likely that each will submit a different figure and that there will be considerable difference of opinion as to what the low bid is going to be. Builders vary in their eagerness for a job and in their methods of calculating costs as well as in their methods of construction.

Cost Estimates

Three procedures may be followed in obtaining cost estimates, although assessors and appraisers are generally concerned with what will be described as the third procedure. The first is to collect and compute detailed cost estimates just as a builder would in preparing a bid. This is too laborious, time-consuming, and expensive for the average valuation by an insurance inspector, assessor, or appraiser. The second procedure is to total all of the unit costs that go into construction. For example an overall figure including labor and materials could be computed for a unit of a certain type of roofing, the unit being 100 square feet. Even this unit procedure, however, requires more detail than most valuation personnel will want to use unless they are interested in obtaining a valuation supported by detailed figures such as might be desired in a court case where the valuation is being contested.

The square foot or cubic foot methods are the third and most common procedure used in cost estimating. Actually comparison is the basic element in both these procedures—comparison between the building to be valued and one similar in type for which costs have been figured per square foot or cubic foot. Value of the building in question is determined by multiplying its size in square or cubic feet by the appropriate square or cubic foot cost figure. Of course many refinements and variations are necessary because buildings are seldom alike. The refinements consist of plus and minus changes to take care of the differences or variations between the bench mark property and the building being valued. As an example, allowance must be made if a special siding or roofing material is used instead of the regulation type on the bench mark property. This allowance or adjustment will be plus or minus depending on whether the item of difference is worth more or less than the item on the standard or bench mark property. This need for adjustment is especially applicable to dwellings which have many variations in such items as porches, fireplaces, flooring, heating, and the like. Size is another factor that must be taken into consideration since cost does not increase as fast as size increases, but size as well as many other variable factors can be readily handled by the use of plus and minus adjustments applied to the standard or key property.

To cube a building, it is necessary to multiply three dimensions, width times length times height. To get the height it is customary to measure from the underside of the basement floor to a point on the roof which makes an even cube. For example with a gable roof, one-half of the vertical distance of the roof story would be used.

To get the square footage of a building, the dimensions of each floor are used. Thus a 20-foot by 30-foot house of two stories would figure 1,200 square feet. Since only two dimensions are involved, this method is not usually as accurate as the cubic foot method, but where buildings closely approach the standard in height the results will be as accurate as necessary. Practice by the appraiser will soon indicate whether to adopt the square foot or the cubic foot procedure. Of course where appraisers, assessors, or insurance inspectors are working under an established system, either the square foot or the cubic foot method will undoubtedly be the accepted procedure.

One of the dangers of both the square foot and cubic foot methods is an apparent accuracy which does not exist. A value carried out to the last cent, such as $7,616.53 to represent the cost of a building, looks accurate but, on account of the large amount of estimating involved, is not accurate at all. Constant vigilance should be exercised to compare overall figures with standard properties of known cost to see that results are consistent and reasonable.

Procedures and Examples

Standard, bench mark, or key buildings—examples to use as models—are the first step in a cost approach to building valuation. Two such examples are presented in Figures 14.4 and 14.5. For a complete set of models the reader is referred to an assessment or building appraisal manual designed specifically for a particular area. However, the examples in Figures 14.4 and 14.5 provide illustrations of the type the appraiser will find useful.

Costs were figured on blueprints and specifications prepared by the Midwest Plan Service for the buildings in Figures 14.4 and 14.5. These costs on both a square foot and cubic foot basis are shown at five-year intervals from 1938 to 1968. The upward trend in building costs is readily apparent. An appraiser can provide his own set of costs for typical buildings in his area by checking with lumber yards, contractors, and farm owners who have recently had farm buildings constructed on their farms.

REPLACEMENT COST LESS DEPRECIATION (OLD BUILDINGS)

The second and more common building valuation problem relates to a set of buildings that is not new. Valuation under these conditions is not so easy. The first step is to estimate the replacement cost of the buildings; the second is to estimate what is called depreciation made up of two factors, physical deterioration and obsolescence.

Replacement Cost

Replacement cost of an old set of buildings presents an interesting problem. Is it desirable to figure the cost-new of an outdated building that would not be built the same today? What should be the policy in the case of a large old farm dwelling or of an obsolete horse barn? If these buildings should be destroyed by fire or windstorm, it is obvious that they would be

Year	Index	Cost per Square Foot	Cost per Cubic Foot	Total Cost
1938	100	$ 3.12	$0.148	$ 3,900
1943	131	4.09	0.194	5,100
1948	210	6.56	0.312	8,200
1953	240	7.50	0.350	9,360
1958	275	8.45	0.400	10,550
1963	290	9.05	0.430	11,310
1968	350	10.85	0.510	13,600

FIG 14.4. *Cost data for five-room farm dwelling, 26' x 48'; square feet 1,250; cubic feet, 26,263. (Courtesy Midwest Plan Service, Iowa State University, Ames, Iowa.)*

Year	Index	Cost per Square Foot	Cost per Cubic Foot	Total Cost
1938	100	$ 3.20	$0.135	$2,765
1943	131	4.19	0.177	3,600
1948	210	6.70	0.283	5,800
1953	240	7.70	0.320	6,635
1958	275	8.80	0.370	7,600
1963	290	9.25	0.390	8,000
1968	350	11.10	0.470	9,600

FIG. 14.5. *Cost data for corncrib and granary, 27' x 32'; square feet, 864; cubic feet, 20,484. Cost includes an inside elevator. (Courtesy Midwest Plan Service, Iowa State University, Ames, Iowa.)*

replaced by different structures. And it is this likely replacement that provides our answer. In order to keep close to reality the appraiser should figure replacement cost-new and not reproduction cost-new. Reproduction would be an exact duplication of the old building by a new one. Replacement, on the other hand, is the substitution of a new building conforming to present needs to take the place of the old one. Consequently the first step in the valuation of an old set of buildings is to figure the cost-new not of an exact reproduction of the present buildings but of a set that would replace them if built today.

The second step, the estimate of depreciation, can be handled best by dividing it into its two components—physical deterioration and obsolescence.

Physical Deterioration

Physical deterioration, the more common aspect of depreciation, means wear and tear on the building which ultimately leads to exhaustion or retirement of the unit from service. What do we mean by wear and tear on a farm building, and how do we measure it? The answers can be given in terms of an examination of the building to detect deterioration in detail. To what extent do the materials show disintegration or wear? Is the paint scaling or worn away? Are boards coming loose and allowing water to get into the structure? Is the foundation cracked? Has the building settled? Are there evidences of dry rot or termites? Does the roof leak? Are the gutters rusted through? These are examples of deterioration which have to be evaluated in formulating an estimate of physical depreciation.

Obsolescence

Obsolescence includes all other factors, except a fall in prices, which cause a building to decline in value. Inventions, styles, and changing economic conditions are typical of the factors which cause a building to become out of date or obsolete. Examples of obsolescence in farm buildings are numerous—a big horse barn where horses are no longer used, a poorly designed central hog house, a silo that is too wide so that the silage spoils, a house that is too large for the average farm family.

Obsolescence can be classified into different types: normal vs. special on the one hand, and internal vs. external on the other. Normal obsolescence is an anticipated rate of going out of date or of losing value other than by physical deterioration. Special obsolescence is something altogether different; it is a large unexpected loss in value due to some event or innovation which makes the old building much less useful than a new one. A new milk ordinance might make many of the dairy barns obsolete because of requirements that these existing barns did not meet.

A comparison of internal and external obsolescence provides another insight into the depreciation problem. Internal obsolescence includes those changes in building styles, layout, materials, and the like which make a new building more efficient and desirable than the old. A new type of insulation, the milking parlor, and prefabricated rafters are illustrations. External obsolescence, on the other hand, includes those changes outside the building which make the building out of date. The new milk ordinance, replacement of horses by tractors, and new regulations on grain storage or fire insurance are the kind of changes which bring about external obsolescence.

Obsolescence can be measured but not with precision. Just as with physical depreciation, it is necessary to study the building in detail to evaluate each and every feature in comparison with similar features in a new structure. This evaluation can be summed up by comparing the present building with the type of building which would be erected in its place if the present one

were destroyed by fire or windstorm. The difference in usefulness of the two buildings, aside from physical deterioration, gives
a good idea of the loss in value through obsolescence. A horse
barn costing $8,000 to reproduce which had a physical deteriora-
tion of $4,000 might have obsolescence of $2,000 based on a
comparison of this type. In this instance the current depreciated
value would be only $2,000. If the horse barn were destroyed by
a windstorm, the new barn built to replace it would probably be
much smaller and might cost only $4,000. In the next chapter
we will have more to say about obsolescence in the discussion
of use value of farm buildings.

Depreciation Procedure

Two general procedures may be followed in handling the
combination of deterioration and obsolescence: the current observation method and the age-life annual rate method. The first
method is simpler and easier to substantiate, but the appraiser
and the building owner cannot rely on it alone because it does
not provide an annual depreciation charge which an appraiser
has to have if he figures building expenses and an owner has to
have in figuring net income and income taxes.

Current observation is the comparison of the building as it
stands today with a new modern building. Both physical deterioration and obsolescence are included in the comparison. Some
appraisers may wish to separate the two while others who are
not interested in as much detail may prefer to lump all the depreciation factors into one figure.

If the appraiser is interested only in the present value of
the buildings and not in the value of the farm as a whole, this
observation method has distinct advantages. Although he may
want to know the date the building was erected and may be interested in estimating the remaining life of the building, neither
of these two items is necessary in his evaluation. All that is required is a mental comparison of the difference between the
present building and a new modern one that would replace it if
the present one were destroyed. In making this comparison, the
appraiser will observe physical deterioration conditions such as
paint, roof, bracing, and the like, and also obsolescence in terms
of features in the present building which are out of date.

The age-life method involves an estimate of the remaining
years of useful life of a building. The depreciation—physical
deterioration and obsolescence—is then calculated on an annual basis as a percentage of the building's life. If we are appraising a well-built, 10-year-old barn which we estimate has 90
years of remaining useful life, then the annual depreciation rate
is one percent on the cost-new or 1.11 percent of the present

value. When the barn is 50 years old, if no unexpected changes have occurred and if the price level is the same, the value should be one-half the cost-new, and the annual depreciation rate would be the same—one percent of cost-new, or in this case two percent of current value which is one-half of the cost-new.

What we have just described is a straight-line depreciation procedure for the age-life method. The annual rate is simply calculated by taking the value when new, subtracting the salvage value, if any, at the end, and dividing this figure by the number of years of estimated life.

Other more complex methods of figuring age-life depreciation include the reducing balances procedure which assumes a building will last indefinitely. In this procedure the rate of depreciation—say one percent—is applied not to the cost or value when new but to each year's depreciated value. For example a new $10,000 building would have $100 of depreciation the first year by the straight-line type if the building were estimated to last 100 years, and it would have $100 of depreciation in each succeeding year unless something unforeseen occurred. But in the reducing balances method, if one percent is the rate, $100 is the depreciation the first year, $99 or one percent of $9,900 the second year, and so on.

The sinking fund, or annuity method, offers a still different procedure to figure depreciation. This is a mathematical plan which provides for the setting aside each year of an amount which during the lifetime of the building will accumulate to an amount equal to the cost-new figure. By this method the owner, if he follows the plan exactly, will have on hand an amount sufficient to reproduce the building when it is anticipated that it will be retired from service because of physical deterioration and obsolescence. This plan assumes no change in price levels.

A major difficulty with the age-life depreciation method and with all the procedures followed in applying it is the establishment of a reliable useful-life figure. How can you decide how long a building will last? Some farm buildings in the eastern part of the United States and Canada are over 100 years old and still in good condition and fully used (see Fig. 14.6). In Europe farm buildings in this category are often even older. This does not necessarily mean that all these old buildings would be replaced by similar buildings if they were destroyed. Obsolescence has taken its toll. But the sobering fact remains that a well-constructed building, when new, does not lend itself to an easy estimate of its useful life (see Figs. 14.7 and 14.8).

One way of avoiding, at least to a limited extent, the pressure for an estimate of useful life is the use of an annual expense figure which the appraiser estimates will maintain the property

FIG. 14.6. *A set of buildings on a Frederick County, Maryland, dairy farm. Buildings of this type illustrate valuation problems faced by appraisers in estimating deterioration and obsolescence. (Courtesy USDA.)*

FIG. 14.7. *End view of a $15,000 poultry building, 40' x 168', on a Midwest farm. Tall feed storage tank used in automatic feeding is located on right. These buildings are examples of specialized types which are becoming more common.*

indefinitely. This procedure will be discussed in greater detail in the next chapter.

An example of replacement cost less depreciation is provided in Table 14.2. A combination of observation and age-life depreciation methods is shown for four buildings. The percentages in Table 14.2 can be interpreted in two ways—as the proportion of the original cost-new value which has been used up or as the proportion of the life span which has occurred. For example, one-fifth of the cost-new value of the dwelling in Table 14.2 is estimated as physical deterioration and the same amount

FIG. 14.8. *Full view of poultry building erected in 1959. (This is same building shown in Figure 14.7.) Barn on the left, which has been converted into use for raising hogs, was built in 1899. This contrast in age indicates building situations which appraisers meet in valuing improvements.*

as obsolescence, a total of two-fifths. From the age-life viewpoint, two-fifths of the dwelling life has occurred with three-fifths of the life left.

Depreciation and Price Level

Neither physical depreciation nor obsolescence covers changes caused by fluctuations in prices. If the price level rises substantially after a building is constructed, the value may increase above the cost, offsetting but not preventing physical

TABLE 14.2. Example of Depreciation, Deterioration, and Obsolescence

	Dwelling	Barn	Crib	Machine Shed	Total
Value new	$9,750	$9,500	$6,900	$1,475	$27,625
Deterioration					
Percent	20	50	50	75	
Amount	$1,950	$4,750	$3,450	$1,105	$11,255
Obsolescence					
Percent	20	25	12½	....	
Amount	$1,950	$2,375	$ 860	....	$ 5,185
Total depreciation					
Percent	40	75	62½	75	
Amount	$3,900	$7,125	$4,310	$1,105	$16,440
Present value	$5,850	$2,375	$2,590	$ 370	$11,185

deterioration and obsolescence. This happened in the years 1938–68 when building costs were increasing faster in many cases than depreciation. It was common in 1968, therefore, to have the reproduction cost of an old building amount to double or almost triple the original cost (Figs. 14.4, 14.5). Of course the reverse situation could happen and has happened with values dropping because of declines in building costs. A decline for this reason should be classed as a price level decrease and not as physical depreciation or obsolescence.

There are two ways to handle price level fluctuations in connection with depreciation. One is to stick with original cost and make no allowance for price level changes; the other is to base the annual depreciation, both deterioration and obsolescence, on the current replacement cost of the building. An illustration of how these two systems would operate using the crib and granary pictured in Figure 14.5 is presented in Table 14.3.

In using the original cost of $2,800 as a base for the first system, both the straight-line and observed depreciation methods are shown; in one case the 1968 value comes out at $1,300 and in the other at $1,260. But in 1968 it makes little sense from a replacement viewpoint to value the building at such a low level. To adjust for this situation both the straight-line and

TABLE 14.3.   Depreciation on Original Value and Adjustments for Price Level for Crib and Granary

1. Year built 1938
    Estimated life ....50 years
    Cost new $2,800
    Salvage value ....$ 300
        Straight-line depreciation on original value
            Original value $2,800
            Annual depreciation rate: 2% × $2,500 = $50 a year
            Depreciation 1938–68: 30 years × $50 ....$1,500
                Estimated 1968 value $1,300
        Observed depreciation on original value
            Original value $2,800
            Deterioration: 30% × $2,800 = $ 840
            Obsolescence: 25% ×   2,800 =   700
                Total $1,540 $1,540
                Estimated 1968 value $1,260
2. Depreciation adjusted for price level
    Replacement cost new of crib and granary, similar capacity $9,600
    Salvage value after 50-year life $600 9,000
    Estimated depreciation: 2% × $9,000 = $180 yearly
    Straight-line depreciation: 30% × $180 = $5,400 ....$5,400
    Straight-line depreciated value, 1968 ....$4,200
    Observed depreciation of present building
        60% × $9,600 = $5,760 $3,840
            Observed depreciated value in 1968 ($9,600 less $5,760) $3,840

observed depreciation methods are based on the replacement cost of $9,600 which brings the estimated 1968 value up between $3,840 and $4,200. If a purchaser bought Farm X in 1968, it would certainly be logical for him to figure on a cost value at this higher level in order to provide for the maintenance, remodeling, and replacement of this crib and granary in line with 1968 prices.

The price level, it is evident, is a significant factor in depreciation if the second system is used. As building costs increase under this second system, depreciation increases correspondingly, but in compensation the value of the building increases also. Conversely, as building costs decrease, depreciation decreases, and the value of the building declines also.

Applications

Insurance values on farm buildings are usually set up on the basis of a fixed percentage ceiling on replacement value less depreciation, and depreciation includes both deterioration and obsolescence. If the figure is 80 percent, which is common, that means the insured value or the amount of maximum insurance is 80 percent of the present cash or market value, if the buildings have a market. Since farm buildings usually do not have a market, it is 80 percent of what they would sell for if they did have a ready market. In the case of the buildings valued in Table 14.3, the maximum insurance would be 80 percent of $11,185, or approximately $8,950. It is not easy to keep fire and windstorm insurance at the proper figure. Insurance companies quite naturally have difficulty in keeping obsolete buildings from being overinsured and in keeping most buildings from being underinsured during periods of rising prices.

Assessments of farm buildings are in a class by themselves. The purpose for which they are made—the levy of taxes—is a major factor. As for assessment procedures, the bench mark or standard building approach presented in this chapter is common, many states having a special system which they follow. These systems usually have detailed tables for figuring costs of construction, for adjustments to standard property values, and for physical deterioration and obsolescence. Some states have developed a system of their own with the assistance of professional appraisal specialists. Other states follow an appraisal system developed by an appraisal company. Still other states follow a variety of practices including the reappraisal of all or part of the properties in a city or county by a professional appraisal service.

In reporting income taxes it is important to depreciate buildings and other property in a systematic manner. To assist taxpayers with this problem a detailed statement on depreciation has been prepared and made available by the United States Treasury Department.[1] Farm buildings in the latest edition of this bulletin are mentioned specifically. Where buildings are set out in a separate account without equipment such as plumbing, elevators, and the like, a total life in years of 60 is set forth as reasonable. This gives a straight-line yearly depreciation rate of 1.66 percent. Where equipment is included the composite rate judged as reasonable is higher, 2 percent for good and average construction and 2½ percent for cheap construction. These percentages include normal obsolescence as well as physical deterioration. One final comment is pertinent: In income tax reporting, depreciation is based on original cost and not on replacement or reproduction cost. Although this basis may not always be realistic for appraisal purposes, at least it is simple and definite.

QUESTIONS

1. Why are farm buildings valued separately from the land in some cases?
2. Explain the principal method used in valuing farm buildings separately from land.
3. Discuss the problems connected with estimating the cost of a new building.
4. Explain the difference between the cubic foot and the square foot methods of estimating building cost.
5. How are standard or bench mark properties used?
6. What is the difference between reproduction cost and replacement cost?
7. How can you tell the difference between physical deterioration and obsolescence?
8. If a building has depreciated 20 percent but building costs have gone up 20 percent, is there any depreciation? Discuss.
9. Explain the observation and age-life methods of depreciation.
10. Apply the depreciation methods explained in this chapter to a set of farm buildings in your community.

REFERENCES

Appraisal of Real Estate, 5th ed., American Institute of Real Estate Appraisers, Chicago, 1967.

Ashby, W., Dodge, J. Robert, and Shedd, C. K., *Modern Farm Buildings,* Prentice-Hall, Englewood Cliffs, N.J., 1959.

Ashby, Wallace, and Lindsey, M. M., "Obsolescence of Buildings," *Power to Produce. The 1960 Yearbook of Agriculture,* USDA.

Barre and Sammet, *Farm Structures,* John Wiley and Sons, Inc., New York, 1950.

---

1. *Income Tax Depreciation and Obsolescence, Estimated Useful Lives, and Depreciation Rates,* Bull. F, U.S. Treasury Department.

Carter, Deane G., *Farm Buildings,* 4th ed., John Wiley and Sons, Inc., New York, 1954.

Hansen, E. L., *Farm Building Appraisal,* Voc. Agr. Serv., Univ. of Ill., Urbana, VAS 3017, revised annually.

Income Tax Depreciation and Obsolescence, Estimated Useful Lives, and Depreciation Rates. Bulletin F, Bureau of Internal Revenue, U.S. Treasury Department, Washington, D.C.

Neubauer, Loren W., and Walker, Harry B., *Farm Building Design,* Prentice-Hall, Englewood Cliffs, N.J., 1961.

Pos, Jacob, "Farm Buildings Appraisal, Old and New Construction," in *1960 Appraisal and Valuation Manual,* American Society of Appraisers, Washington, D.C., pp. 259–78.

Singley, Mark E., "The Dynamics of Structures," *Power to Produce. The 1960 Yearbook of Agriculture,* USDA.

Van Arsdall, Roy N., and Ashby, Wallace, "Costs of Farm Buildings," *Power to Produce. The 1960 Yearbook of Agriculture,* USDA.

15

VALUE OF FARMSTEAD ALONE AND AS PART OF FARM

THE VALUE of a set of farm buildings can be obtained either by separating the farmstead from the rest of the farm or by valuing the farm as a whole and dividing the total between land and buildings. These two procedures as considered in this chapter are entirely different from the cost approach for the valuation of individual buildings discussed in the preceding chapter. Buildings in this chapter will be treated as a farmstead unit, either as the farmstead alone or as part of the whole farm.

Separate farmstead appraisal has distinct advantages in those areas where there is a market for farmsteads among non-farmers who may or may not work in the city but like to live in the country. The whole-farm appraisal also has its advantages, particularly in preventing the overvaluation of buildings. One of the serious errors that is sometimes made in valuing farms is placing a separate value on the buildings, another separate value on the land, and then adding the two together to get the value of the whole farm. The total value of the two separate appraisals is likely to equal more than the value of the farm when appraised as a total unit. Especially dangerous is the use of cost estimates for individual buildings, because when they are added together

223

and combined with a land value figure, the result may be unrealistic and unreliable.

A major problem which plagues the appraiser in valuing farm buildings is surplus farmsteads, especially in areas where pressure for farm enlargement is resulting in larger farms and no market for the vacant farmstead buildings. This building surplus was illustrated recently by a farm which was offered for sale for $350 an acre. There were only a few bids and the best was for $300 an acre. The farm had only a fair set of buildings but they were insured for $9,000. The owner turned down the $300 bid. Later the man who had bid $300 said he would raise his bid to $325 an acre if the owner would remove all the buildings and fix the land where the buildings were located so he could raise corn on it. The question immediately arose as to what the buildings were worth in this situation. Were they worth the $9,000 shown in the insurance policy or were they worth less than zero by the amount it would take to tear them down and prepare the area for crops? Situations like this may appear to be insoluble, but there are straightforward and satisfactory methods that can be used to give reasonable estimates of farm building values.

VALUATION OF FARMSTEADS SEPARATELY

The main reason for appraising the farmstead separately as a unit is that there are areas, especially around cities and industrial plants, where farmsteads are being sold separately to people who are not farmers but want to live in the country. Where this practice exists the appraiser will be negligent if he does not include a separate valuation of the farmstead in the appraisal. The rapid increase in farmstead purchase by non-farmers has been accelerated by the major improvements in transportation which have made commuting to work much easier than it used to be. With radio, television, and the inclusion of rural areas in large consolidated school systems, the advantages of country living—including the saddle horse and plenty of space—have become much greater and more attractive.

On the other side of the coin, widespread pressure for larger farms has resulted in a decrease in the number of farmsteads used by operating farmers. As a consequence, surplus farmsteads have come on the market. Those near cities and within commuting distance of industrial plants have been bought by city people; those outside the commuting areas have in many instances been allowed to stand empty and deteriorate.

Appraisers have a difficult assignment in valuing separate farmsteads in the gray areas where there is occasionally a farmstead sold separately but where usually the surplus farmsteads stand empty. In one such area a farm was sold as a whole, land and buildings together, with no thought by the seller that the farm might bring more if the farmstead were sold separately. Not long after the sale the buyer was approached by a person who wanted to buy just the farmstead. In this instance the farm buyer, who owned land adjoining, had no use for the farmstead and was glad to sell it. In fact, the sale of the farmstead was a windfall. The amount he received for the buildings and farmstead acreage reduced the price he paid for the farm by $70 an acre, bringing it well under the market price he had expected to pay.

The farmstead on Farm *A*, as the appraisal in Chapter 1 indicated, was valued both as a separate unit and as part of the whole farm. In 1968 the tenant was occupying the dwelling and using the farmstead, but this situation might conceivably change if the next tenant were a farmer who lived nearby and wanted to expand the size of his farm by renting more land but did not want the farmstead.

Depreciation on the buildings in Farm *A* was figured as a cash cost in Chapter 1. A slightly different procedure is to figure depreciation as a percentage of the resulting value. In doing this a fixed percentage depreciation is estimated and this percentage is added to the capitalization rate in the income valuation. In this procedure depreciation is recaptured as part of the annual return. Recapture through the higher capitalization rate is frequently used in the appraisal of urban buildings, but it is not used as much in farm appraisals because farm buildings usually make up a small proportion of total value and in addition are made up of not one building but several of different ages and condition. Using a 2½% recapture figure for depreciation, and using all of the other figures for building expense as shown in Chapter 1, the results for the two different procedures for Farm *A* come out as follows:

1. Depreciation as an Expense			2. Depreciation as Percent Recaptured	
Income		$ 1,800	Income	$ 1,800
Expenses				
Taxes, repairs $450			Expenses	
Depreciation 450		900	Taxes, repairs $450	450
		$ 900		$ 1,350
Capitalized at 5%		$18,000	Capitalized at 7½%	$18,000

In case 1 the depreciation is estimated at so many dollars each year. In case 2 the depreciation varies directly with the value.

In this instance the estimated dollar amount and the percentage of value were estimated to give the same resulting dollar value.

On a sale value basis the farmstead on Farm A was valued at $18,000 after comparing the buildings and location with other sales in the neighborhood. A group of farm managers and rural appraisers on a sale value tour in August 1968 valued the farmstead on Farm A with three acres of land at an average of $18,650.

One of the fascinating aspects of appraising the separate farmstead is the likelihood that in the future there will be more developments in this area. With the depreciation schedules used in income tax accounting and with the heavy property tax and repair expense on buildings, a landlord has reason to find ways of escaping the ownership of a farmstead. An actual example is a landlord who sold the six acres in his farmstead to his tenant on a long-term contract. The landlord in this instance eliminated the property tax, insurance, and maintenance expense on the buildings. He had nearly exhausted his depreciation on the buildings so he lost little there. Best of all his crop share rent brought in almost as much as he had received before. The tenant did not fare badly either, because he achieved ownership of his own home so that he could make his own improvements and was able to start taking a substantial depreciation allowance on a new higher building value base.

VALUATION OF BUILDINGS AS PART OF WHOLE FARM

Where the farmstead is not sold or valued separately, the reasons for valuing the farm as a unit outweigh the reasons for valuing land and buildings separately. This is true for both the income and sale value methods of valuation. With the income method the chief argument against valuing land and buildings separately is the difficulty in arriving at a satisfactory estimate of building income. The task of estimating building depreciation and repairs presents problems enough, but these are small compared to the puzzling question of how much of the income received by the landlord should be attributed to buildings and how much to land.

Customary rental practice generally makes no provision for dividing total rental between land and buildings. The total cash rent may be figured as so many dollars an acre, or so many dollars an acre for cropland and so many for pasture, or perhaps as a lump sum for the farm as a whole. Rent paid on a combination cash and crop share basis usually requires a payment of a share of the crop plus a cash rent per acre for pasture. A cash building rental, if added, does not necessarily equal the full return for the buildings, because in the fixing of rentals there is no attempt to fix the rental for land and buildings separately. In reality buildings tend to raise the returns from cropland and pasture. The tenant can afford to pay a larger share of the crop or a larger cash rental per acre if he has the buildings which enable him to turn feed into livestock products—assuming that on the average the farmer finds it worth his while to convert feed into meat. The pasture rental in practice includes a portion of the building rental because the tenant could not pay for the pasture unless he had the buildings to house the livestock grazing on the pasture. It must not be overlooked that the tenant and landlord both receive an income from grain and hay storage which, if not provided, may cause spoilage or forced sales at harvest time. The tenant and his family also have the use of the house, a fact which is sometimes not fully appreciated until a house is rented in town. These considerations make it evident that although buildings do contribute to income, the separation of building income from total income is so difficult it should be avoided if possible.

Although it is difficult to value land and buildings separately on an income basis, there is still an opportunity to value them separately on a sale value basis. But even here, estimates in most cases have to be rough because there is little if any market for individual farm buildings. Whenever farm buildings are sold they are not sold separately from the land; they are usually sold as part of the farm as a whole or as a separate farmstead as discussed earlier. Consequently the reliable practice is to determine the total farm value first and then split that value between land and buildings. This procedure avoids the serious error of having two parts that may add up to more or less than the value of the whole.

DIVIDING FARM VALUE INTO BUILDING AND LAND VALUE

The problem of splitting total farm value into building value and land value is to find a satisfactory procedure for the division. In one sense it doesn't make much difference since the total of land and buildings is assumed to be satisfactory; in short, the division could be an arbitrary one depending on the particular attitude of the person making the division. But if building value is to have much meaning, a systematic procedure of obtaining it should be used.

BUILDING RESIDUAL METHOD

The building residual method is one where the value of bare land is subtracted from total farm value, leaving building value as the remainder. A fairly satisfactory value for bare land can be obtained by either income or sale value methods, but neither applies easily to buildings. If whole farm value is reasonable and bare land value is reasonable, then the remainder which is building value should also be reasonable.

An example of the building residual method using Farm X may clarify the procedure. Farm X has an income value of $340 an acre; in Table 15.1 it is shown as having an estimated sale value of $500 an acre. The reasons for the excess of sale value over income value include many factors of which buildings are an important one. Once the income and sale values are established as shown in Table 15.1, the next step is a valuation of the bare land on both an income and sale value basis. In this instance it is evident that an owner could get almost as much net rent from the land without buildings as with buildings, the difference being $15 an acre without buildings and $17 with buildings ($300 as compared to $340 when capitalized at 5%). If the farm were closer to an urban area, it would be easier to rent the house separately from the farm which would have given a higher farm unit value and a higher building residual value on an income basis.

On a sale value basis the farm unit in Table 15.1 is appraised at $500 an acre and the land alone at $400 an acre. This leaves $100 an acre or $12,000 for the buildings. Even this figure of $12,000 is low compared to a replacement cost less depreciation estimate of $18,000 for the buildings as they stood on the appraisal date. In this connection it is interesting to note in Table 15.2 the result when the cost estimate of $18,000 for Farm X buildings is subtracted from the farm unit value. In this procedure the bare land becomes the residual. Bare land in this instance on a residual plan amounted to only $190 an acre on the income method and to $350 an acre on the sale value method, in both cases below what it was appraised.

There are cases where cost less depreciation estimates may be fully justified; that is, where the building residual figure is equal to the cost less depreciation estimate, or in case the land residual plan is followed where the land residual comes out at a figure which is approximately the appraised value of the bare

TABLE 15.1. Building Residual Method Applied to Farm X

	Income Value Estimate		Sale Value Estimate	
	(*per acre*)	(*total*)	(*per acre*)	(*total*)
Farm unit total	$340	$40,800	$500	$60,000
Bare land	300	36,000	400	48,000
Building residual	$ 40	$ 4,800	$100	$12,000

TABLE 15.2. Bare Land Residual Method Applied to Farm X

	Income Value Estimate		Sale Value Estimate	
	(*per acre*)	(*total*)	(*per acre*)	(*total*)
Farm unit total	$340	$40,800	$500	$60,000
Buildings, cost	150	18,000	150	18,000
Bare land residual	$190	$22,800	$350	$42,000

land. Cases of this kind may be found in areas near cities where buildings can be rented separately from the land or in areas where buildings are more important than land.

BARE LAND VS. LAND WITH BUILDINGS

Appraisers are sometimes puzzled over valuation of bare land. The situation can be especially perplexing if two adjoining tracts are appraised, one with and one without buildings. An appraiser who had such an assignment for two 80-acre tracts came up with the same valuation for the two farms which were about equal in productivity but one had a farmstead while the other did not. The absentee owners of these two tracts were receiving practically the same net income although one had a $10,000 investment in buildings which the other did not have. The crop returns on the bare farm were slightly less but this small difference was more than compensated by lower taxes (since there were no buildings) and by no charge for building maintenance. In fact there were years when the landlord of the bare land received a higher net return than the landlord of the improved farm.

This favorable situation for unimproved tracts, the result of pressure for larger farms, showed up in studies made in the early fifties. A report in 1953 by W. D. Touissant and J. F. Timmons showed an income advantage for the unimproved tract.[1] The results of this study are portrayed in Table 15.3. The amazing situation in this study was the net income difference— $3.76 an acre in favor of the bare land. In percentage return on landlord investment the improved farms were bringing 4.6 while the unimproved tracts were yielding 6.7. At the same time improved farms were selling for $211 an acre compared to $200 an acre for the unimproved tracts.

A study of income and rate of return on bare land and on land with buildings for the years 1965–67 showed results similar to the 1953 study cited above but with not as large a premium for bare land. The results of this study are shown in Table 15.4.

1. *Iowa Farm Science*, 8(1):7–9, July 1953.

TABLE 15.3. Rent and Landlord Costs for Land With and Without Buildings, Iowa

Costs and Rents	With Buildings	Without Buildings	Difference
Gross rent	$18.14	$19.31	$1.17
Cost items:			
Building repairs and insurance	2.24	. . . . .	2.24
Property taxes	1.86	1.61	0.25
Other[a]	4.28	4.18	0.10
Total costs	8.38	5.79	2.59
Net rent	$9.76	$13.52	$3.76

Source: *Iowa Farm Service,* Vol. 8, No. 1, July 1953.
[a] Consists of management fee (10 percent of gross), seed costs, fertilizer costs, and other production costs paid by the landowner.

The estimated value per acre and the gross income per acre were remarkably close for both the improved and unimproved tracts. Expenses, however, as would be expected, were much lower on the bare land, thus increasing the net income by $1 to $3 an acre. The net result is a rate of return advantage for the unimproved tracts which varied from .6% in 1965 to .9% in 1966.

From a strictly economic viewpoint, the situation indicated in these studies is not a stable one. For example, the relatively high returns from unimproved tracts should bring about more demand and push the price up on these units. This has been happening with both neighboring farmers and absentee investors seeking these profitable tracts. But on the other side, it must be borne in mind that unimproved tracts include some risk. Although they may be in strong demand at present, this may not always be the situation.

TABLE 15.4. Income and Rate of Return per Acre on Improved and Unimproved Tracts on Iowa Crop-Share Rented Farms, 1965–67

	Improved			Unimproved		
	1965	1966	1967	1965	1966	1967
No. of Farms	334	284	323	169	137	110
Av. size, acres	251	257	261	170	167	151
Av. market value	$369	$400	$410	$358	$389	$412
Gross income	$ 31.62	$ 33.00	$ 34.12	$ 30.81	$ 34.00	$ 35.17
Operating expense	$ 13.98	$ 15.11	$ 17.16	$ 12.07	$ 13.67	$ 15.96
Capital expend.	$ 1.53	$ 1.62	$ 1.82	$ .78	$ .87	$ .86
Net income	$ 16.11	$ 16.37	$ 15.22	$ 17.96	$ 19.46	$ 18.37
Rate of return	4.4%	4.1%	3.7%	5.0%	5.0%	4.5%

Source: Survey sponsored by Iowa Society of Farm Managers and Rural Appraisers and Agricultural Experiment Station of Iowa State University, Ames.

SURPLUS BUILDINGS

This problem of valuing improved and unimproved land can be handled best by checking up on the building situation in the neighborhood. Such a survey will probably indicate one of three general conditions: a surplus of buildings in relation to cropland and pasture land, an intermediate situation, or a shortage of buildings. The surplus condition is found in many localities where the use of labor-saving machinery has resulted in an expansion in the size of the family farm. The owner of a bare piece of land which can be incorporated with surrounding farms has a distinct competitive advantage where this situation exists. Neighboring farmers anxious to expand their operations will pay the owner of this unimproved tract a rental return which will amount to as much as he could obtain if he built a set of buildings on the farm. The average farmer would not be interested in this farm if it had buildings because there would not be enough cropland associated with the buildings and it might be difficult to rent additional land nearby. The premium which the landlord receives for bare land rests entirely on the present surplus of farm buildings which leads farmers to compete with each other to rent adjoining bare land. The extra value of the bare land is represented by the unused building capacity on the neighboring farms. To go back one more step, the surplus of buildings is caused by a surplus or an unused capacity of power and machinery. Actually there are no unimproved or bare tracts in most areas. Tracts which appear to be unimproved are in reality improved in the sense that a nearby farmer with a surplus of buildings is willing to rent the tract and thus provide a set of buildings to go with the tract. If this farmer buys the unimproved tract, then the tract becomes improved automatically without any buildings being erected.

The value of buildings on Farm X is affected by a surplus building situation in the neighborhood. This explains the difference between the $18,000 replacement value of buildings and the $4,800 income value which was finally estimated. If there were no buildings on this 120-acre farm, the land could be rented out readily to adjoining farmers on favorable terms. Landlord gross income would be approximately the same with or without buildings, but net income would probably be higher without buildings—the same situation as shown in Table 15.3. From a strictly short-time view, it could be argued that the buildings have no value, that the farm would be worth more—at least that the landlord would net more income—if the buildings were not there. But it must be emphasized that this is a short view and that the abnormal factor is a heavy temporary demand for additional units of unimproved land that can be added to undersized improved units. A movement on the part of some owners to remove poor buildings and rent or sell the farm in units to neighboring farmers could change this situation in a short period. However, as long as this demand for unimproved units persists, this abnormally high rental income and value will attach to these units.

The second situation is the intermediate one where there is neither a surplus nor a deficit of farm buildings in the community. In this case the landlord of the bare 80-acre tract will have to bargain with neighboring operators to interest them in renting it. He may or may not have difficulty in renting it, but in any event he will receive a smaller income than under the surplus condition because most neighboring farmers will have all the land they want to operate.

Finally, there is the unusual situation of deficit buildings. Here the value of bare land will be below the value of improved land by an amount approximately equal to the cost of improvements. A good illustration of this situation exists in new areas before development of farm units has been completed. The landlord of bare land may not even be able to rent his land under these conditions because there are no nearby buildings which can be used in connection with the operation of his land. Nevertheless, the value cannot get much lower than the value of improved land less the cost of buildings; if it did a profit would be realized by anyone buying up land and improving it. Under these circumstances landlords would be tempted to put up a set of buildings themselves rather than sell at such a low figure.

BUILDING VALUE ESTIMATES BY FARMERS

Another useful building value estimate is that made by farmers—the persons most concerned with the day-to-day service provided by farm structures. The USDA has brought together figures, beginning with the Federal Census of 1940, in which farm operators were asked: What is the value of all buildings on the farm? This record has been maintained through an annual series of farm building estimates included in questionnaires to crop reporters (see Table 15.5). The resulting estimates, which are not cost less depreciation nor building residual figures, are a combination of the two in the sense that farm operators in giving their answers had both factors in mind—depreciated cost and the difference between land with and land without buildings. Whatever may have been in the farmers' minds in answering the building value question, the answers shown in Table 15.5 are highly illuminating both as to the trend over the years and the wide variation that exists throughout the country.

TABLE 15.5   Farm Real Estate: Market Values of Land, Farm Buildings, and Dwellings, 48 States, 1950–67

Years	Land in Farms[a]	Value of Land and Buildings		Value of Land Only		Value of Farm Service Buildings and Dwellings	Value of Farm Dwellings Only	Value of All Farm Buildings as Percent of Total Real Estate Value
		Total	Per acre	Total	Per acre			
	(billion acres)	(billion dollars)	(dollars)	(billion dollars)	(dollars)	(billion dollars)	(billion dollars)	(percent)
1950	1.16	75.3	65	53.1	46	22.2	11.7	29.5
1951	1.16	86.6	75	61.5	53	25.1	13.2	28.9
1952	1.16	95.1	82	67.9	59	27.2	14.3	28.6
1953	1.16	96.5	83	69.2	60	27.4	14.4	28.3
1954	1.16	95.0	82	68.5	59	26.6	14.0	28.0
1955	1.15	98.2	85	71.4	62	26.7	14.1	27.2
1956	1.14	102.9	90	75.6	66	27.3	14.4	26.6
1957	1.14	110.4	97	81.9	72	28.5	15.0	25.8
1958	1.13	115.9	103	86.8	77	29.1	15.3	25.1
1959	1.12	124.4	111	94.1	84	30.3	15.9	24.3
1960	1.12	130.2	116	99.5	89	30.7	16.1	23.6
1961	1.11	131.8	118	101.6	91	30.1	15.8	22.9
1962	1.11	138.0	124	107.2	96	30.7	16.2	22.3
1963	1.11	143.8	130	112.8	102	31.0	16.3	21.6
1964	1.11	152.1	138	120.2	109	31.9	16.8	21.0
1965	1.10	160.9	146	128.0	116	32.9	17.4	20.5
1966	1.10	172.5	157	138.1	126	34.5	18.2	20.0
1967	1.09	182.5	167	146.6	134	35.9	18.9	19.7

Source: *Farm Real Estate Market Developments,* USDA, CD-70, Apr. 1968, p. 27.
[a] Land in farms based on census of agriculture. Acreages for intercensal years are obtained by straight-line interpolations. Acreages for 1965 and later years are projected from the 1964 census by the change in the annual USDA estimates of land in farms.

BUILDING VALUE AS A PERCENTAGE OF TOTAL FARM VALUE

Farmers have estimated buildings to be a declining percentage of whole-farm value since 1950, the percentage declining, as shown in Table 15.5, from 29.5 in 1950 to 19.7 in 1967. During this period, however, the average value of the buildings on a farmstead rose from $3,900 to $12,450, an increase of over three times. The decline in building value as a percentage of whole-farm value was not caused by a drop in the value of farm buildings; instead it was caused by the increasing size of farms and fewer farmsteads.

That a wide variation in farm building values existed over the country is shown in Table 15.6. Total value of all farm building was highest in the Pacific region with an average farmstead building value of $20,650, followed by the Northeast with

TABLE 15.6. Value of Farm Buildings by States, Mar. 1, 1960, 1964, and 1967

State and Region	Buildings as Percentage of Land and Buildings (percent)			Total Value of Buildings[a] (million dollars)			Average Value per Farm, 1967[b] (dollars)	
	1960	1964	1967	1960	1964	1967	All buildings	Dwellings
Maine	50.5	46.8	44.1	126.6	116.4	113.2	10,900	5,550
New Hampshire	54.0	47.8	45.0	60.9	54.1	51.8	13,800	7,250
Vermont	46.9	41.4	38.4	108.3	107.9	119.3	15,450	7,000
Massachusetts	46.4	42.5	39.8	159.6	143.4	134.2	20,950	10,850
Rhode Island	47.7	42.0	39.0	23.7	20.6	20.4	21,900	10,150
Connecticut	41.7	38.5	35.4	158.3	146.8	140.2	28,800	13,550
New York	54.6	48.2	44.0	1,046.4	1,013.2	1,042.0	18,100	9,350
New Jersey	34.3	30.0	27.4	241.8	217.2	199.0	23,600	12,500
Pennsylvania	46.5	42.6	41.4	1,020.3	1,002.8	1,130.3	14,950	7,650
Delaware	38.0	30.0	28.2	69.5	65.4	75.2	20,000	10,500
Maryland	33.9	28.0	26.5	334.5	352.8	425.0	23,050	11,550
Northeast	45.6	40.3	37.7	3,349.9	3,240.6	3,450.6	17,450	8,850
Michigan	39.1	35.5	31.9	1,104.1	1,098.2	1,136.8	13,500	6,800
Wisconsin	42.7	39.2	36.0	1,193.4	1,195.6	1,289.1	11,700	5,200
Minnesota	37.0	33.0	32.1	1,769.7	1,649.6	1,864.7	15,400	7,100
Lake States	39.1	35.4	33.1	4,067.2	3,943.4	4,290.6	13,600	6,350
Ohio	30.4	28.5	25.6	1,382.4	1,439.3	1,515.1	14,000	7,400
Indiana	25.8	22.8	21.0	1,260.8	1,219.3	1,437.1	14,800	7,700
Illinois	18.9	17.5	16.2	1,813.9	1,827.8	2,144.8	17,500	7,850
Iowa	20.0	19.4	17.9	1,737.3	1,737.5	2,106.6	14,750	6,350
Missouri	28.7	25.3	23.1	1,091.1	1,196.4	1,409.3	10,200	5,400
Corn Belt	23.1	21.5	19.6	7,285.5	7,420.3	8,612.9	14,150	6,850
North Dakota	21.5	17.7	15.8	477.4	483.7	522.0	11,500	5,300
South Dakota	22.1	19.3	17.3	510.3	542.5	566.2	12,200	5,400
Nebraska	19.6	17.0	15.0	836.7	851.6	924.8	12,400	5,700
Kansas	16.8	15.0	13.2	850.0	868.5	943.8	11,000	5,650
Northern Plains	19.3	16.8	14.9	2,674.4	2,746.3	2,956.8	11,700	5,550

Virginia	36.0	32.2	31.6	648.2	684.2	816.3	11,450	6,700
West Virginia	47.6	45.6	45.3	211.3	209.2	203.7	6,800	4,100
North Carolina	37.7	34.0	31.9	1,095.0	1,179.9	1,308.3	9,450	5,600
Kentucky	32.9	32.2	31.3	758.4	915.0	1,021.6	8,000	4,650
Tennessee	38.4	34.5	33.9	809.2	879.9	1,029.6	8,400	5,150
Appalachian	36.8	33.8	32.6	3,522.1	3,868.2	4,379.5	8,950	5,300
South Carolina	38.5	33.6	31.6	470.4	445.8	499.6	10,250	6,400
Georgia	35.6	30.2	29.3	683.1	692.2	881.9	11,350	6,450
Florida	9.6	9.2	9.0	317.6	390.1	403.4	11,550	6,500
Alabama	39.8	34.0	34.0	586.7	608.4	774.6	8,900	5,350
Southeast	26.0	22.1	22.5	2,057.8	2,136.5	2,559.5	10,300	6,050
Mississippi	38.3	32.1	30.6	754.2	800.0	982.2	10,150	6,250
Arkansas	24.2	20.5	19.3	450.7	563.4	690.0	9,600	5,800
Louisiana	30.9	25.9	25.3	553.4	582.3	817.1	15,050	9,500
Delta States	31.3	26.0	24.9	1,758.3	1,945.7	2,489.3	11,200	6,900
Oklahoma	15.9	14.5	13.6	490.1	594.6	700.1	8,150	5,000
Texas	11.6	10.7	9.6	1,417.1	1,603.7	1,657.2	8,700	5,200
Southern Plains	12.5	11.5	10.5	1,907.2	2,198.3	2,357.3	8,500	5,150
Montana	13.0	11.9	11.1	296.1	320.1	351.4	13,600	7,150
Idaho	20.6	18.2	16.8	352.5	360.4	383.6	13,850	7,750
Wyoming	13.8	12.7	11.8	109.3	128.0	137.5	15,900	8,600
Colorado	16.3	14.9	14.0	339.5	388.2	423.5	14,900	7,850
New Mexico	11.8	10.8	9.9	129.8	167.4	186.6	14,150	8,000
Arizona	5.6	5.1	4.8	108.3	108.0	108.1	18,550	10,300
Utah	23.7	20.6	19.1	179.8	180.8	187.2	12,150	7,250
Nevada	9.2	8.2	7.7	31.1	30.9	33.6	16,300	8,450
Mountain	14.1	12.8	11.9	1,546.4	1,683.8	1,811.5	14,350	7,800
Washington	28.1	25.0	23.3	701.9	700.8	775.9	18,500	9,850
Oregon	21.4	20.0	18.5	397.3	441.9	480.2	12,800	6,950
California	10.5	10.0	9.4	1,394.7	1,617.0	1,714.8	26,650	14,050
Pacific	14.1	13.0	12.3	2,493.9	2,759.7	2,970.9	20,650	10,950
48 States[c]	23.6	21.0	19.7	30,662.7	31,942.8	35,878.9	12,450	6,550

Source: *Farm Real Estate Market Developments*, USDA, CD-70, Apr. 1968, pp. 28–29.
[a] Includes both farm dwellings and service buildings. Based on building percentages by size class as reported by crop reporters.
[b] Values are rounded to nearest $50 and are calculated using estimated farm numbers projected from the 1964 agricultural census.
[c] Regional and national totals derived from unrounded state values.

235

$17,450. At the other extreme were the Southern Plains and the Appalachian regions with average values of $8,500 and $8,950. From the standpoint of buildings compared to land, the Northeast region stood out with buildings representing 37.7 percent of the farm value in 1967. This was in marked contrast to the Pacific region where buildings amounted to only 12.3 percent of total farm value in 1967.

A study of building values compared to total farm values for individual states showed wide variations. For instance, according to Table 15.6, buildings made up 45 percent of the value of West Virginia farms but only 9 percent of the value of Florida and California farms. In the Corn Belt the variation was from 16 percent in Illinois to 25.6 in Ohio. In Ohio the appraiser could expect one-fourth of the farm's value to be in the buildings while in Illinois the percentage would drop to one-sixth. On Farm A the estimated $18,000 value for the farmstead buildings represented 15.4 percent of the total farm's value. In the ranching areas of the mountain region the percentage dropped to 4.8 in Arizona and 7.7 in Nevada.

BUILDINGS IN THE INCOME VALUE APPRAISAL

The first time a student in appraisal looks at an income value statement, such as the one for Farm A in Chapter 1, he is likely to be puzzled by the absence of any reference to building income and value. Income for all of the crops and pasture is shown and a small farmstead rental may be indicated, but no income is listed which reflects the value of the dwelling and service buildings. On the other hand, a heavy expense is shown for repairs, maintenance, insurance, and depreciation of buildings. As a consequence the tendency of beginning appraisers is to add a replacement cost less depreciation value of the buildings to the income value of the farm obtained by capitalizing net income. To do this, however, would be incorrect because all the returns on buildings are included in gross crop and pasture rental income. In brief, gross rental income shows a full estimate of what the landlord can expect from the farm as a whole—land and buildings.

The treatment of buildings in an income appraisal requires special consideration because buildings are different from land; buildings depreciate and are replaceable, whereas land does not depreciate in the same way and is, for all practical purposes, not reproducible. The following discussion based on the capitalization of net returns explains how buildings can be handled under this method.

"CONSTANT BUILDING INVESTMENT" PLAN

The procedure followed in appraising a farm unit is simplified if the buildings and land are treated as a unit; otherwise, building life and depreciation rates have to be estimated. Such computations may be avoided in some cases by assuming a set of buildings, similar to those now present, that are maintained indefinitely just like the soil. This means that an annual allowance, like a tax on the land, will have to be provided for maintenance. This allowance will take care of reshingling a roof when necessary, applying paint, replacing certain timbers, or pointing up the foundation. In any one year some buildings will depreciate while others are being improved, but on the average the buildings will remain at the same condition level. The roof on the barn, for example, may need reshingling one year, while the next year some foundation repairs will take all of the annual allowance. The expense actually incurred may be more than the allowance one year but less the following year; the point being that the annual allowance would represent average anticipated expenses in keeping the buildings in their present condition.

There may be some question as to how a building can be maintained indefinitely. A partial answer is supplied by reference to many of the farmhouses and barns found in the eastern part of this country and in Europe that are well over 100 years old. It may well be that these examples are similar to the proverbial axe which, although of a ripe old age, was as good as new because it had been maintained by repairs consisting of four new handles and three new heads during its long history. Nevertheless the proper allowance for new axe handles and heads will answer the question because it will maintain the axe indefinitely.

An objection may be raised on the grounds that buildings become obsolete. This is undoubtedly true but is not as important on a farm as in a city where residence and industrial districts may change rapidly. The farmhouse and service buildings being tied to a given tract of soil gain a permanence from this association. Obsolescence, moreover, can be provided for by increasing the annual allowance and by using the allowance to remodel and rearrange existing buildings.

The advantage of treating buildings as an integral part of the farm can now be visualized. A permanent set of improvements may be considered as though they were part of the soil. As long as the annual allowance is spent, these buildings can be expected to house the livestock, provide storage for the grain, and furnish living quarters for the operator. Returns from buildings can thus be linked with income from the soil, making it possible to capitalize total net income to obtain the value of the farm as a whole. It may be argued further that this method is a close approximation to reality in those instances where buildings are well constructed and may last for many years if properly maintained. It must be borne in mind, however, that there is an annual maintenance charge, and that this charge has to keep the buildings in good condition, replacing those parts that wear out or become obsolete and maintaining the building investment at approximately the same level.

NEW AND OLD BUILDINGS

Adjustments can be made to the constant building investment plan to allow for a new set of buildings or for one that is very old. The procedure is to add or subtract from the income value an amount which represents the difference between a constant building investment and the replacement cost less depreciation value of the buildings that exist.

If a new set of buildings exists, the appraiser follows the same procedure as in the case where the buildings are about one-half worn out. But at the end of the calculations an amount is added which equals the difference between the half-worn-out buildings and replacement cost of the new set (Table 15.7). It is important to keep in mind that it is the replacement cost and not just the new cost of the buildings.

If a very old set of worn-out buildings exists the same procedure is followed except the difference is subtracted. Furthermore, care must be exercised to measure the difference between the replacement less depreciation value of the existing buildings on the one hand and the average value of a set of half-worn-out buildings on the other. Another way to look at it is to estimate the amount of money it would take to put the existing buildings into shape so they could be maintained indefinitely with an ade-

TABLE 15.7. Income Valuation of Farm X Assuming Different Building Situations

	Average Buildings (half worn out)	New Buildings	Almost-Worn-Out Buildings
Gross landlord income	$ 4,020	$ 4,020	$ 4,020
Less expenses, taxes, building repairs, and depreciation	1,980	1,980	1,980
Net income to the landlord	2,040	2,040	2,040
Value of farm, capitalizing at 5 %	40,800	40,800	40,800
Building adjustment	0	+ 8,000	− 8,000
Final income value	$40,800	$48,800	$32,800

quate building expense figure. It would be similar to the purchase of a farm on which one of the buildings was in such poor repair that the appraiser estimated $2,000 would have to be spent immediately to put this building in usable condition. In estimating income value for this farm the appraiser would figure as though all the buildings were in usable condition but would deduct $2,000 from this income value to take care of the building needing attention. In short, if the buildings on Farm X were almost worn out, an expenditure of $8,000 would put them in shape to be maintained with the annual expense provided. On the other hand, as indicated in Table 15.7, if the buildings on Farm X were new, $8,000 would be added because this amount represents the difference between the replacement value of a new set and an average set partially worn out.

USE VALUE OF BUILDINGS

Another way to size up buildings is to estimate their use value. Use value is the value which is left after obsolescence has been deducted from their replacement cost and physical depreciation has been allowed. In the preceding chapter obsolescence was discussed in detail. Here the emphasis is not so much on obsolescence as on what is left after obsolescence has been subtracted and allowance made for physical depreciation. This is use value. Oftentimes not enough deduction for obsolescence is made to bring building values into line with use values. In brief, obsolescence has been larger than estimated. Since there are those who prefer the use value approach, it will be well to discuss these use values in more detail.

Use value may be determined by figuring how much of a loss it would be if a given building or part of it were taken away or burned to the ground. The loss could be measured by the reduction in amount of cash rent that would be paid for the farm, or by the drop in rental income that would follow leasing to a less capable tenant, or by the loss in income to the owner-operator either in lower returns on feed fed to livestock, spoilage of feed, or in less satisfactory living quarters. Another way of measuring the loss would be to estimate the cost of the building which would replace the present one. The large horse barn with a big mow for loose hay is a good example. If such a building is destroyed by a windstorm, the replacement would probably be an entirely different type of structure. Of course some use can be found for the present building, but it will not be an entirely satisfactory solution. The use value of the building in an instance of this kind will be much lower than reproduction cost less physical depreciation. In short, obsolescence has been extensive.

Another example of use value below cost is overinvestment. A barn, although well adapted to the farm's needs, may have cost twice as much as one that would serve the same purpose. The extra cost may be justified to some extent because depreciation and repairs will probably be less on the more expensive structure, but low maintenance will not always compensate entirely for the initial high cost. Or again, if the barn is the right type but twice as large as necessary for the farm, the cost will have to be cut drastically in order to arrive at use value.

The house should be included in this discussion of use value. The large farmhouse with 10 or 12 rooms is not uncommon. The farmer with a family of five or six children doubtless felt the need for a house of this size when he built it. The present operator, with a smaller family and some of the hired help eliminated by the increased use of machinery, might prefer a smaller house because it would not require as much upkeep expense. Therefore, a house smaller than the present one should be estimated in figuring a replacement.

It sometimes happens that out of surplus funds an owner builds himself an expensive house costing far more than the average farmer can afford to pay for or maintain from the earnings on his farm. The appraiser valuing this house will figure on replacing it, not with a similar structure but with one in line with the earning level of the farm. Consequently the use value of the house will be much lower than reproduction cost less physical depreciation. It is recognized, of course, that if the house is attractive and well built, it may draw to the farm an outstanding tenant or a farm purchaser interested in obtaining living quarters above the average. Thus it will be reasonable to add some value over and above that for the average house on this type of farm.

Use value enters into the valuation of the farm through the building expense item. No expense will be figured on a building not in use, and expense will be low in estimating the value of a new set of improvements. An unusually large house requiring heavy upkeep expense will increase the building expense item and reduce the value of the farm as a result. If, on the other hand, the farm is without a building which is needed for proper operation of the farm, the cost of such a building should be subtracted from income value. In figuring income value in this instance, it would be assumed that the building existed.

BUILDINGS AND SALE VALUES

Little has been said about the effect of buildings on sale values, yet it is well known that buildings do have an important effect on prices paid for farms. Buildings in the sale value of Farm X, as indicated in Table 15.1, had a residual value of $12,000—in this case they made up 20 percent of the market value of the farm. But on an income value basis the residual value of the buildings was only $4,800, indicating that buildings contributed $7,200 to the market value of Farm X over and above income value. Buyers and sellers, it is evident, place a value on buildings which is not represented by income. This is particularly true of farm dwellings which have little effect on money income but contribute a great deal to living satisfaction.

The value of farm buildings when sold separately, as a farmstead with a small acreage, provides an excellent opportunity to use the sale value approach on buildings. The main difficulty, as discussed earlier, is measuring the demand for these units, especially where the demand is sporadic. There is a ceiling, however, in the cost that would be incurred in building a new set of buildings. No one is likely to pay much more than cost because under these circumstances the potential buyer would probably prefer to buy bare acreage and build a new set of buildings. Sale value then is likely to be somewhere between a top represented by cost and a bottom equal to zero, the value in any individual case depending on the demand for the specific farmstead that is up for sale or is being valued.

OVERALL EVALUATION OF BUILDINGS

Several complex, difficult, and even confusing aspects of building value have been presented in this chapter and the one preceding. One could easily say that it would have been much simpler, and certainly more comforting to the mind, to have omitted these chapters and in place of them to have stated merely that buildings are included in an income valuation along with the land, and that since no separate income is available for the buildings, no separate valuation can be placed on the buildings. But this would be sticking our heads in the sand, for farm buildings present appraisers with difficult questions for which farmers, lenders, tax officials, and insurance executives want answers.

Valuation of the farmstead as a separate unit is a must for the appraiser in areas where there is a demand for these units by nonfarmers interested in living in the country. Sale and income values will vary widely on these units depending on the demand for them. However, with a cost value the new cost is a ceiling because potential buyers would generally prefer to build rather than to pay more than what the buildings would cost.

SUMMARY

In valuing the farm as a whole, we have seen that it is fallacious to value the land and the buildings separately and then add them together. The correct procedure is to value the farm as a unit and then, if desired, split the resulting figure between land and buildings, using the building residual method to estimate building value.

In the process of splitting total farm value between land and buildings we may run into abnormal situations—situations that do not fit into a neat pattern. From an income standpoint, bare land in certain cases may actually be worth more than the same land with buildings on it. This is the result of a surplus building situation.

A feasible procedure in valuing a farm as a whole is to assume that the buildings, like the land, will be maintained indefinitely and that the expense item for building repairs and maintenance will do this job. For new buildings or for very old buildings an adjustment has to be made to fit these situations.

Use values and sale values are other aspects which throw light on farm building values. An appraiser needs to distinguish between a building investment that is justified by use and one that is not; and between building value represented by income and building value represented not by income but by market value.

Finally, recognition should be given to the existence of more than one building value. It is common to think of a tax assessment value as different from a sale value. It should also be common to think of building values for assessment, insurance, and bookkeeping as distinct from building residual values obtained to show the difference between land with and land without buildings.

QUESTIONS

1. Take an example of a farm near a city or town where there is a demand by nonfarmers for a separate farmstead in the country. Value the farmstead and the farmland separately, then compare the combined total of these two values with the value of the farm as a whole. Discuss the result.
2. For a farm with which you are familiar, figure the income value and estimate the sale value. Then figure (a) the building value by subtracting the bare land value from the total (building residual method), and (b) the bare land value by subtracting a replacement cost less depreciation value of the buildings from the total (bare land residual method). See Tables 15.1 and 15.2 for example. Discuss results.
3. For a community where you have appraised one or more farms, explain whether a surplus or deficit of farm building units exists.

4. Analyze two sets of buildings as to their use value. What type and investment in buildings would be justified in both cases if the present units were to be replaced?

REFERENCES

Barre, H. J., and Sammet, L. L., *Farm Structures,* John Wiley and Sons, Inc., New York, 1950.

Pos, Jacob, "Farm Buildings Appraisal, Old and New Construction," in *1960 Appraisal and Valuation Manual,* American Society of Appraisers, Washington, D.C., pp. 259–78.

Wills, J. E., "Economics of Farm Buildings," *Journal of American Society of Farm Managers and Rural Appraisers,* 28(1):27–32, Apr. 1964.

5

SALE VALUE
VS. INCOME VALUE

Down through the years farm appraisal has been a battleground between two forces, one advocating the sale value approach and the other the income approach. Both of these approaches, as the discussion will show, are necessary parts of a logical and scientific system of farm valuation.

The difference between sale value and income value approaches rests chiefly on those aspects of a farm that are of a nonincome and intangible character. Measuring the effect of these features requires judgment based on sale value information.

16

INTANGIBLE AND
NONINCOME FEATURES

WHEN THE SALE VALUE ESTIMATE in an appraisal is above the
income value, which is the usual case, the difference represents
intangible and nonincome features. To find the dollar amount
of these features, all that is necessary is to subtract income
value from sale value. The objective of this chapter is to study
the individual factors which make up this dollar amount and to
indicate how the appraiser can evaluate them.

Intangible and nonincome features may be divided into
what will be called the amenities on the one hand and anticipated
income and value increases on the other. The amenities can be
broken down into locational factors such as distance from town
and road type; community factors such as nationality groups,
schools, and churches; home attractions such as farmstead site
and landscaping; and buildings to the extent they are not reflected in income value. The anticipated increases do not include
any part of the estimated income represented in income value.

DIFFERENCE BETWEEN SALE VALUE AND INCOME VALUE

A fundamental distinction in farm appraisal is the difference between the appraiser's estimates of sale value and of income value. The sale value estimate is made by using comparable sales, which are checked and brought up to date. The
income value estimate is based on soils, yields, and net returns

to the landlord capitalized at an arbitrary investment rate of interest. These two value estimates are arrived at independently of each other using different source materials. Since income value is based solely on estimated earnings while sale value includes all value factors of which income is only one (though the major one), it is clear that sale value would usually be higher than income value and that the difference would represent all value factors other than income.

One way to indicate the distinction between sale value and income value is the equation which was used in the discussion of capitalization. By designating all intangible and nonincome features as "I" the equation reads as follows:

$$\text{Sale Value} = \frac{a \text{ (annual income)}}{r \text{ (capitalization rate)}} \pm I \text{ (Intangibles)}$$

When the data for Farm *A* are used in the equation, the result is as follows:

$$\text{Sale Value} = \frac{\$27}{.05} + \$190 = \$730 \text{ per acre}$$

AMENITIES

Those features which are clearly of a nonincome nature can be grouped in various combinations. One combination made up of four major groups—location, community, home attractions, and buildings above average—is shown in Table 16.1. A rating scale is added for the appraiser's use while he is inspecting the farmstead. Unless some list is used which can be systematically checked in the field, the appraiser is liable to overlook some important features during the inspection. In addition, the rating scale requires the appraiser to make definite decisions on the various nonincome features while he is on the spot.

It is possible to carry this appraisal of amenities too far, as is illustrated by some of the attempts to assign numerical weights to each intangible factor and then, by a summation process, obtain an index figure which gives the effect of these factors on value. The danger in this procedure is that those not familiar with the method may infer an accuracy that obviously does not exist. Another difficulty with a numerical weighting scheme is that the weights should vary, but this is seldom provided for because it would supposedly give too much leeway to the appraiser. If the weights set were as follows: location 50, community 20, home attractions 20, and buildings above average 10, this dis-

TABLE 16.1   Appraisal Form for Rating Intangible Features

Features	Rating Scale
I. Location Distance to town. *2 miles* Road type. *paving* Electric service. *no* Mail. *yes* Telephone. *yes* Hazards, nuisances. *none*	√ (Road type) A B C D E
II. Community Nationality. *mixed* Tenancy. *30 percent* Debt record. *fair* Schools. *good* Churches. *several* Employment opportunities. —— Recreation. *good*	√ (Debt record, C) A B C D E
III. Home Attractions Home site. *fair* Yards. *poor* House architecture. *fair* Landscaping. *none* Windbreak. *none*	√ (Yards, D) A B C D E
IV. Buildings Above Average House. *no* Barn. *no* Others. *no*	√ (House, D) A B C D E

tribution would be suitable for one territory and not for another. The home features in one area might have more relative importance in sale value than location, while in another area the weights as given above would reflect the correct situation.

Why does income value fail to reflect the amenities in the competitive bidding among tenants for a desirable farm? In order to reflect amenities in income value the landlord would have to find a tenant willing and able to pay additional rent for the amenities. But tenants are not always willing to bid against each other for the intangible features. They are often prevented from such bidding because the income which they can get from the land will not be sufficient to justify their bid. It is the farmer who has accumulated reserves who is willing to bid on land having these amenities.

It is entirely possible, though not as likely, for income value to be above sale value because of certain unattractive features. An excellent piece of bottomland may justify an income value above the price set by buyers and sellers because of a very unattractive homestead, poor drainage through the lots, and a poor road to the farm. Another example would be a farm located near a refuse dump or sewage disposal plant. The conditions expected in such cases are the reverse of those where sale value exceeds income value.

DISTANCE TO TOWN

Among the locational features of a farm, distance to town or to other important centers is often of vital significance as a value factor. A farm located three miles from the outskirts of a growing city certainly would have a higher valuation than an equally productive farm located three miles from a small town with a declining population. Also there is a value difference between a three- or eight-mile distance from either the growing city or the small town.

Other factors being equal, the closer the farm is to a city or town the higher is the value. This fact has been brought out in numerous studies: by Hammar in a study near Kansas City, Missouri; by Welborn in an Iowa study; by Parcher in an Oklahoma study;[1] and by others. Parcher's study in Oklahoma, which included 2,000 sales, produced the following conclusions on distance to town:

> Farms within a mile of a market sold for about one-third more on the average than those three to five miles away.
> Farms within five miles of a principal city sold for about 30 percent more than farms 10 to 15 miles away.[2]

Since there is always the possibility that farms at a distance have poorer soil, this soil factor was held constant in a portion of the Oklahoma study by comparing farms with a certain kind of soil, i.e. good soil, at varying distances from town. The results of this special comparison where soil was held constant were the same, indicating that distance was an effective factor; the shorter the distance from a town or city, the higher the value.

In a later study in Oklahoma covering sales in Logan County in 1960–61, Parcher found the same difference in the effect of location factors on land values.[3] Even though land values had increased, the percentage relationships with respect to location stayed the same. Parcher's conclusion was: "The evidence is that as all land prices rise the premiums (or discounts) for location

---

1. C. H. Hammar, *Factors Affecting Farm Land Values in Missouri*, Mo. Agr. Exp. Sta., Res. Bull. 229; L. A. Parcher, *Influence of Location on Farm Land Prices*, Okla. Agr. Exp. Sta., Bull. 417; W. R. Welborn, *Equitable Allocation of Highway Costs with Respect to Agricultural Land Values*, M. S. thesis, Iowa State University Library.

2. Parcher, *Influence of Location*, pp. 5, 6, 10–13.

3. L. A. Parcher, "Location as a Factor in Land Prices," *Oklahoma Current Farm Economics*, 36(1):8–16, Mar. 1963.

TABLE 16.2   Average Value per Acre of Farm Real Estate in Metropolitan and Non-metropolitan Counties in Selected States, 1959

State	No. of Metropolitan Counties	Average Value per Acre	
		Metropolitan counties[a]	Nonmetropolitan counties
New York	17	$242	$119
New Jersey	11	621	577
Pennsylvania	21	309	119
Michigan	14	281	169
Ohio	19	373	224
Indiana	10	346	262
Illinois	15	468	287
Virginia	10	224	134
Georgia	12	207	93
Florida	10	514	192
Texas	28	161	77
California	16	693	262

Source: USDA, *Farm Real Estate Market,* May 1961, p. 15.

[a] Counties were classified as metropolitan if they contained at least one city, or "twin cities," of 50,000 inhabitants or more. Only states with 10 or more metropolitan counties were selected.

relative to a principal market tend to rise at about the same rate."[4]

The influence of large cities on farm values in surrounding areas is well recognized. How large a factor this can be is seen in Table 16.2, a list of the states with ten or more metropolitan counties showing the average value per acre of farms in the metropolitan counties and in other counties. In all of the states listed, farm values in the metropolitan counties were higher than in the nonmetropolitan counties. In one-half of the states, values in the metropolitan counties were more than twice the values in the other counties. The area where this was not true was in the Midwest, a region in which soil productivity is more important compared to city influence than in other parts of the country.

ROAD TYPE

Paved, gravel, and dirt roads rank in this order, as would be expected, in their effect on farm values. The points which concern appraisers are how much premium a paved road has over gravel and dirt, and how much premium a gravel road has over dirt. A comprehensive study covering these points was con-

---

4. Ibid., p. 15.

ducted in all parts of the country by the USDA, using a question-naire sent out to real estate brokers and others familiar with farm values. The persons questioned were asked to list bona fide sales giving the sale prices and their opinion of what the farm in each case would have sold for if it had been located on a lower or higher type of road. The results of this study for sales of farms on gravel roads are shown in Table 16.3. On the average, a gravel road adds around eight percent to the value of a farm as compared to a farm with a dirt road, and a paved or hard-surfaced road adds about seven percent to the value of a farm on a gravel road.

Location is usually not static but subject to change—a condition which applies especially to roads. The unimproved road by a farm cannot be assumed to be permanent, particularly in areas where improvement in road surfacing is under way. Main highways are often rerouted or new highways constructed to eliminate sharp corners and to avoid going directly through the center of towns. A comparison of present-day highway maps with those of 20 years ago brings sharply to mind the effect of the changing road system.

Another illustration of the changing aspect of location is method of transportation which, although closely associated with type of road, deserves separate treatment because of its importance. The statement "8 miles from town" has various meanings depending on whether the farmer is driving to town

TABLE 16.3   Sale Prices for Farms on Gravel Roads, with Estimates for Dirt and Hard-Surfaced Roads, 1958

| Type-of-farming Area | Sales[a] | Average Sale Price[a] | | Premium or Discount, as a Percentage of Sale Price for Location on | |
		Per acre	Per farm	Dirt road	Hard-surfaced road
	(*number*)	(*dollars*)	(*dollars*)	(*percent*)	(*percent*)
Lake States dairy	107	141.75	18,375	− 7	+ 5
General farming	141	101.95	15,290	−17[b]	+13
Eastern Corn Belt	136	278.00	42,285	− 8	+ 3
Western Corn Belt	423	170.10	29,090	− 8[b]	+ 6
Spring wheat	88	40.65	23,015	− 5	+ 4
Winter wheat	52	99.80	22,825	− 4	+ 5
Central cotton	56	116.65	20,025	−11	+ 7
Western cotton	31	78.00	19,300	−16	+20
Northern range livestock	81	44.90	31,280	− 3	+ 4

Source: USDA, *Farm Real Estate Market*, Oct. 1958, p. 27.

[a] Most of the sales probably occurred in the preceding 6 months.

[b] Areas in which the estimated discount was statistically significant at the 5 percent level.

in a buggy or an automobile, in a wagon or a truck. The revolution in transportation has made distance from town no longer as much of a handicap as formerly on the basis of time saved.

OTHER AMENITIES

Distance to town and road type, only two of a large number of nonincome factors that influence farm value, have received special attention because they are at least partially susceptible to measurement and are important throughout the country. Other factors are just as important in certain areas. For instance, a nationality group may exert a strong influence on farm values where members of the group pay high prices to obtain farms in the area so their children may be raised with others of their kind. Although an appraiser will not find this factor a common one, he should recognize it when he appraises farms in one of these areas.

Community features cover a large number of factors which vary in their importance. Similarity in natural resources among the surrounding farms has a significant influence on value. One is attracted more to a productive, well-improved farm located in a community made up of such farms than to the same farm situated among unproductive, poorly improved units. Schools, churches, recreational facilities, and the type of people living in the community are other characteristics that need to be evaluated Schools and churches are very important to some prospective farm purchasers.

There are negative community characteristics which require special attention. Taverns, roadhouses, or the location of a garbage dump or a sewage disposal plant near a farm can lessen the farm's attractiveness and value. Certain industries, a heavily traveled highway, a railroad, huge power lines, and similar factors can lower the value of a farm, not so much because of income effects—although it may be difficult to get a good tenant —but more because of the intangible effects of these factors on making the farm a less desirable place to live.

Utilities such as electricity, natural gas, telephone, fire protection, mail, and delivery service from town are features that enter into the attractiveness as well as the income-producing power of a farm. Cheap electricity and natural gas, for example, have favorable income effects and also make the farm more attractive as a place to live. Some of these utilities, like electricity and mail service, are so nearly universal that the appraiser's task is not to add value when they are present but to subtract value when they are absent. On the other hand, if a more unusual feature such as cheap natural gas is available, this can be the basis for added value.

Buildings above the average require special consideration by the appraiser. A new attractive home may add a sizable amount to sale value over and above any representation of the home in income value. In making his sale value estimate of a farm with such a dwelling the appraiser will have to decide how much effect this dwelling would have in the bid of a prospective purchaser. Knowledge gained from what other purchasers have paid for buildings above average is the best aid available in making these decisions. Expensive silos, new dairy barn equipment, and automatic feeding installations are other examples of buildings and improvements above average which the appraiser has to evaluate. And in these as well as in all others the test of what purchasers are willing to pay is the best procedure to follow in reaching decisions on adjustments to sale value for these factors.

NONFARM USES

A nationwide USDA study of nonfarm uses and land values showed residential and development uses as the most important types in every region of the country (see Table 16.4). Residential included both suburban and rural, and development included purchases of land for future development. Next in order came highways, industrial and commercial use, public and private recreation, and oil, gas, and mineral use. The demand for farmland for these various uses pushes the price above what the land would bring for strictly agricultural use.

The amount of overvaluation caused by nonfarm uses, according to the USDA study, amounted to 17 percent on the average. This was the figure reported by two-thirds of the reporters; the other one-third said there was no overvaluation in their areas (see Table 16.5). But this 17 percent average overvaluation did not indicate the wide variation in different parts of the 48 states. The high areas were the Pacific region with 35 percent overvaluation and the Appalachian region with 32 percent. At the other extreme were the Mountain and Corn Belt regions with percentages of 6 and 7 respectively. When the word "overvaluation" is used in this connection it should be clear that it means the amount over and above farm market value for which the land either sells or is valued because of these nonfarm uses. This "overvaluation" is included in the intangible and nonincome features discussed earlier in this chapter.

ANTICIPATED VALUE INCREASES

A difficult intangible to measure is the farm buyer's anticipation of increased farm values in the future. This can be the buy-

TABLE 16.4 Nonagricultural Uses for Land Contributing to Market Prices in Excess of Agricultural Value, October 1965[a]

Production Region	Number of Replies	Total Number of Reasons Given	Specified Nonfarm Uses								
			Residential		Industrial and commercial	Public		Private recreation	Oil, gas and minerals	Land held or acquired for development	Other[b]
			Suburban	Rural		Highways and other public works	Recreation				
						(percent)					
Northeast	198	462	18	21	8	8	9	11	1	19	5
Lake States	190	265	17	21	6	6	6	11	1	17	15
Corn Belt	456	725	17	19	8	11	6	4	3	17	15
Northern Plains	259	265	11	13	3	7	4	2	3	16	41
Appalachian	321	610	18	19	8	13	6	4	2	20	10
Southeast	183	406	18	18	13	11	3	4	3	21	9
Delta	148	257	13	17	7	8	6	3	10	23	13
Southern Plains	293	490	15	17	3	5	4	6	12	17	21
Mountain	205	338	14	14	3	6	9	6	4	21	23
Pacific	176	338	16	16	6	5	6	6	1	27	17
48 States	2,429	4,156	16	18	7	9	6	6	4	19	15

Source: *Farm Real Estate Market Developments*, USDA, CD-68, July 1966, p. 27.
[a] Reporters who considered market prices for farmland in their communities to be above agricultural values were asked to check one or more nonagricultural uses they believed were responsible.
[b] About half of these replies related to farmer demand for land to expand existing farms.

TABLE 16.5　Indicated Overvaluation of Current Market Prices in Relation to Agricultural Values, October 1965

Production Region	Market Prices in Line with Agricultural Values	Market Prices Exceeding Agricultural Values[a]
	(percent)	*(percent)*
Northeast	31	23
Lake States	49	22
Corn Belt	37	7
Northern Plains	43	6
Appalachian	26	32
Southeast	36	22
Delta States	29	27
Southern Plains	19	9
Mountain	32	18
Pacific	22	35
48 States	34	17

Source: *Farm Real Estate Market Development,* USDA, July 1966, p. 26.
[a] Contains the average prices given by a few reporters who indicated market prices were below agricultural value.

er's forecast of city expansion, a new highway, an airport project that may affect the farm, prospective oil or other mineral developments, or simply a belief that inflation is ahead. Inflation means higher farm product prices, which in turn means higher farm values. In countries where inflation has prevailed, farm values have increased rapidly.

In the early twenties farm value analysts pointed out that anticipated increases were in all probability the explanation of the long continued rise in farm values which culminated in the land boom of 1919–20 in the Corn Belt. The following equation was devised to represent this idea of a continuous anticipated increase in earnings and in values.

$$\text{Value} = \frac{a}{r} + \frac{i}{r^2}$$

The "i" in this equation is the annual increase in income that was anticipated. With a continuous increase each year of 25 cents an acre, the value of Farm A would come out as follows:

$$\text{Value} = \frac{\$27}{.05} + \frac{\$.25}{(.05)^2} = \$640 \text{ per acre}$$

The assumption behind this equation was that income was going to increase every year at this same rate. Although there is a possibility of income increasing as it has done for periods in the past, there is no likelihood nor reason to substantiate a continuous rise indefinitely.

A more reasonable assumption on anticipated value increase would be a flat increase in the estimated income. For Farm *A* an increase of $5 in annual income per acre might be made by some buyer who felt that inflation to this level or above was likely in the near future. This would add to the formula an anticipated income factor as follows:

$$\text{Value} = \frac{a}{r} \pm \frac{c}{r} \pm I$$

In this new equation "c" indicates a change in income that is anticipated. Using the estimates for Farm *A* in the formula we get the following:

$$\text{Value} = \frac{\$27}{.05} + \frac{\$5}{.05} + \$90 = \$730 \text{ per acre}$$

On the basis of a 5% investment return the current income on Farm *A* gives a value of $540 to which we add $100 because of anticipated increases in income and $90 because of intangible and nonincome features made up of nearness to town and the development of a recreational area nearby.

In contrast to the income approach in the formula above, a sale capitalization formula could be used. In this case the capitalization rate would be 3.7%, a rate which would include all three factors—current income, anticipated income, and nonincome features. In short the anticipated income increase and the nonincome features are represented in the reduction in the rate from 5% to 3.7%.

In conclusion it is evident that intangibles and nonincome features, as their name implies, are difficult to measure. Fortunately their total can be estimated by subtracting income value from sale value. How much of the difference is due to closeness to town, a paved highway in front of the farmstead, an attractive home, or anticipated increase in values due to forecasts of inflation has to be left largely to the appraiser's judgment. This judgment, on the other hand, can be reinforced by a systematic rating of the various intangible features and by information from buyers on what it was that led them to pay the price they did for the farms they bought.

QUESTIONS

1. Discuss the effect of road type and distance to town on the sale value of one or more farms you have appraised.
2. What was the effect of closeness to a city on the value of Farm A at the end of Chapter 1?
3. Are the benefits of a hard-surfaced road and a short distance from town measurable in income? If they are, should these features be included as intangibles?
4. How much more rent can a landlord get from a farm close to town than from one a long distance from town? How much more rent on a farm with a paved road than from one with a gravel road?
5. Is inflation a factor in the sale values of farms in your community? Discuss.

REFERENCES

Appraisal of Real Estate, 5th ed., American Institute of Real Estate Appraisers, Chicago, 1967.

Holm, Paul L., and Scofield, William H., The Market for Farm Real Estate," *Land: The 1958 Yearbook of Agriculture*, USDA.

Wendt, Paul F., *Real Estate Appraisal*, Holt, Hinehart, & Winston, New York, 1956.

17

LONG VIEW ON
SALE VALUE AND
INCOME VALUE

THE SCENE CHANGES in this chapter from current appraisal of an individual farm to issues of a wider scope. First is a long view of sale values showing the wide fluctuations which have brought financial tragedy as well as prosperity to many farm owners. Second is a controversy over sale value versus income value methods of appraisal which has flared up from time to time in both Europe and North America. The question to be answered is whether sale value or income value is a better approach when considering the evidence over a long period.

LONG VIEW OF SALE VALUES

One of the advantages of the farm appraisal profession is the long-range perspective of sale values it provides those engaged in making appraisals. Not all appraisers avail themselves of the opportunity to study the past record of sale values to obtain a better understanding of what lies ahead, but the record is there for all who wish to see it. Moreover, appraisers can render a useful service to buyers, lenders, and others interested in farm real estate by bringing to their attention this long view of sale values which is so often overlooked to the detriment of agriculture and society generally.

An analysis of sale value fluctuations over a long period leads to an understanding of what is behind these ups and downs in the land market. To be meaningful such an analysis should cover at least fifty years. The home county is the place to start, then the home state can be added, and finally the country as a whole. When the figures are all brought together they will look something like those in Table 17.1 which tell a long and interesting story covering both success and tragedy in farm ownership over a hundred years.

With only minor exceptions the long continuous rise in farm values from 1850 to 1920 followed the same pattern in Story County, Iowa, and the United States. But the land boom of 1919–20 which ended this long rise was much more severe in some areas than in others. For instance, the value rise in Story County was much sharper than in Iowa as a whole, and much sharper in Iowa than in the United States as a whole. A few quotations from the current writings in the 1919–20 boom period indicate the prevailing psychology and attitude of the time. In September 1919, an officer of a large Iowa bank in addressing the Farm Mortgage Bankers Association stated:

I believe from personal experience farming Iowa land, from observation, from contact with farmers all over the coun-

TABLE 17.1   Value per Acre of Farmland and Buildings for Farm *A*, Story County, Iowa, and United States, Census and Selected Years

Year	Farm *A*	Story County	Iowa	United States
		(*dollars*)		
1850	160.00	...	...	...
1854	...	310	227	69
1860	...	154	124	49
1870	100.00	89	72	31
1880	...	109	79	32
1890	...	224	161	65
1900	...	351	254	111
1910	...	518	362	167
1913	730.00	546	375	178
1920	...	...	6	11
1930	1.25[a]	...	...	...
1935	...	9	12	16
1940	...	25	20	18
1950	...	21	23	19
1959	...	20	28	21
1967[b]	47.50	52	43	20
1968[b]	...	119	96	40

Source: Federal census for all figures except for Farm *A* and for 1967 which are estimates.

[a] The 1854 figure is sale of public land to first owner.

[b] Estimates based on federal census for 1964 and annual survey by Iowa State University.

try, from recent investigation with this address on my mind, and from a lifelong study of farming and banking as correlated subjects, that the present land prices are warranted and that we shall see no appreciable decline for many years to come.[1]

In Tama County, two counties east of Story County, the editor of the Traer *Star Clipper,* a weekly newspaper, sent out a questionnaire in July 1919 to leading members of the community asking their views on the land boom. Here are some of the replies he received and published:

A young farmer: ". . . my opinion is that Tama County land will never be as cheap again as it was six months ago, and I look for the best to reach $700 (an acre) if not more."
Real estate broker: ". . . within the next ten years no good farm in Tama County can be bought for less than $500 to $800 per acre."
A manufacturer: "This should not be called a land boom. We are simply getting to the actual value of good Iowa land. It will stay around the $500 and $600 mark."
A banker: "It (Tama County land) will never be worth any less and the tendency will be for higher prices from now on, as land will be the safest investment in the world."[2]

Almost all of the statements which were returned to the newspaper and published were optimistic regarding farmland values in the future. There was one notable exception, however—a banker who had this to say:

No one seems to take into account the readjustments of the future, makes no allowance for crop failure, nor seemingly, will admit any fear of the tremendous mortgage indebtedness being created (on an inflated dollar basis) by the interselling of these farms which necessitates the carrying charge of a heavy annual interest payment. . . .
I have talked with many who think a new price level is permanent. For my part, I cannot believe it will be for more than the period of another favorable crop year, or two at the most, and thereafter the prices will be downward. Europe is hungry and our salable products will find a ready market, but we must bear in mind that their needs will soon be supplied.
I, therefore, believe it behooves us all to go cautiously and

1. John A. Cavanagh, *The Future Valuation of Farm Lands in the United States,* address before the Farm Mortgage Bankers Association of America, Chicago, Sept. 23, 1919 (private print), p. 9. See also William G. Murray, "Iowa Land Values, 1803–1967," *The Palimpsest,* State Historical Society of Iowa, Vol. 48, No. 10, Oct. 1967.
2. From various issues of the Traer *Star Clipper* published in Traer, Iowa, during the fall of 1919. The questionnaire was mailed July 26, 1919, and the replies were published in August, September, and October of the same year.

that instead of contracting heavy future obligations we should be utilizing these high prices to free ourselves of debt.

During this 1919–20 land boom what, if anything, could the appraiser do to prevent the skyrocketing sale prices and assumption of heavy mortgage debt obligations? Some appraisers, like the banker quoted last, did contribute in this period by reporting income values based on farm product prices at a lower level than those prevailing, but the public generally was not in a mood to listen or to study these income value appraisals. And unfortunately there were lenders who would neither listen nor study, because it was second mortgages and other easy credit granted by these lenders which made the excesses of the boom possible.

Farm values in 1968–69, almost fifty years later, were much higher than in 1919–20, but the recent period was not referred to as a boom. The big difference was that purchases in 1968–69 were not being made for a quick profit based on easy credit. Instead many of these purchases were made for farm enlargement and much of the land was yielding a far greater return per acre than in 1919–20, thanks to improved technology. Nevertheless, in the late sixties farms were selling at a price which did not, on the average, yield a net return equal to the mortgage rate of interest. Unless a purchaser made a substantial down payment or had other resources, he faced the grave danger of not making enough from the farm to pay taxes, interest, and the installment on the principal. The danger, evident from a study of the long period, is that land prices go too high in prosperous periods and fall too low in depressed periods.

The recent period differed from the twenties in another respect—that of loan policy. While many lenders in 1919–20 gave scant attention to income value appraisals, the lenders in the recent period gave income top priority in determining their lending policy. In consequence the future of farm values has much more stability built into it than was true of the situation in 1919–20.

INCOME VALUE VS. SALE VALUE METHODS

Little if any disagreement exists regarding the appraisal of farm productivity, but there is a divergence of opinion on the best method to follow in converting production estimates into value. At one extreme are those who favor the use of sale prices or sale values as "bench marks" or standards of comparison. This sale price comparison method uses as a base in valuation of the farm in question the actual sale prices of farms whose productivity is known. At the other extreme are those who favor the determination of value by the capitalization of net rental income. This method, called the net income capitalization method, requires (1) an estimate of productivity and net rental income and (2) the choice of two capitalization rates, one an investment rate which gives a value independent of sale value and the other a sale capitalization rate representing the return being received on farms that are selling. This second rate—a sale value rate—reflects intangible nonincome features as well as income and gives an estimated sale value as a result.

Both of these methods have definite advantages; indeed, the use of both appears to be a desirable solution. Yet there are those who favor one method to the exclusion of the other, thus setting the stage for the controversy as to which method is better. This controversy, which developed during the 1930's in the United States, can be traced to a much older and sharper division in Germany which broke out in 1912 with the publication of Aereboe's treatise on land appraisal.[3]

Aereboe took violent exception to the income methods of valuation. In an article in English, Aereboe says:

> This so-called valuation according to revenue (Ertragswert) is impracticable, unscientific and indefensible. . . . A mistake of only one bushel per acre of wheat in the crop valuation makes, under the assumption of the same cost of cultivation, such a difference in the net returns as to reduce when capitalized the whole valuation *ad absurdum*. . . .
>
> If, however, a valuation of net returns were practicable, the determination of the capital value of this revenue would not yet be complete, for the rate of interest at which the net revenue is to be capitalized remains to be settled. Who will decide this question?
>
> In fixing the rate of interest at a half percent more or less, a very considerable difference in the amount of the value follows. . . .
>
> In short, the purchasing power of the same net revenue in money varies immensely with the different localities in which farms are situated and consequently the amount of enjoyment which may be procured with the same amount of money. If this is true, how can the same net returns of properties situated in different localities correspond to equal values? It is preposterous.[4]

3. Friedrich Aereboe, *Die beurteilung von landgütern und grundstücken*, 2nd ed., P. Parey, Berlin, 1924.

4. Friedrich Aereboe, *The Value of Landed Property Based on Its Net Revenue, Its Purchase Price, and the Credit That Commands*, International Institute of Agriculture, Bulletin of the Bureau of Agricultural Intelligence and of Plant Diseases, Nov. 1912, pp. 2344–45.

DIFFERENT VIEWPOINTS

A vigorous exposition of the sale value approach was made in the United States by G. C. Haas in a bulletin published in 1922.

> Appraising land means forecasting or predicting what it would sell for on the basis of the present market. The figure sought is probable market price, and not what any person, no matter how good his judgment, thinks the land should be worth. Market prices are the results of the judgments of the land market composed of buyers and sellers of the general order of intelligence.
>
> Many persons consciously or unconsciously assume that the only scientific basis for land appraisal is the productivity of the land. While it is true that land derives its value solely from its products, and, therefore, its value must be proportional to the value of its product, nevertheless its productivity cannot be made the basis for its appraisal for several reasons, as follows: (1) The product of land is perpetual and no one can forecast the amount or value of it or determine the present worth of future products. To determine the present worth, one must know the rate at which land income is capitalized now and in the future. (2) The product required is net product, and to obtain net product, one must have costs of production. Several of the important costs of production, such as value of family labor and wages of management and responsibility-taking, cannot even be estimated. There is a very wide range in net produce on different farms if the profits of farmers are included in net product. Also many costs of production, such as taxes, fertilizers, wages, are likely to be different in the future. (3) It is impossible to determine the income from farm land from its use as residence and all the elements of psychic income associated with this. For all these reasons, no reputable scientific method of appraisal can be based on productivity.[5]

Another statement in defense of sale values is the following taken from an article by Karl Brandt:

> In my own judgment based upon experience in the practice of appraisal, and that of a large staff of college-trained appraisers who cooperated with me, this method (capital value or sale value method) of land valuation is the only one that is fit for quick application on large scale and productive of reliable results without excessive costs. Theoretically it appears to me the only method that can claim to operate on a scientific basis.[6]

-----

5. G. C. Haas, *Sale Prices as a Basis for Farm Land Appraisal*, Minn. Agr. Exp. Sta., Tech. Bull. 9, Nov. 1922, pp. 3, 4; A. G. Black and J. D. Black, "The Principles Involved in Farm-Land Appraisal Procedure for Loan Purposes," *J. Land and Publ. Utility Econ.*, II, No. 4, Oct. 1926.
6. Karl Brandt, "Land Valuation in Germany," *J. Farm Econ.*, 19(1):178, Feb. 1937. Words in parentheses are the author's.

A statement of Farm Credit Administration appraisal policy by P. L. Gaddis, former head of the FCA appraisal division, indicated that the method of this agency occupied a middle position between the net income method at one extreme and the sale value method at the other extreme. Gaddis did not favor the direct capitalization of income as a method but admitted its usefulness as a guide to acre values and as a check.

> Let us first differentiate between two methods of using the capitalization principle. First is the direct method under which an effort is made to determine the net income of an individual property and build values by a capitalization of that income. Second is the checking or corroborative method under which values are first estimated on the basis of comparative productivity and other factors and then the capitalization method is used as a check in determining the soundness of those values. The Farm Credit Administration believes that serious errors will be more easily avoided if the latter method is used. It has carefully weighed the direct capitalization method and has not adopted it, believing that it is undependable, subject to serious error, and that it assumes a scientific accuracy that it does not possess under farm conditions in America.
>
> . . . I am hopeful that we can avoid the two extreme positions that have caused disagreement in Germany for several decades, that is, the direct capitalization approach and the sales approach. The method I have attempted to describe may be considered as one which, in my opinion, avoids the evils of both and retains the good points. It is a method which employs analysis, comparison, and capitalization, in the order mentioned.[7]

The issue in this country was brought to the surface by a committee which endeavored to establish a system of appraisal that would be generally recognized and adopted. This committee, with representatives from lending agencies, agricultural colleges, and various agricultural interests, could not agree entirely on one appraisal approach and so finally suggested two methods, one with major emphasis on net returns and the other based on comparisons. As an outgrowth of this committee's work, the American Society of Farm Managers and Rural Appraisers adopted a set of "cardinal principles" in which capitalization of income had a prominent place.

True D. Morse, then with the Doane Agricultural Service, endorsed the capitalization method:

> Income capitalization is the direct and primary approach to appraisal value and one that lends itself to exhaustive mathe-

---

7. P. L. Gaddis, "The Appraisal of Farm Lands," *J. Farm Econ.*, 19(2):404, 415, May 1937.

matical analysis. Because of the fact that it can be developed to
such a large extent from definite mathematical data and formulas, it has appeared logical to use it as the foundation on
which to build.

The income capitalization approach to value is the determination and capitalization of the annual net money income.
It is based upon the monetary returns that may reasonably be
anticipated from the future operation of the property.[8]

There are situations, moreover, where sales are so scarce or
so unreliable that competent authorities prefer to use values
based on income rather than to attempt to use market values.
N. Westermarck of the University of Helsinki, Finland, after
noting the position of Aereboe referred to earlier in the chapter,
made the following statement in 1952:

> . . . even if, in principle, one is in favour of valuation ac-
> cording to market values, in most cases one meets with insu-
> perable difficulties when trying to construct a system having this
> basis, simply for lack of sufficiently comprehensive statistical
> data. In most European countries land is not a marketable
> commodity, and free supply and demand is a very limited phe-
> nomenon.
>
> There is only one way out of this dilemma—the net output
> of the field must be taken as a basis. The fields are grouped in
> such a way that both the gross output and the costs involved
> in producing it—rent and interest excluded—are taken into
> account. The difference, viz. the net output, is taken as index
> of value. In other words, we apply the principle of output
> value.[9]

Westermarck takes the firm position that in his country
sale values are not frequent enough nor sufficiently reliable to
use as a value base. He prefers what he calls "output value"
which can be likened to what in this book has been called productivity or income value.

The income value method, which had its heyday in the
1930s and gave ground to the sale value method in the forties
and fifties, was fighting for its life in some quarters during the
sixties. Part of the shift in emphasis stemmed from the extensive highway and airport takings which were necessarily based

---

8. True D. Morse, "The American Rural Appraisal System," *J. Amer.
Soc. of Farm Managers and Rural Appraisers*, 2(2):98, Nov. 1938. See also
a later statement of procedure and principles of "The American Rural Appraisal System," ibid., 10(2):84–99, Oct. 1946. For an able analysis of
this appraisal system see Robert R. Hudelson, "The American System of
Farm Appraisal in Theory and Practice," ibid., 9(1), Apr. 1945.

9. N. Westermarck, "Economic Problems in the Classification of
Land," *Proceedings of the International Conference of Agricultural Economists*, Oxford Univ. Press., New York, 1953, p. 254.

on sale values because farmers whose land was purchased or condemned were entitled to the going market price to replace the land taken. Part was based on the attitude of the courts which favored sales and looked with suspicion on income capitalization because of the difficulty of substantiating the figures. Appraisers from opposing sides came to court too often with income valuations which were "miles" apart with no objective test, such as sales, which could be applied. And finally part of the sales emphasis was due to the rise of sale values themselves during the forties and fifties to a level where they provided a higher value than income would justify. On the other hand, during the thirties when sales were few and extremely low, it was possible by assuming normal prices of farm products above the existing level to obtain income values well above the existing sale value level.

It is unfortunate to have this changing emphasis on income and sales over the years, especially on loan appraisals. In the case of these appraisals it is important at all times to show estimated income and income value because of the significance these figures have to a buyer and lender who are counting on the farm to earn the interest and principal payments on a mortgage loan, with the farm as the major security.

DIFFICULTIES WITH BOTH METHODS

Handicaps to Income Value Method

The most serious of all the handicaps to the income method of appraisal is the absence of rental agreements in certain areas. Success of the income estimate, as shown previously, depends on calculation of net income to the land. Estimation of owner-operator expenses and receipts constitutes an almost hopeless task, not only because of the many items that enter in but also because with each item the amount will vary widely according to the type of manager on the farm. An attempt to estimate landlord income in a territory where there is little or no renting will be unsatisfactory because the estimates will be entirely unrelated to fact. The income method consequently is at a distinct disadvantage in those areas such as New England and the East generally where tenancy is uncommon.

A high proportion of land value made up of intangible factors, the so-called amenities, constitutes another drawback to the income approach. Here estimates of income can account for only a portion of the answer at best. Examples of this situation are common in the same areas where tenancy is practically non-existent. Farms in connection with summer homes, lake frontage, and suburban areas where other employment may be had are good illustrations. The intangible elements in the value of such cases may be as important as farm income from the property.

Valuation of a farm's intangible features actually rests in the final analysis on sale values. Since there is no income base, or at least no reliable income base, these features cannot be included in the income calculations. If an appraiser attempts to put arbitrary values on farmstead attractiveness, community, and the like, he can be challenged to produce a basis for his intangible values. His only logical reply is some form of sale value. Where sale values do not exist the answer is that intangibles are worth what the appraisers estimate they would be worth if there were some farm sales. This makes the valuations highly subjective but the best that can be done in this situation.

Another handicap to the income method exists where a very small proportion of the annual return comes from the soil. This applies particularly to those farms which closely approximate manufacturing plants, such as poultry farms, mushroom farms, and even dairy farms where most of the feed is purchased. The common characteristic to be noted in these examples is the small part which the soil contributes. Buildings, location, and typical management are likely to be the central issues in valuing these farms, and the soil is a minor item. Still another application of this point may be made to those farms where large annual purchases of fertilizer are common and necessary for crop production. The soil itself may not be very productive, but location near a large consuming center may be the key factor which supports the value because, with use of fertilizer, crops can be grown and marketed profitably.

Farms which produce a wide variety of crops are difficult to appraise by the income method. Similar difficulties are met in areas where no two farms are operated in the same way; that is, where the emergence of any standard practice is prevented by the wide variety of crops and methods of operation practiced. Examples can be found around cities, particularly where truck gardening, orchards, nurseries, apiaries, and many other types and combinations may be found. A multitude of detail will be encountered in cases of this kind if a typical operation, including crop rotation, landlord share, and prices of produce, is invoked.

The income method is at a disadvantage when the farm value is very low, as in areas of marginal farms where average expenses and average returns nearly balance each other. A slight change in the income figures may mean as much as 100 percent rise or fall in land value. An appraisal with gross rental returns of $2 an acre and expenses of $1.50 an acre leaves $.50 an acre net return which, capitalized at 5%, gives a value of $10 an acre. But if one of the expense items is varied slightly so that the total expense is $1.75 an acre, the $.25 an acre left capitalizes at only $5 an acre.

A final handicap to the net income method is the choice of capitalization rate. A change in rate has an important effect on value; a shift from 5% to 4½% adds 11+ percent to value, raising a value of $100 an acre to $111 an acre. The difference in itself would not constitute a difficulty if the shift in rate could be justified.

Handicaps to Sale Value Method

The use of the sale value method also has its handicaps, the greatest of which is the lack of enough sales to establish reliable sale value estimates. A scarcity of sales is found in those areas where owner operation is common and farms are handed down from father to son, generation after generation. If a farm is sold in these areas, its location, its buildings, and certain other characteristics or amenities attached to it may be as important as the earnings. The sale value per acre in such cases has a limited use in making comparisons with other farms. True, it may be the only index of comparison available, but its limitations must nevertheless be recognized. The lack of bona fide sales is felt during periods of depression when sale values are so unrepresentative and so badly mixed with forced sales that they are not reliable as an index.

Another drawback to the use of sale values is the lack of standardization in the land market. That the Smith farm sells for $150 an acre does not indicate precisely the quality of the land on the Smith farm nor the knowledge which the buyer or seller had relative to the Smith farm at the time of the sale. In contrast, a quotation on a certain grade of wheat indicates that a buyer and seller have arrived at a price for a commodity which is accurately described and labeled according to definite standards. With land it is extremely difficult to be sure that a given purchase price represents a going market value. The sale price, for instance, may be unduly low because the two parties are related or unduly high because the buyer is making an extremely small down payment. Sales prices in this unorganized condition of the land market may be misleading unless unusual care is taken to make sure what was in the minds of the buyers and sellers.

THE SOLUTION

The answer to the question of which method to use is to use both methods in all cases. This means a sale comparison value and two income capitalization values, one an arbitrary investment value and the other an estimated sale value based on income and selling price of nearby farms. Each provides significant information bearing on value, and each method serves as a check on the other. Income stresses estimates of productivity made by the appraiser. Sales place emphasis on the actions of buyers and sellers of comparable farms. Income gives information on what can be expected in earnings from the farm; sales cover amenities and anticipated increases in value not represented in earnings. The appraiser who includes both methods in his appraisal can use one to check the other, thus building into his final valuation an added test of reliability.

In using both methods the appraiser will usually vary the emphasis according to the situation. In a condemnation case, sales will be stressed; in a loan appraisal, income will be stressed; but in both cases the two methods can be used to supplement and strengthen each other and the final value figure. Even where income is estimated with difficulty, the estimate should be made because it will serve as a useful check on sale value. Even where there are few or no sales in the immediate locality, the appraiser should reach out as far as is necessary to get some sales which along with market value data will serve as a basis for sale value estimates; this basis in turn will be a useful check on the income value estimate. In short, the appraisal should be a combination of income and sale, with the emphasis depending on the kind of appraisal and the underlying conditions which favor one method or the other.

QUESTIONS

1. From federal census reports available prepare a table similar to Table 17.1 using average value per acre of farmland and buildings for your home county, home state, and the United States. Explain the situation revealed by this table.
2. Discuss the advantages and disadvantages of the income-capitalization method of farm appraisal. Do the same for the sale value-comparison method.
3. For your territory, what do you consider the most desirable method of appraisal? Give reasons.

REFERENCES

Appraisal of Real Estate, 5th ed., American Institute of Real Estate Appraisers, Chicago, 1967.

Scofield, William H., "How Do You Put a Value on Land?" *Land: The 1958 Yearbook of Agriculture*, USDA.

Wendt, Paul F., *Real Estate Appraisal*, Holt, Rinehart & Winston, New York, 1956.

6

STATISTICAL APPROACH
TO APPRAISAL

THE ADVENT of the computer has quickened activity among those who see a new opportunity to develop a statistical approach to farm valuation. A review of a selected list of studies in the field is presented with comments on the advantages and limitations of this approach to farm appraisal.

18

STATISTICAL APPROACH
TO FARM VALUE

THE APPLICATION of statistics to appraisal has been a potent challenge since 1920—a challenge that has resulted in a large and growing number of studies designed to find a new statistical or econometric approach to value. Regression analysis—the establishment of relationships between variable factors—has been the main procedure used in these studies. Regression is a means of handling statistically the different variable factors, such as soil productivity, quality of buildings, road type, distance to town, and the like, all of which determine value. The object of the statistical method is to analyze a group of actual sales or sale estimates in an area to determine for each factor its contribution to value. Soil productivity in the area of study may be responsible for a certain portion of farm value, road type for another portion, and so on. Once an equation has been established which accounts for the major variable factors affecting value, it can be used to estimate the value of a farm in this same area. All that is necessary is to ascertain the numerical quality of each factor for the farm to be appraised and place it in the equation. Solving the equation gives an estimated sale value of the individual farm. Thus the procedure gives each major value factor a specific numerical rating to obtain the farm's estimated sale value by using an equation which includes each of these factors in terms of its contribution to sale value in the area.

In order to see how this statistical approach to farm value has emerged and developed, a review is presented of a selected list of studies. Since these experiments were undertaken with different specific objects in mind they indicate a wide range of possible applications. In going through the list the reader can see how the statistical technique may be used to advantage in many areas. He can also see some limitations of the method.

EARLY STUDIES

G. C. HAAS

Haas conducted a survey in Minnesota using individual farms as his basic unit. His purpose and procedure are stated in the report in these words:

> The method of analysis has been to correlate the sale prices of 160 farms in Blue Earth County, Minnesota, sold in 1916, 1917, 1918, and 1919, with the factors influencing land prices, namely, value of buildings per acre, type of land, crop yields, distance from market, size of adjacent city or village, and type of road upon which located, and to derive from this correlation an equation from which the probable sale price of any other farm land in the same general territory may be determined.[1]

An illustration of the use of the regression equation developed by Haas was presented in this form:

> Farm sold in 1919 for \$135 per acre—dirt road, Class II town:
> X_2 = 1919 depreciated cost of buildings per acre . \$ 12.47
> X_3 = Land classification index 75.62
> X_4 = Soil productivity index 103.7
> X_5 = Distance to market .... 3.5 miles
> $X_1 = + 57.785 + 1.067X_2 + .7279X_3 + .1658X_4 - 3.4219X_5$
> $X_1 = 131.35$ = estimate. \$135 = actual sale price.[2]

In this multiple correlation, four independent factors are included, and adjustments are made for two others—type of road and class of town adjoining the farm. Using the four factors as independent variables Haas obtains a multiple correlation coefficient of .81.

H. A. WALLACE

Wallace singled out four factors and correlated them with the 1925 federal census estimates of land values without build-

    1. G. C. Haas, *Sale Prices as a Basis for Farm Land Appraisal*, Minn. Agr. Exp. Sta., Tech. Bull. 9, 1922, p. 3.
    2. Ibid., p. 22.

ings for the 99 counties in Iowa. The four factors were 10-year
average corn yield per acre, percentage of land in corn, percentage of land in small grain, and percentage of land not plow-
able. Together these four factors gave a multiple correlation
coefficient of .9166. An application is given by Wallace as follows:

> The multiple regression line is predicted X equals 3.4A plus
> 1.8B plus .6D minus .6E minus $74.72.
> Apply this formula to Adair County, for instance, where the
> 10-year average acre yield of corn is 36 bushels, the percentage
> of land in corn is 32, the percentage of land in small grain is 16,
> and the percentage of land not plowable is 19. Thirty-six times
> 3.4 gives $122.40 an acre credit because of the 10-year average
> acre yield of corn; 1.8 times 32 gives $57.60 an acre credit; 16
> times .6 gives $9.60 an acre credit; 19 times minus .6 gives
> minus $11.40; these figures added together with the constant
> factor of minus $74.72; give a net of $103.48, as the predicted
> value of farm land in Adair County on the basis of the four
> independent variables. The actual value is $93, or there is a
> difference of $10 an acre.[3]

Further on in this same article, Wallace comments on the
use of this method of making value comparisons in these words:

> The writer does not care to defend this formula as the last
> word in scientific accuracy. It is his belief, however, that land
> appraisers in Iowa who are willing to accept the Federal census
> values as of 1925 will find the formula of some use. In applying
> it, however, they should keep in mind just how this formula was
> derived and make their own corrections, so as to fit most accu-
> rately the specific time and place.[4]

M. EZEKIEL

A third example of correlation analysis in comparing land
valuations is contained in a farm management study by Mordecai
Ezekiel. Ezekiel attempts to find those factors which account for
the differences in the farm values estimated by farmers whose
farm management records were obtained in the survey. A multiple correlation of eight factors gave R = 0.64 as a result. Ezekiel
found, however, that a multiple curvilinear correlation was more
appropriate, the result being P = 0.772, which means that about

---

3. Henry A. Wallace, "Comparative Farm-Land Values in Iowa," *J.
Land and Publ. Utility Econ.*, 2(4):390, Oct. 1926.
4. Ibid., p. 391.

60 percent of the land value variation is accounted for by this method. The following is taken from the report:

> The relative importance of the several factors in determining farm value per acre, as shown by their coefficients of net determination, was as follows:

Factor	Percentage
A. Dwelling value	11.95
B. Dairy buildings value	12.45
C. Other buildings value	19.21
D. Crop index	4.55
E. Percentage of area tillable	2.81
F. Percentage of area level	6.16
G. Type of road	0.47
H. Distance to town	2.08

> Of the causal factors included, buildings as a whole are responsible for 44 percent of the variations in farm value; the land, including fertility, topography, and proportion usable, 13½ percent; and general farm factors of location and road, 2½ percent. These conclusions, however, are reached in a study where 40 percent of the variation in value is still left to be accounted for.[5]

EDWARD F. RENSHAW

In 1958 Renshaw wrote a comprehensive article in which he reviewed the work that had been done up to this time including one of his studies.[6] He also commented on the present and future of the statistical approach to appraisal. In his own study on appraising land for reclamation investment he used four factors with a multiple-correlation coefficient squared of .88, indicating an explanation of 88 percent of the value variation.[7]

Other studies mentioned by Renshaw besides the three described above included one by H. E. Selby in 1945 on irrigated land values.[8] A multiple-correlation coefficient of .826 was obtained. Another study was one by D. M. Warren of the Canadian Department of Agriculture. This study involved lands in the proposed South Saskatchewan River Development.[9] Using four independent variables, Warren obtained multiple-correlation coefficients ranging from .46 to .88.

5. Mordecai Ezekiel, *Factors Affecting Farmers' Earnings in Southeastern Pennsylvania*, USDA, Dept. Bull. 1400, 1936, p. 49.

6. Edward F. Renshaw, "Scientific Appraisal," *Nat. Tax J.*, 11(4):314–22, Dec. 1958.

7. Edward F. Renshaw, "Cross-Sectional Pricing in the Market for Irrigated Land," *Agr. Econ. Res.*, Jan. 1958, pp. 14–19.

8. H. E. Selby, "Factors Affecting the Value of Land and Water in Irrigated Land," *J. Land and Publ. Utility Econ.*, 21:250, Aug. 1945.

9. D. M. Warren, *A Study of Factors Associated with Land Values in the Proposed South Saskatchewan River Development*, Ottawa, Canada, Dept. Agr., Dec. 1954.

Three obstacles, according to Renshaw, stood in the way of progress in statistical appraisal: (1) lack of available data, (2) hostility of the appraisal profession, and (3) lack of financial support. He pointed out that more progress had been made in farm appraisal than in other areas because more data were available.

Renshaw suggested the formation of an institute to promote statistical appraisal by collecting and analyzing data and by making the results of experiments in this field available to interested persons and agencies.

RECENT STUDIES

M. M. A. AHMED AND L. A. PARCHER

Two significant studies by Ahmed and Parcher from Oklahoma deserve attention.[10] Both studies were aimed at using "actual farmland market prices of recently sold tracts of land in developing a systematic technique for estimating the price per acre of unsold tracts."[11]

In the first experiment, an analysis of Woods County farm values, four independent variables were selected—number of acres, productivity, population of nearest town, and distance to principal city in miles. All bona fide sales of unimproved farmland in the county for 1960, 1961, and 1962 were taken, a total of 46 sales. The coefficient squared was .81 indicating 81 percent of the variation in values was explained by the four variables. In concluding their study Ahmed and Parcher stated:

> It was possible, in this study, to develop a statistical tool . . . for estimating farmland market price in Woods County, Oklahoma. The usefulness of the empirical results of this study is limited to Woods County. Furthermore, since the coefficient of determination of this equation is about 81 percent, the real estate assessor or the real estate appraiser must utilize his own knowledge in making the necessary adjustments for attaining the ultimate value of the land.[12]

The authors also point out that the equation does not include farm buildings so that they too must be taken care of by the assessor or appraiser.

---

10. Mohammed M. A. Ahmed and L. A. Parcher, "Farmland Market Analysis in an Agricultural Area: Woods County, Oklahoma," *Oklahoma Current Farm Economics,* 37(2):29–35, June 1964; "Equitability of Farm Real Estate Assessment for Tax Purposes," ibid., 37(4):87–95, Dec. 1964.
11. Woods County study, p. 29.
12. Ibid., p. 35.

TABLE 18.1   Regression Estimate of the Price of 160-Acre Vineyard Using a 30-Variable Multiple Linear Regression Equation Developed from 110 Irrigated Farm Sales in the Madera Irrigation District, California

Index	Variables Used	Units	Coeff.	Data	Estimate
02	Size of farm	acres	—.33	160	—53.16
03	Date of sale	months	—1.09	42	—45.75
12	Down payment	percentage	—1.38	29	—40.10
33	Paved road	10ths of mi.	—6.44	0	0
34	Intersection	10ths of mi.	—1.22	0	0
35	Shipping point	miles	6.79	4	27.16
39	Metropolitan city	miles	—14.84	5	—74.19
41	Subdivision	10ths of mi.	—.66	20	—13.23
43	Recreation area	10ths of mi.	—.06	100	—6.01
45	Soil profile	percentage	2.76	87	239.72
46	Soil texture	percentage	—2.81	93	—261.25
47	Storie index	percentage	—.42	67	—28.47
48	Land capability	tenths	8.02	20	160.32
49	Soil moisture	10th of in.	.25	19	4.71
50	pH factor	tenths	6.89	76	523.37
51	Weeds	1, 0	—73.28	1	—73.28
52	Land versatility	1, 0	—128.04	1	—128.04
56	Land slope	percentage	29.15	1	29.15
57	Drainage	1, 0	—33.13	1	—33.13
59	Growing season	days	1.00	285	285.00
60	Thermal belt	3, 2	—177.73	3	—533.19
62	Water supply	1, 0	—15.45	1	—15.45
65	Gravity water	1, 0	75.73	1	75.73
66	Percent gravity	percentage	—1.80	55	—99.14
71	Well water	1, 0	44.19	1	44.19
77	Cultivated land	percentage	—1.32	100	—132.34
78	Perm. planting	percentage	8.09	100	808.74
79	Irrigated land	percentage	4.15	100	415.28
83	Allotment	percentage	—1.19	0	0
100	Home	1, 0	199.65	0	0
	Constant term		598.68	1	598.68
	Estimated price per acre				$1675.32
	Standard error of estimate				±494.34
	R squared				.67

Source: Irving F. Davis, Jr., *A Statistical Approach to Real Estate Value with Applications to Farm Appraisal*, Study No. 12, School of Business, Fresno State College, Fresno, Calif., 1963, p. 98.

The second Oklahoma study, concerned especially with farm assessments, concentrated on farm tracts near the city of Tulsa. Ahmed and Parcher used six variables—number of acres, productivity, distance to Tulsa, distance to paved road, mineral rights transferred, and road type. The area was divided into regions with the R^2 for one equal to .89 and for the other .85. The authors conclude their report with this statement:

Two equations were estimated. . . . However, these equations appear not to be perfect tools for estimating land price, and the assessor must utilize his experience in making the necessary adjustments in the estimated prices. This means that while the problem of nonuniformity of real estate assessment may

TABLE 18.2 Regression Estimate of the Price of 160-Acre Vineyard Using a 7-Variable Multiple Linear Regression Equation Developed from 110 Irrigated Farm Sales in the Madera Irrigation District, California

Index	Variables Used	Units	Coeff.	Data	Estimate
02	Size of farm	acres	—.18	160	—28.31
35	Shipping point	miles	6.50	4	25.99
39	Metropolitan city	miles	—15.77	5	—78.86
50	pH factor	tenths	5.45	76	414.33
78	Perm. planting	percentage	8.11	100	810.56
79	Irrigated land	percentage	3.87	100	386.51
100	Home	yes:1, no:0	220.15	0	0
	Constant term		292.41	1	292.41
	Estimated price per acre				$1822.63
	Standard error of estimate				±458.01
	R squared, unadjusted for degrees of freedom				.64

Source: Davis, *Statistical Approach*, p. 99.

not be achieved, the degree of nonuniformity in assessment rates will be reduced by utilizing these equations.[13]

IRVING F. DAVIS, JR.

A study in book form by Davis in 1965 bore the title, *A Statistical Approach to Real Estate Value with Applications to Farm Appraisal*.[14] The project, conducted at Fresno State College, was financed by the California State Division of Real Estate.

After explaining in detail the multiple linear regression analysis, Davis gives several examples to show how the procedure can be used in actual appraisal work. The first example presented was a 160-acre vineyard valued by two appraisers and the owner at $1,800 an acre. Using a 30-variable equation based on 110 farm sales, Davis calculated a statistical value of $1,675 an acre with a standard error of plus or minus $494 an acre and an R^2 of .67 (Table 18.1). When a 7-variable equation was used the results were fairly close to those for the 30-variable equation (Table 18.2). When an 18-variable equation was used based on 38 vineyard sales, the result was a higher price and R^2 of .92 (Table 18.3).

Davis stresses the following cautions in the use of regression analysis:

1. Sampling errors. These can be large because only a small number of sales is usually available.
2. Errors of observation. Errors of this type may be made in collecting the data in the field. These can be serious in a

13. Assessment study, p. 95. See also by the same authors, *Assessing Farmland in a Metropolitan Area*, Dept. Agr. Econ., Okla. State Univ., Stillwater, P-503, Apr. 1965.

14. Study No. 12, School of Business, Fresno State College, Fresno, Calif., Jan. 1965.

TABLE 18.3   Estimated Price per Acre of 160-Acre Vineyard Using a Multiple Linear
Regression Equation Developed from Observations of 38 Vineyard
Farm Sales in California

Index	Variables Used	Units	Coeff.	Data	Estimate
02	Size of farm	acres	−2.30	160	−368.00
12	Down payment	percentage	−5.66	29	−164.14
37	Distance to town	miles	14.35	4	57.40
38	Town population	thousands	−38.69	1	−38.69
41	Subdivision	miles	31.34	2	62.68
43	Recreation area	miles	72.00	10	720.00
47	Storie index	percentage	−15.81	77	−1217.37
48	Land capability	number	−607.91	2	−1215.82
49	Soil moisture	inches	−665.18	1.9	−1263.84
50	pH factor	number	−193.55	7.6	−1470.98
56	Land slope	percentage	134.94	1	134.94
57	Drainage	1, 0	627.53	1	627.53
65	Gravity water	1, 0	501.75	1	501.75
78	Perm. planting	percentage	11.79	100	1179.00
100	Home	1, 0	−562.84	0	0
102	Cost of buildings	dollars/acre	1.49	0	0
173	Land capability III	1, 0	1373.04	0	0
174	Land capability IV	1, 0	734.93	0	0
	Constant term		4777.11	1	4777.11
	Estimated price per acre				$2321.57
	Standard error of estimate				±234.32
	Average absolute residual				139.72
	R squared				.92

Source: Davis, *Statistical Approach,* p. 100.

regression analysis because the computer does not exercise
any judgment in accepting or rejecting the data.
3. Errors of omitted data. These are unavoidable because there
are likely to be too many factors in an actual sale to include
all of them in a regression equation.
4. Errors of interpretation. A regression analysis can give
inaccurate results if the sample data used do not represent
the property being appraised. Good judgment should be
exercised in interpreting the regression estimate.[15]

In closing, Davis emphasizes the importance of supplying
accurate data for the computer and in interpreting and converting the regression estimate into a final estimate of value. He
also points out that a coordinator is necessary to prepare the
data for the regression analysis, and finally, computer and technical staffs are required to process the data.

ROBERT J. REMER

This study was conducted by Remer for the Montana Highway Commission to show how statistical and computer techniques might be used by right-of-way personnel and others in

15. Ibid., pp. 108–10.

evaluating factors affecting farm and ranch values.[16] The report was based on 309 sales from 15 areas in Montana.

In developing his estimating equation Remer used up to 11 variables, the number varying in different areas. The multiple regression equation that was developed accounted typically for 60 to 80 percent of the variation in sale price. In one special case the equation accounted for 95 percent of the sale price variation.

Remer in commenting on the limitations of the multiple regression approach wrote:

> There are some very definite limitations to the use of statistical methods. It is difficult if not impossible to find support for special variable items that affect only an occasional subject property or sale. Many sales have special attributes, each of which may be different. A small sample limits the potential usefulness of multiple regression. In this study, where the number of variables used was 8 or less, sample size appears to be a limiting factor when the groups of sales were less than 20 in size. Sale groups of 30 sales or more appear to offer better results.[17]

ABDEL-BADIE AND PARCHER

Another Oklahoma study of the land value problem was made by this research team using two methods—multiple regression and discriminate analysis.[18] In their study Abdel-Badie and Parcher had 293 bona fide land sales for a ten-county area in western Oklahoma. They selected 15 variables as determinants of value: number of acres, quality of land, type of land, productivity index, mineral rights conveyed, road type, distance to paved road, distance to nearest town, population of nearest town, distance to nearest town of at least 1,500, distance to town of at least 50,000, distance to Oklahoma City, and acres in wheat, cotton, and peanut allotments.

The correlation coefficient squared in the regression analysis was .51, indicating that only half of the variation among the sales was explained. Of the 15 variables the following appeared to be the most significant: size in acres, quality, type of land, productivity index, mineral rights, distance to Oklahoma City, and wheat allotments.

In the discriminant analysis, which assigns individual sales to price groups on the basis of the specific characteristics of each

16. Robert J. Remer, *Correlation of Rural Land Factors in Montana,* Dept. Agr. Econ., Mont. State Univ., Bozeman, Sept. 1967.

17. Ibid., p. xii.

18. Farid Abdel-Badie and L. A. Parcher, *Regression and Discriminant Analyses of Agricultural Land Prices,* Okla. State Univ., Stillwater, P-579, Dec. 1967.

sale, the results were similar to those with the regression. Using the same 293 sales and the same 15 variables, the authors found 56 percent of the sales classified in their proper price range.

> In view of the fact that neither of the approaches yields more than about 50 percent accuracy in results, it would appear that a different set of explanatory variables should be selected.
> There is little doubt that certain factors which cannot be measured nor even identified influence the price paid for land.[19]

PENN, BOLTON, AND WOOLF

Multiple regression was used in a study by these authors to explain varying prices of farmland in the cotton area of the Mississippi River Delta.[20] A total of 1,378 valid land transfers for 1964–65 from 17 counties in Arkansas, Louisiana, and Mississippi were used. The object of the regression analysis was to determine prevailing land prices for various categories of land. The results gave a price of $231 an acre for open land, $85 an acre for woodland, and $316 an acre for cotton allotments. The correlation squared was high—.95 for the whole area with the exception of one county in Arkansas. For this Delta cotton area it is evident that the three land conditions of open land, woodland, and acres of cotton allotment explain most of the price variation that exists between these land conditions. But the explanation is for these land conditions, or land components as they are called in this study, on the average and not for a specific sale. On this point the authors state:

> There were wide variations in purchase prices among tracts that appeared to be similar with respect to land components. There were indications that non-measurable factors such as intensity of desire on the part of buyers, lack of knowledge on the part of buyers and sellers, and qualitative considerations such as topography, drainage, and estimated productivity, played some part in these individual value variations. There also appeared to be some difference in average purchase price based on other factors, such as type of road and race of seller. However, differences in mean values associated with these various characteristics, although frequently logical and substantial, were seldom statistically significant. Differences in land components accounted for the major portion of the variation in purchase prices.[21]

---

19. Ibid., p. 22.
20. J. B. Penn, Bill Bolton, and Willard F. Woolf, *The Farm Land Market in the Mississippi River Delta Cotton Region*, La. State Univ., Baton Rouge, D.A.E. Rept. 372, Apr. 1968 (in cooperation with the USDA, Washington, D.C.).
21. Ibid., p. 2.

EVALUATION SUMMARY

Until the computer entered the picture the mechanical hand computations required in the statistical approach (multiple correlation analysis) took so much time and effort that little progress was made. Nevertheless, early studies beginning in 1920 indicated clearly the kind of results which could be obtained. One major limitation of the statistical approach other than the difficulties of computation became apparent—the existence of more variables or factors affecting land value than could be sorted out and included in an estimating equation. In short, the statistical approach could explain about three-fourths of the variation in values, and a little more in some cases, but there was always a significant portion that was not explained by statistical analysis.

With the computer the problem of computations has been solved. But the problem of explaining that last significant portion of value variation by the statistical process still remains a challenge.

Probably the most promising field for the statistical approach is assessment. Here mass or wholesale appraising is required. Every farm tract in a county, for example, has to be reassessed at periodic intervals. In this process the use of regression analysis can be helpful not only as a check but also as a means of sorting out the important factors to be considered. The studies by Ahmed and Parcher in Oklahoma and by Penn, Bolton, and Woolf in the Mississippi Delta indicate the kind of progress that may be made in this area. Average conditions can be explained and handled statistically but the individual sale often needs a special explanation.

The studies by Davis and Remer are also suggestive of future progress. The statistical approach to farm value has definite limitations, as indicated, but at the same time it provides powerful precise tools which will undoubtedly be used increasingly along with human judgment in improving the farm appraisal process.

QUESTIONS

For those able to do regression analysis, develop an estimating equation based on a group of sales from some district in your state. Instead of a group of sales, a group of county land value averages as reported in the federal census may be used. Select one or more studies mentioned in this chapter for suggested procedure. If an estimating equation is already available, use it with information from one or more farms from which you can obtain required data.

REFERENCES

Studies cited in this chapter can be used for further study, especially those by Ahmed and Parcher, Irving F. Davis, Jr., and Robert J. Remer. For information on regression analysis see:

Ezekiel, Mordecai, and Fox, Karl A., *Methods of Correlation and Regression Analysis*, John Wiley and Sons, Inc., New York, 1959.

7

FARM PRODUCTIVITY APPRAISAL

THE SIX CHAPTERS which follow deal specifically with evaluation of physical productivity on the farm. Determining and mapping soil productivity, rating soils, and estimating crop yields make up the first four chapters; typical management, highest and best use, and rating whole farms are discussed in the last two chapters. An understanding of farm productivity as set forth in these chapters is an essential element in the appraisal process; it provides the factual background for the critical judgments made by the appraiser in sale value comparisons and in estimates of farm income.

19

APPRAISAL MAP AND
SOIL INVENTORY

ONCE A FARM has been definitely located, its setting in the community evaluated as to highways, schools, and the like, and its boundaries carefully noted, the next task is a detailed inventory —a picture of what lies within the boundaries. This appraisal inventory includes a description of physical resources such as soils, topography, drainage, and climate.[1] Productivity represents the main purpose behind the inventory because it brings together the various contributions of different physical resources to land value. A productivity inventory, therefore, is a major objective whether grazing land, irrigated land, wheat land, or truck-crop land is being appraised. It is true that other features apart from productivity (such as farmstead attractiveness and location) contribute to land value, but these other features were discussed earlier. Productivity will be considered for the time being as the one factor above all others which makes one farm more valuable than another.

---

1. This chapter and the two following are not intended as a complete discussion of soils, drainage, erosion, and the like as they relate to appraisal. On the contrary, these chapters are designed to stress the major contributions which soil and other agricultural sciences are making in the appraisal field and to outline the type of work that an appraiser should master. The reader not familiar with the subject matter treated in this part of the book should consult the standard text and reference books covering the subjects discussed.

IMPORTANCE OF ACCURACY

Because of its significance the appraisal map should be signed and dated by the person or persons preparing it. The appraisal map may one day be considered as important in the transfer of farm property as an abstract of title. Nor is it strange to think of this comparison, because the abstract gives the chain or evidence of title while the appraisal map testifies to what is contained within the legal boundaries described in the abstract. These inventory maps, however, must be prepared in a manner that will merit respect.

The expense of an appraisal map is generally warranted. Numerous instances may be cited where foreclosure of a farm mortgage occurred as a result of inadequate information on the producing ability of a farm. A few more dollars spent in a more thorough examination of the land might have averted a loss of a thousand dollars or more. Usually the added expense of an appraisal map represents only a small fraction of the value of the land.

An appraisal map in a report is the best method of presenting the physical resources of a farm as they relate to productivity. Such an inventory map will show soil areas, drainage, erosion, slope, and permanent pasture distinguished from cropland. Many other minor items like stones, trees, and weeds also may be shown on the map. The appraiser prepares a field copy of the appraisal map as he walks over the farm. He indicates drainage by appropriate lines showing streams and dry runs. He indicates soils by drawing dotted boundary lines around areas of soil of similar producing potential as indicated by physical properties. And in the same manner he maps other factors according to any one of a number of different systems of symbols used in making farm maps. Appraisal maps of three Corn Belt farms designated as X, Y, and Z are presented in Figures 19.1, 19.3, and 19.5. The aerial maps of these same farms, presented in Figures 19.2, 19.4, and 19.6, indicate that these aerial pictures are most useful in preparing appraisal maps for farms like Farm Z where the field boundaries are irregular and the topography is rough.

An outstanding advantage of the appraisal map is the systematic inspection of the whole farm which is necessary in the preparation of the map. As much attention is given to the "back forty" as to any other part of the farm. Furthermore, the appraiser records on the map, as he traverses the farm, the information which applies to a certain part of the farm while he is actually standing on or in the vicinity of that part. If no map is prepared at the time, the appraiser may not remember certain points correctly or may locate them inaccurately or forget them altogether when he writes his report later.

The amount of detail placed on an appraisal map will vary not only with the importance of the different features of the farm but also with the use to be made of the map and the appraisal. It will not pay to map as much information where land is unproductive and low in value as where land is productive and high in value. In drawing the map the question of what scale to use will arise. A scale of 4 or 8 inches to the mile is recommended as well adapted to farms from 40 to 640 acres. The 8-inch scale, which coincides with USDA aerial photo enlargements, is easy to use because 1 inch equals 40 rods.

SOIL AND THE APPRAISAL MAP

The aim of the appraiser as he sets out to make a physical examination of a farm should be to determine variations in the soil which indicate differences in crop-producing ability. The appraiser will want to measure and to record in some manner the productivity differences inherent in soil areas within the farm, often within a single field. A desirable method of making the soil inventory is to divide a farm on the appraisal map into soil areas of uniform crop-producing ability. Each of the areas may then be evaluated by rating it in terms of the most important crop. For instance, the appraiser, after examining a farm in the Corn Belt, decides that of two soil areas on the farm the first is capable of yielding 115 bushels to the acre while the second will probably produce only 95. He not only maps these areas as he passes from one to the other on the farm, but at the same time he compares their producing ability. Two steps are involved in this procedure: (1) determining the boundaries of the soil areas and (2) estimating the yields for the different soil areas. The remainder of the chapter will be limited to a discussion of the first step—determination of the boundaries between soil areas.

In plotting the soil areas on an appraisal map it appears reasonable to use as far as possible the methods and naming system of the USDA. This does not mean that an appraisal map need be the same kind of soil map that a soil scientist might prepare or that it need be similar to a soil survey map on an enlarged scale. The appraiser is interested almost exclusively in soil productivity, not in soils as an end in themselves. While the soil scientist may be concerned with the difference between two soils of equal productivity, the appraiser may justifiably consider two such soils as only one soil grouping and not bother to separate them on the appraisal map. Moreover, the appraiser is interested in certain soil yield factors which the soil scientist may not care to show on his soil map. Generally the appraiser uses the soil-naming system followed in surveys conducted jointly by the USDA and state agricultural experiment stations.

Appraisal maps and aerial photographs of three farms are presented on the pages which follow. Location and legal descriptions of these farms are:

Farm X

    Location: 5 miles north of Ames, Story County, central Iowa.

    Legal description: SW¼ NW¼ and W½ SW¼ of sec. 35, T 85 N R 24 W of 5th p.m., containing 120 acres more or less.

Farm Y

    Location: 3½ miles east and 4 miles north of Hartley; O'Brien County, northwest Iowa.

    Legal description: NW¼ of sec. 12, T 97 N   R 39 W of 5th p.m., containing 160 acres more or less.

Farm Z

    Location: 3½ miles west of Albia, Monroe County, southern Iowa.

    Legal description: E½ SE¼ of sec. 24, T 72 N   R 18 W and SW¼ of sec. 19, T 72 N   R 17 W of 5th p.m., containing 235 acres more or less.

LEGENDS FOR APPRAISAL MAPS OF FARMS X, Y, AND Z

1. Legend for Soil Types (Number above line):

Symbol	Soil Type
26	— Nodaway silt loam.
26a	— Olmitz-Wabash complex.
55	— Nicollet loam.
65	— Lindley loam.
80	— Weller silt loam.
81	— Jackson silt loam.
107	— Webster silty clay loam.
111	— Lamoure silty clay loam.
111a	— Lamoure silty clay loam (gravelly phase).
138	— Clarion loam.
147	— Lakeville loam.
149	— Clarion fine sandy loam.
x	— Alkali.

2. Legend for Slope (First letter below line):

Symbol	Percentage
A —	0 – 3
B —	4 – 6
C —	7 – 10
D —	11 – 15
E —	16 –

3. Legend for Surface Soil Depth (Number after dash below line):

Symbol	Depth		
1 —	Over 12 inches	3 —	5 – 8 inches
2 —	9 – 12 inches	4 —	0 – 4 inches

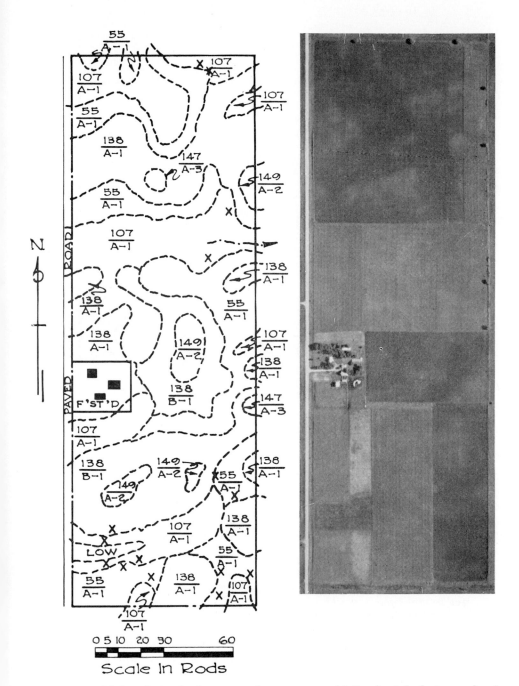

FIG. 19.1. *Appraisal map of Farm X. Farm contains approximately 120 acres. (Map made by the author.) See explanation of map symbols on opposite page.*

FIG. 19.2. *Aerial photograph of Farm X. Photographs from the air are particularly helpful in checking boundaries and acreage. (Courtesy USDA.)*

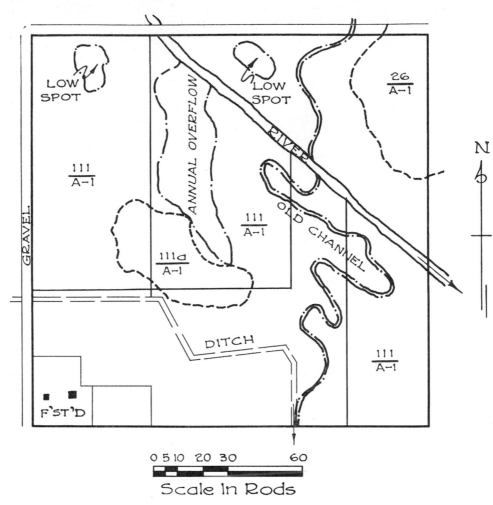

Scale In Rods

FIG. 19.3. *Appraisal map of Farm Y. Farm contains approximately 160 acres. (Map made by Herbert Pike and the author.) The map symbols may be interpreted as follows: The numbers above line refer to soil type, e.g., 111 is Lamoure silty clay loam; the first letter below line indicates topography, e.g., "A" is a slope of 0–3 percent; the number after dash below line indicates surface depth, e.g., 1 is 12+". Alkali is indicated by an "x." (See explanation of legends on page 292.) Aerial map was not available when this appraisal map was made.*

FIG. 19.4. *Aerial photograph of Farm Y. An interesting feature is the old river channel which is a low, wet area on the farm. Comparison of this photograph with Figure 19.3 indicates the advantage of the aerial photograph in showing irregular lines. (Courtesy USDA.)*

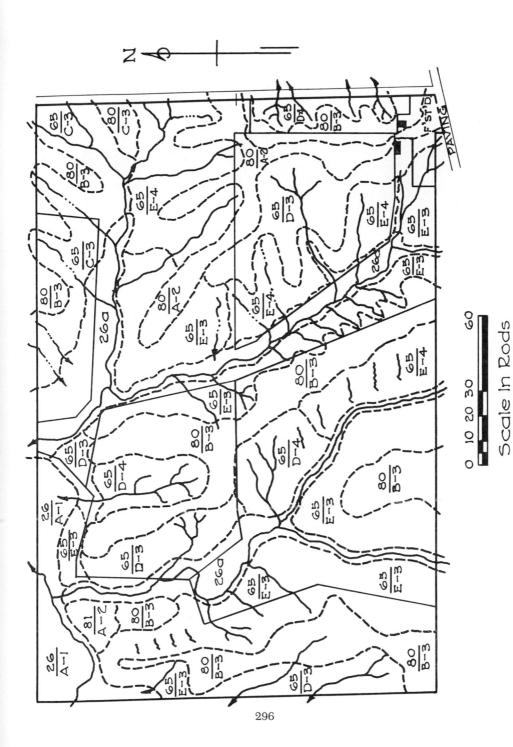

Scale In Rods

0   10 20 30   60

FIG. 19.5. *Appraisal map of Farm Z. Farm contains approximately 235 acres. (Map made by H. R. Meldrum and I. I. Wallace.) Explanation of symbols appears on page 292.*

296

FIG. 19.6. *Aerial photograph of Farm Z. Note small areas of cropland situated on high ground between the ravines. Paved highway crosses southeast boundary of farm. (Courtesy USDA.)*

SOIL TERMS

When such terms as Norfolk fine sand, Ontario loam, Marshall silt loam, and Summit clay appear in print, the appraiser should have no difficulty in knowing to what they refer. These terms describe soil types. The first part of the soil designation, usually a proper name, indicates a *soil series;* the name commonly is taken from the place where the soil was first identified. The second part of the term is the *soil type designation,* a description of the texture of the surface soil. In the soil types named at the beginning of this paragraph the surface soil textures vary from Norfolk fine sand to Summit clay. Texture indicates the size and combination of the mineral particles which make up the soil. In short, texture is the relative proportion of particles of different sizes occurring in the soil. It is evident that the method of separating one surface soil texture from another by such generally accepted terms as "sand," "clay," and "loam" makes interpretation of surface soil texture relatively easy regardless of where the soil is located. In fact, to speak of sand, sandy loam, silt loam, or clay is to use descriptive words that are readily understood.

SERIES

The 1957 Yearbook of Agriculture uses this definition:

> *Soil Series:* A group of soils that have soil horizons similar in their differentiating characteristics and arrangement in the soil profile, except for the texture of the surface soil, and are formed from a particular type of parent material. Soil series is an important category in detailed soil classification. Individual series are given proper names from the place near the first recorded occurrence. Thus names like Houston, Cecil, Barnes, and Miami are names of soil series that appear on soil maps and each connotes a unique combination of many soil characteristics.[2]

Reduced to common terms, this definition means that soils having the same origin and similar in all other respects except surface soil texture are given a name, this name being the series name. Wherever soils of the same general characteristics occur, they receive the same series name. If the makeup of Norfolk soil is known, then the name "Norfolk" may be attached to any soil which answers the Norfolk description. With the series designation, however, the use of proper names has not been entirely satisfactory because the name in itself gives no clue to the characteristics which distinguish soil series from

---

2. USDA, *Soil: The 1957 Yearbook of Agriculture,* Washington, D.C., p. 766.

each other. Furthermore, soil series are more difficult to comprehend than texture, because the series refer to the soil as a whole—to subsoil and subsurface soil as well as to surface soil, and to a number of characteristics rather than just one like texture.

TEXTURE

Within a given soil series there may be a number of soil types representing different surface soil textures. These texture differences, as previously explained, refer to the size and combination of the mineral particles which make up the soil. Many do not realize what a great difference there is between the size of large and small soil particles. The difference, on an enlarged scale, is graphically portrayed in Figure 19.7. A given soil type is usually a combination of several sizes—a loam, for example, being a mixture of sand, silt, and clay. Clarion loam is a soil in the Clarion series with a loam-textured surface soil. For the appraiser, then, the soil series name is identified by examining the soil as a whole, subsoil as well as surface soil, while the texture, which indicates the soil type, is determined by examining the surface soil only.

PHASE

The term "phase" is a means of denoting variations in any soil class but chiefly variations or subdivisions within a soil type. For example, differences in slope, in susceptibility to erosion, in surface soil depth, in stoniness, and in drainage frequently provide for the use of phases. We may find on a given farm one soil type that has several phases, such as Fayette silt loam, slightly eroded phase, and Fayette silt loam, severely eroded phase. Again we may find one soil type subdivided into shallow phase and deep phase. Or we may encounter a single soil type with wide differences in stoniness which can be handled by subdividing the area and calling the stonier area "very stony phase." The term phase is recommended wherever a significant difference in productivity is indicated within a given soil type. Such cases will probably be most common where the land has considerable slope, because erosion and surface soil depth often vary widely under such conditions.

A systematic method of identifying soils by symbols, if it could be developed on a nationwide uniform basis, would be of great assistance to appraisers. The symbols, once they had been learned by the appraiser, would make it possible for him to recognize the characteristics of a soil any place in the country.

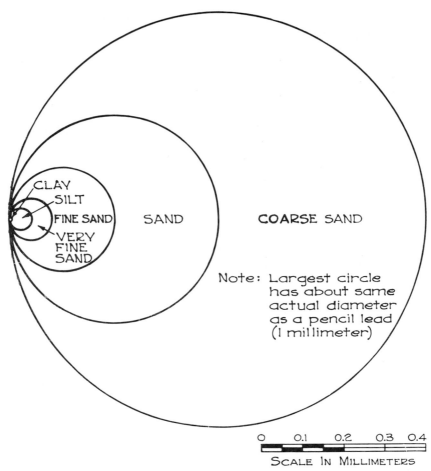

FIG. 19.7. *Relative size of soil particles.* (*Courtesy Herbert Pike.*)

Appraisers would find it distinctly helpful if, instead of having to learn the names and associated characteristics of a hundred or more soil types, they could, by means of a short list of symbols, learn how to identify the characteristics of any soil type. A 5-unit "formula" system could be followed: the first symbol representing origin and formation; the second symbol, color; the third, texture; the fourth, depth of surface soil; and the fifth, reaction. Thus the symbol "A-5-H-4-C" might represent a soil described as follows: "A" indicates an upland glacial prairie soil; "5" indicates a chocolate brown surface soil and light brown subsoil; "H" indicates a silt loam surface texture and a loam subsoil texture with small pebbles and pockets of sand; "4" indicates a surface soil depth of from 0 to 4 inches; and "C" indicates the soil is slightly acid. Another soil designation "A-5-H-2-D" would indicate the same soil as the one above except a deeper surface soil ("2" instead of "4" for the fourth symbol) and a stronger acid reaction ("D" instead of "C" for the fifth symbol). The advantage of a system of this kind is that it brings out the points of difference and similarity between soils. With the present soil-naming system, two soil series like Webster and Floyd may be similar, but this similarity is not at all apparent from the names.

A symbol system somewhat similar to the one described has been proposed by W. M. LeVee and H. E. Dregne;[3] as an illustration they give a certain New Mexico soil the following description: 3231 (which indicates the soil profile) over 2-17V (which indicates the slope and erosion). The first numeral of the profile, 3, is surface texture; the second, 2, is subsoil permeability; the third, 3, is substratum permeability; and the fourth, 1, is the presence or lack of geologic materials which restrict crop production such as loose sand. The symbols 3231, according to the code, indicate a medium-textured surface, slowly permeable subsoil, moderately permeable substratum, and an effective depth without geologic materials which retard plant growth. As for the bottom symbols, 2 indicates a slope of 2 percent, and 17V is slight water erosion with occasional deep gullies. Further comments on this rating system, including provision for special factors, will be found following the explanation of the Storie system in Chapter 21.

FIELD IDENTIFICATION OF SOILS

Since there are several hundred different soil series in the United States, the reader is referred to the soil bulletins and soil survey reports for his own state where he will find complete descriptions of the soil series and soil types occurring in his territory. A bird's-eye view of soils throughout the country may be obtained from the generalized soil map of the United States shown in Figure 19.8. This map gives the location of the great soil groups or associations. Since each of these great soil groups is a combination of a number of individual soil series, the appraiser will find it useful to become acquainted with these overall groupings before tackling the differences between the large number of soil series and types.

A soil series is distinguished by the following characteristics: origin of the soil, color, acid or alkaline reaction, structure

---

3. W. M. LeVee and H. E. Dregne, *A Method for Rating Land,* N. Mex. Agr. Exp. Sta. in cooperation with U.S. Soil Conservation Service, Bull. 364.

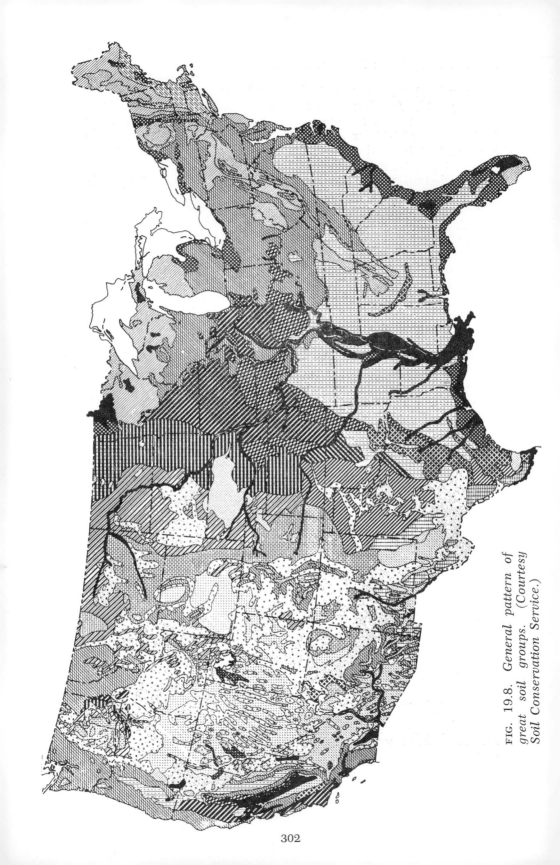

FIG. 19.8. *General pattern of great soil groups. (Courtesy Soil Conservation Service.)*

ZONAL

Great groups of soils with well-developed soil characteristics, reflecting the dominating influence of climate and vegetation. (As shown on the map, many small areas of intrazonal and azonal soils are included.)

PODZOL SOILS
Light-colored leached soils of cool, humid forested regions.

BROWN PODZOLIC SOILS
Brown leached soils of cool-temperate, humid forested regions.

GRAY-BROWN PODZOLIC SOILS
Grayish-brown leached soils of temperate, humid forested regions.

RED OR YELLOW PODZOLIC SOILS
Red or yellow leached soils of warm-temperate, humid forested regions.

PRAIRIE SOILS
Very dark brown soils of cool and temperate, relatively humid grasslands.

REDDISH PRAIRIE SOILS
Dark reddish-brown soils of warm-temperate, relatively humid grasslands.

CHERNOZEM SOILS
Dark-brown to nearly black soils of cool and temperate, subhumid grasslands.

CHESTNUT SOILS
Dark-brown soils of cool and temperate, subhumid to semiarid grasslands.

REDDISH CHESTNUT SOILS
Dark reddish-brown soils of warm-temperate, semiarid regions under mixed shrub and grass vegetation.

BROWN SOILS
Brown soils of cool and temperate, semiarid grass lands.

REDDISH BROWN SOILS
Reddish-brown soils of warm-temperate to hot, semiarid to arid regions, under mixed shrub and grass vegetation.

NONCALCIC BROWN SOILS
Brown or light reddish-brown soils of warm-temperate, wet-dry, semiarid regions, under mixed forest, shrub, and grass vegetation.

SIEROZEM OR GRAY DESERT SOILS
Gray soils of cool to temperate, arid regions, under shrub and grass vegetation.

RED DESERT SOILS
Light reddish-brown soils of warm-temperate to hot, arid regions, under shrub vegetation.

INTRAZONAL

Great groups of soils with more or less well-developed soil characteristics reflecting the dominating influence of some local factor of relief, parent material, or age over the normal effect of climate and vegetation. (Many areas of these soils are included with zonal groups on the map.)

PLANOSOLS
Soils with strongly leached surface horizons over claypans on nearly flat land in cool to warm, humid to subhumid regions, under grass or forest vegetation.

RENDZINA SOILS
Dark grayish-brown to black soils developed from soft limy materials in cool to warm, humid to subhumid regions, mostly under grass vegetation.

SOLONCHAK (1) AND SOLONETZ (2) SOILS
(1) Light-colored soils with high concentration of soluble salts, in subhumid to arid regions, under salt-loving plants.
(2) Dark-colored soils with hard prismatic subsoils, usually strongly alkaline, in subhumid or semiarid regions under grass or shrub vegetation.

WIESENBODEN (1), GROUND WATER PODZOL (2), AND HALF-BOG SOILS (3)
(1) Dark-brown to black soils developed with poor drainage under grasses in humid and subhumid regions.
(2) Gray sandy soils with brown cemented sandy subsoils developed under forests from nearly level imperfectly drained sand in humid regions.
(3) Poorly drained, shallow, dark peaty or mucky soils underlain by gray mineral soil, in humid regions, under swamp-forests.

BOG SOILS
Poorly drained dark peat or muck soils underlain by peat, mostly in humid regions, under swamp or marsh types of vegetation.

AZONAL

Soils without well-developed soil characteristics. (Many areas of these soils are included with other groups on the map.)

LITHOSOLS AND SHALLOW SOILS
(ARID–SUBHUMID)
Shallow soils consisting largely of an imperfectly weathered mass of rock fragments, largely but not exclusively on steep slopes.

(HUMID)

SANDS (DRY)
Very sandy soils.

ALLUVIAL SOILS
Soils developing from recently deposited alluvium that have had little or no modification by processes of soil formation.

The areas of each great soil group shown on the map include areas of other groups too small to be shown separately. Especially are there small areas of the azonal and intrazonal groups included in the areas of zonal groups

303

or granular arrangement, subsoil texture, and a combination of all these factors in a vertical cross section or profile of the soil. It is obvious from this array of distinguishing marks that the description of several hundred soil series is a huge task. The individual appraiser finds the task much simpler because he is dealing with relatively few soil series in a territory. Moreover, many soil series occur so infrequently that they may be combined with the more common soils of the same productive capacity.

An illustration of different soil profiles in the area in which Farms A and X are located is provided in Figure 19.9. There is a striking contrast, it will be noted, between the profile of the Glencoe soil with its deep black surface layer formed under poor drainage conditions and the light brownish surface layer of the Hayden soil formed on timbered slopes.

The explanation of surface soil texture is much easier and more generally applicable than that of soil series. The following list includes the soil textures commonly encountered; they are given in order from coarse to fine:

I. Sand Soils	*II. Loam Soils*	*III. Clay Soils*
1. Coarse sands	5. Sandy loams	10. Sandy clays
2. Fine sands	6. Fine sandy loams	11. Silty clays
3. Very fine sands	7. Loams	12. Clays
4. Loamy fine sands	8. Silt loams	
	9. Silty clay loams	

A systematic chart showing the varying proportions of different-sized particles which make up the various textures is shown in Figure 19.10.

DETERMINING SOIL TEXTURE

The beginner will experience difficulty in determining surface soil texture. It is not a question of telling the difference between a sand and a clay but in being sure that the soil is a loam and not a sandy loam or that it is a silty clay loam and not a silty clay. Rubbing the soil between the fingers is the standard test in the field. The sand particles will be recognized readily because they have a gritty feel. Silt particles and particularly clay particles are so small that the main test is to determine how smooth the soil feels. If with average moisture the soil sample will not fall apart after being squeezed together in the hand, there is considerable clay present. But all this sounds much easier than it actually is. The difficulty of the task may be demonstrated by comparing two samples of the same soil type varying only in the amount of moisture contained in each,

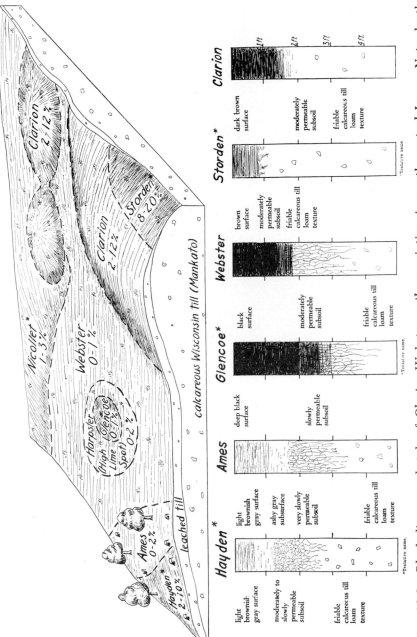

FIG. 19.9. Block-diagram sketch of Clarion-Webster soil association in north central Iowa. Note depth measure in feet identified on right side of Clarion profile. This soil association is the one typical of the area in which Farms A and X are located. These farms are made up chiefly of soils shown in this chart indicating the typical grouping of soils in the area. (Courtesy Iowa Agr. Exp. Sta.)

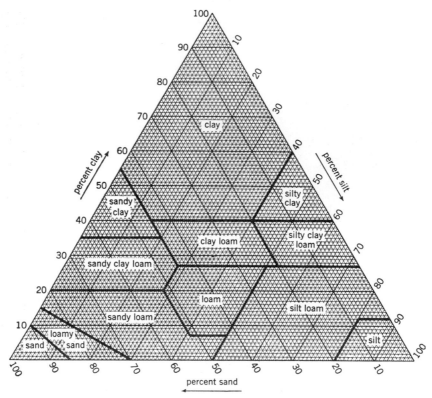

FIG. 19.10. *Chart showing the percentages of clay (below 0.002 mm.), silt (0.002 to 0.05 mm.), and sand (0.05 to 2.0 mm.) in the basic soil textural classes. (Courtesy USDA.)*

or the same test may be made using varying amounts of humus or organic matter instead of moisture.

MAPPING TEXTURE BOUNDARIES

If an SCS soil capability map for a farm showing soil type boundaries is available, much time will be saved for the appraiser, as he will need only to check the SCS map in going over the farm. When such a map is not available he will need to establish and sketch the different soil types as he walks over the farm, particularly the boundaries between types which he judges to be of unequal yielding ability. Even within an area of one soil type, subareas may be outlined which differ in crop-producing ability. One of the principal reasons for these subareas or subtypes is a large amount of variation in surface soil depth within the same soil type. On the appraisal map for

Farm Z (Fig. 19.5), for example, two subareas of Lindley loam are shown, one with from 5 to 8 inches of surface soil, $\dfrac{65}{D \, or \, E\text{--}3}$, and the other with only 0 to 4 inches, $\dfrac{65}{D \, or \, E\text{--}4}$. Since these two subareas of Lindley loam differ greatly in productivity it is essential that such conditions within the same soil type be shown.

The appraiser, when he has finished his appraisal map, will have the farm area divided into those soil conditions or subareas within which the yield will be estimated as uniform. He will subdivide permanent pasture in a similar manner, all pasture within a soil condition or subarea being considered capable of producing approximately the same amount of pasturage.

VARIATIONS WITHIN A SOIL TYPE AREA

One of the difficulties connected with soil mapping is determining where the boundary should be drawn between two soil type areas. In some instances, such as between upland and terrace soil type areas or two soils differing markedly in color, the difficulty may not be great, but where one type changes gradually into that of the other it is almost impossible to find a clear-cut dividing line.

An example of the variation existing within soil types is provided by a study of the appraisal map of Farm X shown in Figure 19.1. Two small elongated areas of Number 149—Clarion fine sandy loam—are located between the farmstead and the south boundary of the farm. This type is surrounded by Number 138—Clarion loam, and the Clarion loam is bordered on the south by Number 107—Webster silty clay loam.

Actually there is little difference between Number 149 and Number 138 just on each side of the arbitrary line drawn to separate these two types, nor is there any marked difference between Number 138 and Number 107 on either side of the line which divides them. If we designate Number 138 where it joins Number 149 as Number 138a, and where it joins Number 107 as Number 138b, then we can explain the situation as follows: Soil Number 138a resembles Number 149 more than it does Number 138b. In a similar manner Number 138b is more like Number 107 than it is like Number 138a.

This situation is one which cannot be avoided where there is gradual variation. It is important, however, that the appraiser recognize these conditions when he sees them. Furthermore, it will be helpful for the appraiser to understand this difficulty when he is faced with the problem of mapping soil type areas that shade gradually from one to the other.

SOIL ASSOCIATIONS, CATENAS OR TOPOSEQUENCES, AND COMPLEXES

Where soil types are difficult to map, and for others as well, the appraiser will find the soil association, the catena, and the complex concepts especially helpful. The 1957 Yearbook of Agriculture defines these terms as follows:

> *Soil association:* A group of defined and named kinds of soil associated together in a characteristic geographic pattern. Except on detailed soil maps, it is not possible to delineate the various kinds of soil so that on all small-scale soil maps the areas shown consist of soil associations of two or more kinds of soil that are geographically associated.[4]
> *Soil catena:* A group of soils, within a specific soil zone, formed from similar parent materials but with unlike soil characteristics because of differences in relief or drainage.[5]
> *Soil complex:* An intimate mixture of tiny areas of different kinds of soil that are too small to be shown separately on a publishable map. The whole group of soils must be shown together as a mapping unit and described as a pattern of soils.[6]

The term *toposequence,* indicating soils developed under different topography conditions, has been gaining in favor over *catena* as the term to use in explaining situations like those in Figures 19.9 and 19.11.

When an appraiser becomes acquainted with the soils of a given territory, he will find that a certain pattern is frequently repeated. These patterns, associations, or groupings of soils cover large areas and can be readily identified by the appraiser with a little practice. Iowa, for example, is made up for the most part of five soil associations. Farms *A*, *X*, and *Y* are in the Clarion-Webster association of north central Iowa, while Farm *Z* is in the Grundy-Weller-Lindley association in southern Iowa.

Within the association slight soil variations caused by differences in topography are called a toposequence or a catena. (The word *catena* comes from the Latin meaning *chain.*) On Farm *X*, for example, a Clarion-Webster toposequence is well marked with the Clarion soils on the higher, well-drained upland and the Webster on the lower, poorly drained upland. A good illustration of the Clarion-Webster soil association and toposequence is provided by Figure 19.9, which shows how the soil types are related to slope, drainage, and vegetation. The catenary chain linking Clarion and Webster soils, for example, is slope— Clarion developed on slopes up to 12 percent, Nicollet on slopes

4. USDA, *Soil: The 1957 Yearbook of Agriculture,* Washington, D.C., p. 767.
5. Ibid., p. 754.
6. Ibid., p. 755.

up to 3 percent, and Webster on the poorly drained level upland.

The toposequence for the Farm Z area is presented in Figure 19.11. On Farm Z, which is hilly and timbered for the most part, only the Weller and Lindley soils in the catena are represented, but Figure 19.11 indicates the typical situation for these two soils found on this farm.

The appraiser's problem of getting acquainted with a large number of soil types can be eased by the use of a toposequence grouping or key. This emphasizes the relation of one soil series to another. For example, in the Clarion-Webster toposequence the Clarion is on the higher sloping ground, the Webster on the lower level or depressed area. Another example, this one from New York State, is shown in Table 19.1, a grouping of soils according to lime content, origin, and drainage. An understanding of the underlying relationships between soil series, as provided by the catena idea, is an essential tool in the appraiser's kit.

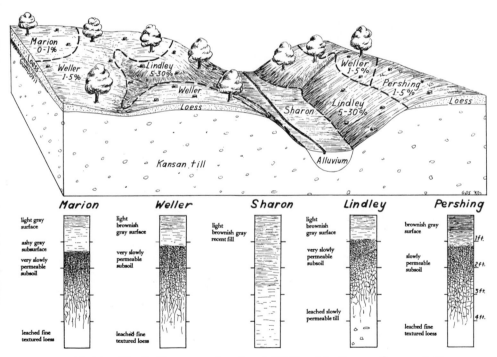

FIG. 19.11. *Block-diagram sketch of Weller-Lindley soil association in southern Iowa. This soil is typical in the area in which Farm Z is located. Appraisal map of Farm Z is shown in Figure 19.5, and aerial map in Figure 19.6. (Courtesy Iowa Agr. Exp. Sta.)*

TABLE 19.1. Guide to Some Central New York Soils

Lime Content	Origin	Drainage				
		Good or excessive	Moderately good	Somewhat poor	Poor	Very poor
		Soil Names				
High	Till	Honeoye	Lima	Kendaia		Lyons / Sloan
High	Alluvial	Genesee	Eel		Wayland	
High	Outwash	Palmyra	Phelps			
High	Lakelaid	Schoharie	Odessa		Lakemont	Poygan
Medium	Till	Lansing	Conesus	Kendaia		Lyons / Sloan
Medium	Alluvial	Chagrin	Lobdell		Wayland	
Medium	Outwash	Howard	Phelps			
Low to medium	Till with pan	Valois	Langford	Erie	Ellery	Alden
Low	Till with pan	Bath	Mardin	Volusia		Chippewa
Low	Till with no pan	Woostern				
Low	Shallow till	Lordstown	Arnot			
Low	Alluvial	Tioga	Middlebury		Holly	Papakating
Low	Outwash	Chenango	Braceville	Red Hook		Atherton

Source: M. G. Cline. *Soils and Soil Associations of New York*, Cornell Ext. Bull. 930, May 1955.

A soil complex often develops in a river bottom which has been subjected to frequent overflows. Areas of heavy and light soils may exist in all sorts of irregular patterns. In situations like this the appraiser can designate such an area as a soil complex and at the same time give the reader of the appraisal map an indication of the soil condition on the river bottom. The soil type as well as boundaries may be altered with each overflow, making previous maps and appraisals subject to revision.

NOTES ON MAPPING

The heart of the appraisal is usually the inspection trip around the farm. Plenty of time should be allowed so that a thorough job is done. An example of an inspection trip is shown in Figure 19.12. The author has found that considerable time can be spent to advantage in the first part of the inspection, making certain beyond question the kinds of soil conditions present; once these have been determined the remainder of the trip over the farm goes much faster. It is always well, however, to examine any possible variation in soil, topography, drainage, or similar features that might have some effect on the land value.

In making appraisal maps in the field the appraiser may find it desirable to use a sharp 4-H pencil, glazed durable map paper, a clipboard, and a small ruler that is adapted to scale of map. The appraiser will often save time in orienting himself at the start by beginning at a corner of the farm. In estimating distances, rough methods such as counting fence posts, pacing, and dividing fractionally are more practical usually than actual measurements with chain or tape. If an aerial map is available, it will serve as an excellent guide. One practice that may prove helpful is to use a red pencil to indicate drainage ways: a dot-dash red line for intermittent drains and a solid red line for flowing streams. This makes the drainage stand out in marked contrast to the soil boundaries.

A definite marking system should be used to indicate permanent pasture so that its separation from tillable land can be seen at a glance.

The finished appraisal map may be traced from the field map. Here again a fixed series of colored pencils or crayons is advisable in showing different grades of productive soil on a farm map; for example blue areas for excellent soils, green for very good, brown for good, red for fair, and yellow for poor soils. It is possible with such a system to tell at a glance the general productivity rating of a farm.

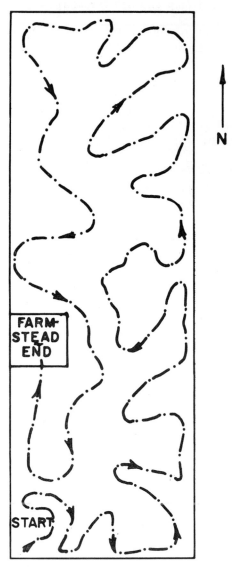

FIG. 19.12. *Route taken by author in making appraisal map of Farm X. Objective in inspecting a farm is to examine all the different soil conditions and other features that bear on the farm's value and make an accurate map of these soil conditions and other features while going over the farm.*

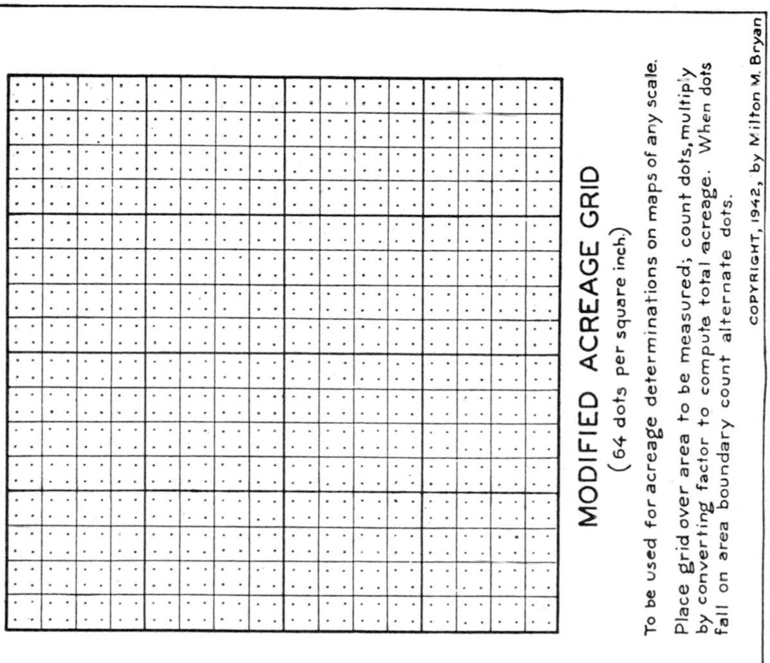

MODIFIED ACREAGE GRID
(64 dots per square inch)

To be used for acreage determinations on maps of any scale.

Place grid over area to be measured; count dots, multiply by converting factor to compute total acreage. When dots fall on area boundary count alternate dots.

COPYRIGHT, 1942, by Milton M. Bryan

MAP SCALES AND EQUIVALENTS

Fractional Scale	Inches Per Mile	Acres Per Square Inch	Converting Factor Each dot equals:	
1″ = 7,920″	8.00	10.000	0.156	Acres
1″ = 9,600″	6.60	14.692	0.230	Acres
1″ = 15,840″	4.00	40.000	0.625	Acres
1″ = 20,000″	3.168	63.769	0.996	Acres
1″ = 31,680″	2.00	160.000	2.500	Acres
1″ = 63,360″	1.00	640.000	10.000	Acres
1″ = 125,000″	0.507	2,490.980	38.922	Acres
1″ = 250,000″	0.253	9,963.906	155.686	Acres
1″ = 500,000″	0.127	39,855.627	622.744	Acres

FIG. 19.13. *Modified acreage grid. (Courtesy Milton M. Bryan.)*

313

ACREAGE MEASURE

A special lined paper for appraisal maps is recommended for beginning appraisers because it makes estimation of acreages of the different soil conditions relatively easy. Some appraisers use this map paper regularly. A common type, often referred to as "engineering paper," has heavy green lines marking off one-inch squares with each of these squares marked off with faint green lines into 25 small squares. With an 8-inch-to-the-mile scale, one inch on the map equals 40 rods, each one-inch square contains 10 acres, and each small square contains .4 of an acre. By counting the small squares in each soil condition area, it is a simple matter to estimate the acreage in each soil area and to make adjustments so the total acreage in the farm is properly allocated.

A separate grid can also be used to measure acreage. An example of such a grid, which is placed over the completed appraisal map, is shown in Figure 19.13.

QUESTIONS

1. Soil identification

   If possible, obtain a county soil survey map for an area which can be reached readily. Make a trip to this area to check the soil survey on one or more farms by locating boundaries between soil areas as indicated on soil survey map. Note the difficulties probably encountered by soil surveyors in drawing some of the boundary lines. Use soil auger in determination of soil profiles, or if an auger is not available, use spade. Compare different textures, such as sand, loam, and clay.

2. Appraisal maps

   After you have checked one or more farms with use of published soil maps, make an appraisal map by yourself without the aid of a published map. From legal description draw farm boundaries, then starting in one corner of farm, work back and forth determining variation in soils as you go. Note relation between topography and soil. Indicate boundaries between soil conditions on the map.

   (These problems may be used in connection with the next three chapters by requiring additional material on physical features and estimated yields. Skill in soil identification and making of appraisal maps may develop slowly. But since this skill is basic for succeeding steps, practice should be continued until proficiency is attained.)

REFERENCES

County soil survey reports. These may be obtained from the state agricultural experiment station in each state or from the USDA,

Washington, D.C. Not all counties are surveyed; a list of the counties for which surveys are available will be furnished by the state agricultural experiment station.

Baver, L. D., *Soil Physics*, 3rd ed., John Wiley and Sons, Inc., New York, 1956.

Buckman, H. O., and Brady, N. C., *The Nature and Properties of Soils*, 6th ed., Macmillan, New York, 1960.

Millar, C. E., Turk, L. M., and Foth, II. D., *Fundamentals of Soil Science*, 4th ed., John Wiley and Sons, Inc., New York, 1965.

Soils of the North Central Region of the United States, N. Cent. Reg. Publ. No. 76, Agr. Exp. Sta., Univ. Wis., Madison, Bull. 544, June 1960.

Thompson, L. M., *Soils and Soil Fertility*, 2nd ed., McGraw-Hill, New York, 1957.

USDA, *Soil: The 1957 Yearbook of Agriculture*, Washington, D.C. See especially sections on "Principles," "Systems," and "Regions."

———, *Soil Survey Manual*, Handbook No. 18, Washington, D.C., 1951. This gives a detailed description of the methods followed in the preparation of county soil surveys maps and reports.

———, *Soils and Men: The 1938 Yearbook of Agriculture*, Washington, D.C. See especially Part IV, "Fundamentals of Soils Science," and Part V, "Soils of the United States." An excellent reference for the appraisal student. Note especially the generalized soil map of the United States found on the inside back cover.

20

SOIL PRODUCTIVITY FACTORS

AN ESTIMATE of productivity goes hand in hand with the preparation of the appraisal map because each soil condition indicated on the map has meaning chiefly in terms of yield. It is this yield index which justifies the time taken in making the map. Productivity as measured in yield is not estimated easily, however; it is the result not of one factor but of many. Hence the following discussion will include the effect on yield of soil texture, depth of surface soil, character of subsoil, fertility, topography, erosion, drainage, and climate.

SOIL TEXTURE

The extent to which texture influences crop yield may be illustrated by assuming an experiment set up in the Corn Belt with three widely differing kinds of soil texture—a sand, a loam, and a clay. A crop of corn on sand gives a low yield in normal weather because the large size of the soil particles allows the water to seep away and is not conducive to holding organic matter. Sands are often called drouthy soils. A field of clay under the same climatic conditions is difficult to prepare for planting and still more difficult to cultivate if normal rainfall prevails. Clay is so compact, so given to puddling and running together when wet, that the corn yield will be low. A loam soil is the happy medium, consisting as it does of sand, silt, and clay, with enough clay to keep the soil from being drouthy and enough sand so the land can be worked advantageously. Corn yields on the loam texture soil in the prairie regions are much higher than on the sand at one extreme and the clay at the other. Most of the important crop-producing soils are some type of loam. Loam soils, as noted in the previous chapter, include many combinations of sand, silt, and clay which shade into one another like colors in the spectrum.

How yields vary with surface soil texture is shown by a study of the productivity ratings given soils by various state agricultural experiment stations. A rating system devised by R. E. Storie for California gives each texture a weight, with 100 as the best. The most productive textures are a straight loam along with fine sandy loam on one side and silt loam on the other. (see Table 21.1). The productivity rating decreases as the soil becomes progressively more sandy or composed of more clay.

Another method of rating surface soil texture is to classify the different soil types according to their corn yield potential. When this was done for Iowa soil types the resulting classification, presented in Table 20.1, showed loams on top with silty clay loams and silt loams predominant and with a few straight loams in the top group. No sandy loams or clay loams were rated in the select group producing above 90 bushels to the acre. Since corn yields have been rising rapidly in recent years, this 1965 classification is used to show relative ratings, not absolute corn yields for the different surface soil textures. A third classification by texture is presented in Table 20.2 for the southern tidewater area of North Carolina. In this region where annual precipitation averages between 45 and 60 inches, sandy loams are more productive than the heavier clay loams. There is a

TABLE 20.1. Surface Soil Textures of Iowa Soil Types Classified by Corn Yield Potential

Texture of Surface Soil	Corn Yield Potential in Bu/Acre						
	30–40	41–50	51–60	61–70	71–80	81–90	90+
Loamy sand	2						
Loamy fine sand	1		1				
Sandy loam	1				1		
Fine sandy loam		1		2			
Loam		1	7	5	10	19	4
Silt loam		2	10	7	19	21	13
Silty clay loam				3	16	16	21
Clay loam	3	7				1	
Silty clay	1	1	2	2	1		
Peat and muck			1	1			

Source: *Principal Soils of Iowa,* Spec. Rept. No. 42, Iowa Agr. Exp. Sta., Ames, 1965.

TABLE 20.2. Surface Soil Textures of North Carolina Soils in Southern Tidewater Area Classified by Productivity Rating

Texture of Surface Soil	\multicolumn{10}{c}{Number of Soil Types Rated[a]}									
	10	9	8	7	6	5	4	3	2	1
Sand	3			3	2					
Fine sand	3	3		2	2	1				
Loamy fine sand					1	1				
Sandy loam					2		1	1	3	2
Fine sandy loam	1	1	1		3	3	3	2	4	4
Very fine sandy loam					1	2			2	1
Loam		1		2	1		3	1		
Silt loam				2	1					
Clay loam			1			1	1			
Clay					1					

Source: Prepared from data in *Agricultural Classification and Evaluation of North Carolina Soils,* N.C. Agr. Sta., Bull. 293.

[a] Soil ratings range from 10 for poorest soil to 1 for best soil.

difference in the type of loam which rates at the top in California, Iowa, and North Carolina, although the loams are more productive than the sands or clays in all three states. In Iowa the silt loams and silty clay loams outrank the sandy loams, while in California and North Carolina the advantage lies with the sandy loams. This situation can be explained by moisture, temperature, and kinds of crops grown. The heavier soils are preferable where moisture is not always adequate and temperature is relatively high as in Iowa. A sandier soil is more desirable where irrigation is practiced or rainfall is abundant. Moreover, some crops do better on heavy soil, while others require soils with more sand in them.

In the dry wheat areas a heavy soil texture is especially desirable. This is illustrated by a study of wheat yields in Saskatchewan in western Canada where yields per acre over a period of years for typical soils averaged as follows:[1]

Heavy clay . . . . 16.2 bu. Loams—clay loams . . . . 11.2 bu.
Clay . . . . 14.1 Loams—light loams . . . 10.6
Clay loams . . . . 12.3 Fine sandy loams 10.1

Another example of the effect of soil texture is presented in Table 20.3. In the northern New York dairy region, hay yields were higher on loam soils than on clays and sands. The farmers on loam soils needed less cropland to keep a cow and also had available more roughage for each cow. ("Hay equivalent" is the tons of hay plus one-third of the tons of silage.) The amount

1. J. Zeman, *An Economic Classification of Land and a Study of Farm Organization of the Biggar-Kerrobert-Unity Area, Saskatchewan, 1948,* Canada Dept. of Agr. Marketing Service, Ottawa, 1953.

TABLE 20.3. Hay Yields and Related Items on Different Soils, 556 Farms, North
             Country Region, New York, 1955

Item	Clay Soils	Loam Soils	Sandy Soils
Number of farms	239	143	174
Tons of hay per acre	1.4	1.8	1.5
Crop acres per cow	3.7	2.9	3.2
Tons of hay equivalent harvested per cow	4.5	5.0	4.5
Pounds of milk sold per cow	6,500	7,000	6,200

Source: L. C. Cunningham, *North Country Dairy Farming,* Mimeo. Rept. A.E.
1084, Dept. of Agr. Econ., Cornell Univ.

of milk sold per cow was higher on the loam soils largely because
of the more adequate amount of roughage available.

DEPTH OF SURFACE SOIL

Yield is often dependent on depth of surface soil. The
deeper the surface layer the higher will be the resulting crop
yields up to a certain point. Beyond that point, naturally, additional depth is of no real benefit to yield because the plant roots
have all the surface soil they need in which to work. This critical
point will vary according to whether the plant is shallow or deep
rooted. Since the importance of the depth factor cannot be overemphasized, an accurate inventory of a farm should include a
detailed description of the surface soil. This measurement of
surface soil means that a third dimension of depth is added to
the usual concept of a farm as having length and breadth.

A vertical cross section of soil may be divided for our purposes into two parts, surface soil and subsoil, the subsoil including what is referred to as subsurface as well as subsoil. The soil
scientist divides the soil profile into horizons indicated by the
letters A, B, and C. The main distinction for the appraiser, however, is between surface soil and subsoil. In prairie soils the surface layer can be recognized as the plant root-growing area in
which organic matter is present. A change of color is usually
the most common characteristic distinguishing this surface layer
from the subsoil, the reason being that plant roots and organic
matter close to the top of the soil produce a different colored soil,
usually darker than that below where conditions are entirely
different.

A few experimental figures are available showing in relative
terms the increase in yield that accompanies increased surface
soil depth. Corn yields for two years on varying depths of Tama
silt loam in Tama County, Iowa, are presented in Table 20.4.
Depth up to 10 inches is apparently extremely important in its
effect on yield of corn. Additional surface soil over 10 inches,

TABLE 20.4. Average Corn Yields on Tama Silt Loam According to Depth of Surface Soil (Yields on Basis of Perfect Stand)

Approximate Depth of Surface Soil in Inches	Poor Year (Dry)		Good Year	
	Number of samples	Average yield	Number of samples	Average yield
		(*bushels*)		(*bushels*)
0 – 2	4	31	7	47
3 – 4	8	28	10	69
5 – 6	30	39	19	77
7 – 8	39	45	33	82
9 – 10	23	50	19	88
11 – 12	12	50	25	82
Over 12	11	53	19	88

Source: W. G. Murray, A. J. Englehorn, and R. A. Griffin, *Yield Tests and Land Valuation,* Iowa Agr. Exp. Sta., Res. Bull. 252, 1939, p. 66.

at least as far as this example shows, has no influence on yield. A more recent rating of Iowa soil types on slope and depth in terms of corn yield potential indicates a substantial decline in yield as the slope increases and the surface soil depth becomes thinner. The estimated drop in corn yield as slope increases and depth decreases is shown for four soil types in Table 20.5.

Depth of surface soil should be a part of the appraisal report, the chief concern of which is to estimate the value of the land. If depth does not actually appear as it should in the report itself, at least the influence of depth on yield should be considered by the appraiser in arriving at his estimates of the crop-yielding ability of the soil. If 80 acres of 4-inch Tama silty clay loam yields a certain amount, and 80 acres of 8-inch Tama yields 15 percent more (in line with the results obtained in the

TABLE 20.5. Corn Yield Potential of Selected Iowa Soil Types Classified by Slope and Depth of Surface Soil

Soil Type	Slope	Depth of Surface Soil		
		7–12 in.	3–7 in.	Under 3 in.
	(*percent*)	(*bushels*)		
Clarion loam	2–5	90		
	5–9		85	
Fayette silt	2–5	92		
loam	5–9		87	
	9–14			78
Tama silty clay	2–5	98		
loam	9–5		93	
	9–14		85	
Clarinda clay	5–9	45		
loam	9–14		30	

Source: *Principal Soils of Iowa,* Spec. Rept. No. 42, Iowa Agr. Exp. Sta., Ames, 1965.

example in Table 20.5), the 8-inch soil is worth more than the 4-inch, even though both areas of soil have identical type names.

The effect of soil depth on yield is indicated by the studies of R. T. Odell in Illinois. In these studies a good comparison is the Muscatine-Sable soils with a deep top layer and the Cisne soil with a thin top layer. In the 10 years ending in 1955 the Muscatine-Sable soils averaged 73–80 bushels per acre whereas the Cisne for the same period averaged 51–60 bushels.[2]

SUBSOIL

Subsoil has an important place in the production of a crop but is often ignored because it is seldom seen except in such instances as road cuts, post holes, or soil borings. The appraiser should make sure that this factor is properly weighted in the valuation. This means that when inspecting a farm an appraiser must use a spade or soil auger. This is hard work but there is no substitute for it.

The effect of subsoil on yield can be expressed in terms similar to the discussion on surface soil texture. Three types of subsoil in the Corn Belt may be used as an illustration: (1) a porous, sandy-textured subsoil with practically no water-holding capacity, (2) a mixture of sand, silt, and clay which allows moisture to filter through but has a relatively high capacity for holding water, and (3) a hard, clay-pan layer composed largely of clay with little chance for water to seep through. With the same surface soil and with normal weather conditions assumed for all three cases, crops planted on the porous subsoil will dry up, crops on the second type will produce a good yield, while on the clay pan it will be difficult to even get the planting done because of the wet condition which is likely to prevail. An added disadvantage of the tight subsoil on slopes is its susceptibility to erosion, the topsoil washing readily because the rain, unable to penetrate easily, runs off instead.

These three types of subsoil can be compared under actual conditions. In Iowa, for example, Waukegan loam with a moderately deep surface soil and a porous sandy or gravel subsoil fits the first type; Tama silty clay loam with a favorable mixture in the subsoil fits the second type; and Weller silt loam with a heavy tight subsoil fits the third type. Estimated yields for these soils are 50 to 80 bushels of corn per acre on the Waukegan,

2. R. T. Odell, "Soil Survey Interpretation—Yield Prediction," *Soil Sci. Soc. Amer. Proc.*, 22(2):158, Mar.–Apr. 1958. See also Odell, *Productivity of Soils in the North Central Region*, Univ. of Ill. Agr. Exp. Sta., Bull. 710, May 1965. In this regional bulletin Muscatine is given an estimated corn yield under high management of 106 bushels and Cisne 78 bushels.

85 to 98 bushels per acre on the Tama, and 57 to 65 bushels on the Weller.[3]

An object lesson in the difference between a porous, gravelly, or sandy subsoil and a good, well-mixed subsoil may often be observed at harvest time in the crops growing above these types. The outline of gravelly subsoil areas in a field of small grain will probably be clearly defined by the difference in the appearance and yield of the grain. The appraiser should endeavor to estimate accurately the difference between poor and good subsoil in terms of yield. Continued practice and observation will soon improve the beginning appraiser's estimates.

The difference between the three subsoils cited above is clear, but the measurement of intermediate types presents more of a problem. The beginner should concentrate on the differences between a few easily recognized types, like those mentioned, which are wide apart in their makeup. This plan leaves for the more experienced appraiser the identification and evaluation of the in-between subsoils—the large range between gravels at one end, loams at the center, and clays at the other end. *The student should progress from a simple separation into more complex and difficult separations as fast as, but not any faster than, his experience in the field dictates.*

FERTILITY

Two areas of soil may differ markedly in their productivity because of variation in fertility, even though these two areas may be identical in (1) surface texture, (2) depth of surface soil, and (3) character of subsoil. One area may have been farmed continuously for years with little in the way of plant food put back into the soil. The other area, in contrast, may have considerable organic matter and have been well handled. The difference between these two areas consists mainly in organic matter or plant food; a laboratory analysis in which the amount of organic material is determined will confirm this. What is important of course is the difference in yields, and it is this difference which the appraiser must reflect in the final valuation.

An appraiser may have little or no difficulty in ascertaining texture and soil type, depth of surface soil, and character of subsoil, but he is often baffled when it comes to measuring soil fertility. Fertility depends not only upon the amounts of nitrogen, phosphorus, and other plant nutrients present in the soil but also on whether they are available for the use of plants in the right proportions. What effect would heavy applications of ni-

---

3. *Principal Soils of Iowa,* Spec. Rept. No. 42, Iowa Agr. Exp. Sta., Ames, 1965, pp. 48, 70, 76.

trogen fertilizer have on a given soil? Is calcium or phosphorus or organic matter a limiting factor on a farm being appraised? Since the answers to these questions may have an important bearing on the valuation of the land, the appraiser needs to keep up to date on the latest and most reliable methods of soil testing. An ideal instrument would be one similar to a thermometer which, when inserted in the soil, would indicate the normal crop-producing ability of the soil. A yield thermometer for each important crop would then be to the appraiser what the fever thermometer is to the doctor. There are various methods which have been developed for testing soils to determine the amount of the principal plant food requirements present in the soil, a discussion of which appears later in this chapter.

Actual yield, with consideration given the type of management, is by far the best determinant of soil fertility. Some appraisers claim, however, that they can judge the fertility of a soil by rubbing a few samples of soil between their fingers and feeling the difference. This test, although a common one, is only a rough method of determining humus content. For instance variation in soil moisture is an obstacle to reliable results; a wet soil will appear to have more humus than the same soil when dry. Appraisers in general look for signs of fertility or the lack of it in growing crops, or, if the season is early, they look at the remains of last year's crop, such as the size of the corn stalks. Here again, one year's crop is only a rough check on fertility. An appraiser using such methods must be on his guard lest he underrate the soil because of a poor season or poor management, or overrate it for the opposite reasons. Appraisers may increase their ability to evaluate fertility by using as observation laboratories some fields where they are familiar with the soil variation or where they know something of previous management. Here they may observe how different textures, different depths of surface soil, and different subsoil conditions within the same field influence yield. By studying the same farm through successive years, they may follow the progress of specific management practices such as erosion, pest and weed control, and soil-building programs. Knowledge gained in this way, plus evidence from state experiment stations, may be applied directly in making appraisals in a territory. This practice does not conform to strict scientific calculations, but it is the best method to follow until more material on soil fertility and related yield factors becomes available.

LIME

In many regions the pH reaction of soil is an important factor. A soil so alkaline that crop yields are almost negligible will generally be low in value. An acid soil has to be discounted in raising those crops which are sensitive to the acid condition, because the correction of acidity, as by the application of lime, involves expense. Many farmers do not buy sufficient lime and as a result obtain lower yields than they could obtain with the correct amount of lime applied. It often happens that a particular soil may rate high in all other productivity factors save this one of reaction. Thanks to the soil scientists, relatively easy tests have been devised for obtaining this reaction. A field kit may be used to obtain the pH values. The *Soil Survey Manual* gives a description of soils of various pH values as follows:[4]

	pH		pH
Extremely acid	Below 4.5	Neutral	6.6–7.3
Very strongly acid	4.5–5.0	Mildly alkaline	7.4–7.8
Strongly acid	5.1–5.5	Moderately alkaline	7.9–8.4
Medium acid	5.6–6.0	Strongly alkaline	8.5–9.0
Slightly acid	6.1–6.5	Very strongly alkaline	9.1 and higher

A simple test for calcareous material, or "alkali," is the application of dilute hydrochloric acid to the surface soil. A bubbling action caused by the acid indicates the presence of the calcareous material, the amount being measured by the extent of the bubbling. An appraiser equipped with a small bottle of acid will find this a handy test for an estimate, but a laboratory test will be needed to establish the exact value.

OTHER SOIL CONDITIONS AFFECTING YIELD

There are several soil conditions influencing yield which may be designated as minor, not because they are always unimportant but because, as a whole, they do not affect yield as much as those factors discussed thus far. Among these conditions are color of surface soil and subsoil, origin or parent material, stoniness, structure, and porosity.

Color may or may not be an indication of producing ability. In temperate climates the darker the color of the surface soil, the more productive it is likely to be because organic matter darkens the soil. This condition must not be confused, however, with varying colors dependent not on organic matter but on the origin of the soil.

Subsoil color is likewise significant, revealing the conditions existing in the area. In certain depressed areas, for instance, a gray color tells the appraiser that drainage has not been good, that little air circulation has taken place resulting in a small

4. USDA, *Soil Survey Manual*, Handbook No. 18, Washington, D.C., 1951, p. 235.

amount of oxidation. A brown color, on the other hand, will in-dicate entirely different conditions, particularly good oxidation.

An understanding of the origin of a soil is fundamental, not so much because of the name of the parent material as because of what that original source signifies. A loessial soil may be ex-pected to have a uniform surface texture, usually a silt loam. A glacial soil may have considerable variation including stones and pockets of gravel.

Soil productivity, as presented in the discussion thus far, includes more than merely soil naming or identification. It in-cludes how yield is affected by texture, depth of surface soil, subsoil, fertility, lime, and other factors. Consequently the ap-praiser in making the appraisal map should draw in the soil area boundaries with these productivity factors in mind. The resulting appraisal map may include two or more areas of the same soil, the difference between the areas being depth of sur-face soil. It is essential, therefore, that the appraiser go beyond the naming of soils to the crucial question of what a given area of soil will yield. In fact, *the yield differences within a given soil type caused by depth (and other) variations may be even greater than the yield differences between certain types.*

SOIL SAMPLING AND TESTING

A good soil test is one that is based on a good soil sample. If the sample is improperly taken the test will not be reliable. When the soil samples have been properly taken, they can be sent to a soil testing laboratory, usually located at the state agri-cultural experiment station, where the tests will be run and the results returned for analysis and use.

The following suggestions for taking a soil sample should prove helpful:

1. Each sample should represent a *uniform* soil area with narrow slope range and *similar* past management.
2. Avoid, or sample separately, all odd or dissimilarly treated areas not representative of the uniform soil area.
   a. Odd areas would be: dead furrows, back furrows, old straw-stack bottoms, old fencelines and small field depressions.
   b. Dissimilarly treated areas would be: actual crop rows which may have been fertilized; a 50-pace wide strip along limerock roads; and the same soil type but with only a portion of it limed, fertilized, or manured previously. In the latter case, separate samples would be needed even though only one soil type is involved.
3. Each sample should represent *10 acres or less* of the uniform soil area.

4. Each sample should be made up of 15 to 20 separate cores, borings, or trowel slices taken to plow depth at random from the 10 acres or less of the uniform land area.
5. Place cores, borings, or trowel slices in a *clean* pail, mix thoroughly, and then fill plastic-lined bag half full. Identify and number bag before filling.
6. Grasp bag just above soil level and, with both hands, bring sides of bag together to squeeze out air. Then roll upper portion of bag down to soil level. Clip roll at bag ears so that it will not unravel. The soil sample must *not be allowed to dry out.*
7. Fill out information sheet and map showing sample and field location with reference to buildings.
8. Mail samples to the soil testing laboratory *within 12 hours* after they are taken. If this is impossible, put samples in the refrigerator or deep freeze until they can be mailed.[5]

TOPOGRAPHY

Topography deserves a generous share of the appraiser's time. Besides being a key to the water erosion hazard, topography gives an indication of soil and drainage conditions and often determines the amount of pasture and kind of farming practiced. The appraiser, therefore, will do well to map topography as he covers the farm, examining and mapping the soil and other physical features.

Topography will often help in identifying the soil type. A study of soil survey maps indicates that many soils are formed and exist only under certain topographic conditions. It is usual in any region to find certain soil types associated only with definite topographic features. There are for example soil types that occur only on level upland, and other types that occur only on terraces or bottoms. If the appraiser is on an upland area, he will find that possible types are limited to upland soil types. This linking of soil type with topography not only makes soil mapping easier but also brings the two together in a way which may suggest how the soil originated. Topography can be noted with the eye over wide areas, while determination of soil types by soil borings is a tedious, time-consuming process. It should not be inferred, however, that topography can be substituted for soil types, because many times the same topography represents two entirely different soils, one valuable, the other poor. No general rule applies because conditions vary from region to region, but continued appraisal practice in an area will soon familiarize the appraiser with the correlation between soils and topography. A safe rule is to use the soil auger when in doubt.

---

5. J. A. Stritzel, "Take a Good Soil Sample," Iowa State Univ., Ext. Serv., Pam. 287 (Rev.), July 1963.

It is desirable to follow a definite and uniform system of measuring topography. The usual division into five kinds will serve most purposes: level, undulating to gently rolling, rolling, strongly rolling, and steep. This classification, however, leaves something to be desired because "rolling" to one person may mean the same kind of topography as "gently rolling or undulating" to another, and "strongly rolling" to a third. The use of "percentage of slope" is growing in favor because it eliminates misunderstandings. Percentage of slope is the number of feet rise in 100 feet of horizontal distance, a slope of 5 percent being a 5-foot rise in 100 feet of horizontal distance. A slope of 0 percent is level or horizontal and a 100 percent slope is a 100-foot rise in 100 feet, making a 45° angle (Fig. 20.1). Several scales used in translating slope percentages into general terms are

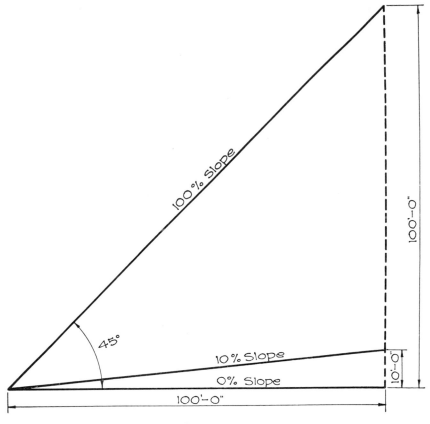

FIG. 20.1. *Measurement of slope in percentage.*

TABLE 20.6. Classification of Topography in Percentage of Slope

A. Adapted From U.S. Soil Survey Manual[a]

Class		Characteristic	Range in Percent
A			0– 3
	Single slopes	Level or nearly level	
	Complex slopes	Level or nearly level	
B			1– 8
	Single slopes	Gently sloping	
	Complex slopes	Undulating	
C			5–16
	Single slopes	Sloping, moderately sloping	
	Complex slopes	Rolling	
D			10–30
	Single slopes	Moderately steep	
	Complex slopes	Hilly	
E			20–65
	Single slopes	Steep	
	Complex slopes	Steep	
F			Over 45
	Single slopes	Very steep	
	Complex slopes	Very steep	

B. West Virginia Classification[b]

Class	Range in Percent
Level to gently rolling	0–12
Gently rolling to rolling	12–25
Rolling to steep	25–40
Steep	Over 40

C. Nebraska Classification by Soil Conservation Service[c]

Class	Range in Percent
A	0– 3
B	3– 7
BB	7–10
C	10–15
D	Over 15

D. California Classification (See Storie Rating, Factor C, p. 349.)

[a] Soil Survey Manual, Handbook No. 18, USDA, pp. 161, 261–64, and 291.
[b] G. G. Pohlman, Land Classification in West Virginia Based on Use and Agricultural Value, W. Va. Agr. Exp. Sta., Bull. 284, pp. 11–12.
[c] Arthur Anderson, A. P. Nelson, F. A. Hayes, and I. D. Wood, A Proposed Method for Classifying and Evaluating Soils on the Basis of Productivity and Use Suitabilities, Nebr. Agr. Exp. Sta., Res. Bull. 98, p. 6.

FIG. 20.2. *An Abney level used in measuring slopes. (Courtesy Keuffel and Esser Co. of New York.)*

presented in Table 20.6 to show the variations and the possibility of error unless slope readings are stated in percentages.

The slope classification used in the appraisal maps for Farms X, Y, and Z is shown with the other data on these farms in Chapter 19. Slope in these examples is indicated by the letter below the line in each soil designation.

Even percentage slope readings do not convey an exact picture because slopes vary in all directions. Near the bottom of the slope the percentage may be 5, farther up it may be 10, and near the top may be 6. Actually the slope percentage is only an overall estimate of the average slope for an area.

Several simple methods are used in measuring slope in percentage. One easily constructed device consists of a scale marking off the degrees in a half circle and expressing them in percentages, plus a plumb bob and a small level. Another instrument which can be carried and operated with ease in the field is an Abney level, which is illustrated in Figure 20.2.

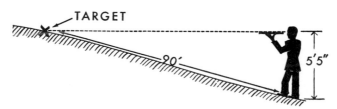

FIG. 20.3. *In this sketch percentage of slope is being estimated with a hand level. This man has established that his eye or line of sight is 5.4 feet above the ground when he stands erect. He sights uphill through the level to an object which he has placed on the ground for a target. He watches the bubble in the level and adjusts his position up- or downhill until he has established a level line of sight. Then he paces to the target and finds the distance to be 90 feet. Then 5.4 ÷ 90 × 100 = 6.0 percent slope. This is an approximate answer because 5.4 feet should be divided by the horizontal distance, not 90 feet. But for most cases the difference is too small to matter.*

A simple level can be used to determine slope, as illustrated in Figure 20.3. This method and the others suggested appear easy, but actually the task is much more complicated because slopes as straight as the one pictured in Figure 20.3 are not common. What you usually find in the field is a complex situation with all kinds of variations either by moving up and down the slope or by moving sideways. For example you can get different slope readings by reading at different points. The answer is to get a bench mark or general idea by one or more slope readings and then select a percentage which in your judgment best fits the particular slope situation you are measuring.

It is possible to overemphasize the actual percentage of slope. The particular soil type or condition must be taken into consideration, some types eroding much more rapidly than others. Rainfall and crop adaptation may permit the cropping of a 10 percent slope in one area and not in another. The appraiser should familiarize himself with the special soil and crop conditions which govern his territory.

EROSION

The change in soil productivity which accompanies erosion ranks this factor as one of the most difficult for the appraiser to evaluate. If it were necessary to estimate only the present producing power of the soil, the task would be relatively easy. But in most cases the buyer or lender on farm security wants to estimate future as well as present yields, and any factor such as erosion which may cause yields to decline must have an important place in the valuation.

One way to consider erosion is to estimate the damage it has done and may do in terms of inches of surface soil lost. Tables 20.4 and 20.5 indicate a decline in crop yield accompanying reduction in topsoil. If the surface soil of a given area has a depth of 8 inches but is subject to excessive erosion, it may in 15 years or so be reduced to an average of 4 inches. In Table 20.5 it is estimated that a reduction in surface soil from 8 to 4 inches would result in a reduction in yield on Tama silty clay loam of approximately 15 percent. Although reduction in yield caused by erosion cannot as yet be estimated accurately, an approximation of the influence of this factor on value should be included. A map showing the extent of erosion in the United States is shown in Figure 20.4.

An accurate appraisal of erosion conditions, their effect upon present and future productiveness, and an estimate of the rate of erosion spread or increase is a vital part of any appraisal of land that will be used for any kind of cultivation. In appraising land that is to be taken out of cultivation—such as for sub-

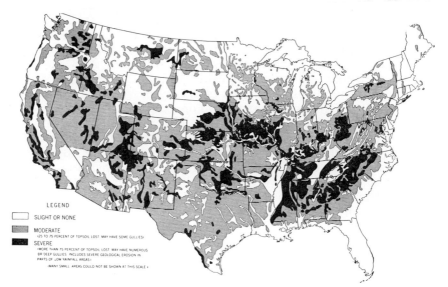

FIG. 20.4. *Generalized erosion map of the United States. (Courtesy USDA.)*

urban or industrial development or highway construction—the amount of time and energy expended on accurate and analytic appraisal of erosion conditions may be considerably reduced.

Two methods may be used to indicate erosion losses. One is based on the amount of surface soil left, the other is based on the percentage of surface soil lost or removed. To use the second method of surface soil lost involves an assumption as to the amount present in the beginning, an assumption that may be exceedingly weak where it is impossible to determine the original depth of surface soil. The appraiser can avoid argument by reporting the surface soil present at the time of appraisal, adding notes on evidences of erosion which he sees.

A breakdown of erosion into types and subtypes has much to commend it to the appraiser. Gullies which cannot be crossed with farm machines should be indicated in the appraisal report. Hillsides where sheet or rill erosion is starting also belong in the records, accompanied when possible with an estimate of the rate of soil loss. The importance of erosion in the appraisal report may be recognized by a comparison of two farms similar in every respect save that one has a rolling topography subject to heavy erosion while the other is level with little danger of erosion. A much less intensive crop program must be followed on the first farm than on the second, or considerable expense will be required for erosion control. The first farm, subject as it is to erosion, should have the lower valuation, even though both farms would yield approximately the same for a year or two if put into similar crops.

Soil phases can be used to indicate the extent of erosion and the hazard that exists of future erosion. When an appraisal map is made the following phases, as described in the *Soil Survey Manual*, provide one method of handling both water and wind erosion:[6]

Water erosion

Slightly eroded phase — management to maintain production has to be slightly different than on soil not subject to erosion.

Moderately eroded phase—requires an entirely different set of management practices than on soil with no erosion hazard.

Severely eroded phase—erosion has changed use that can be made of soil and makes a drastic treatment such as terracing necessary to maintain production.

Gullied land—not suitable for cropping because of numerous gullies.

Wind erosion

Wind-eroded phase—different set of management practices required in comparison to soil not subject to wind erosion. No change in use.

Severely wind-eroded phase — reworking of soil required to use for crops; may be desirable to use for permanent grasses or trees.

Blown-out land — land is practically barren. Trees or grasses are main possibilities.

Instead of using erosion phases, the appraiser may prefer to describe the erosion situation in the notes following the appraisal map. The actual extent of erosion can be indicated by the symbols giving the slope and depth of each soil area on the appraisal map. *The important point is to convey a clear picture of the situation, not only as it exists at present but indicating the erosion hazard of the future.* Since conditions vary widely, a method that is successful in one territory may not be in another. Consequently whether to use erosion phases, or notes plus slope and depth data, or a combination of the two is a matter for the appraiser to decide after trying out various methods and combinations designed to show the effect of erosion on productivity.

DRAINAGE

Topography and erosion fill the eye and mind of the appraiser with vivid impressions the minute he sees the farm. Drainage, although just as significant in many cases, is often

6. USDA, *Soil Survey Manual*, Handbook No. 18, Washington, D.C., 1951, p. 295.

hidden in such a way that it may escape the appraiser's attention altogether. An actual situation illustrating this point proved a bitter experience to some farm mortgage lenders. Certain loan companies were making loans in a region where level to undulating land was situated next to an area of rolling hills. Some of the companies made loans on the level sections but refused to make any loans on the rolling land. A few made loans in both areas. The loans in the hills came through the depression with few foreclosures, while a large percentage of the loans in the level sections were foreclosed. Poor drainage and heavy subsoil on the level area were the reasons for the appraisal error. On the basis of topography, the hills appeared to warrant lower values and loans than the level areas, but on the basis of an intensive examination of the soil profile, the two areas should have been given almost the same value.

An appraiser's first problem when considering drainage is to determine approximately how much water the farm is likely to receive. He must take into account the amount of rainfall during the entire year as well as that likely to come in any one month. He must include information on the amount of water that might pour onto the farm from neighboring farms. He will need to estimate the overflow hazard if there are streams passing across or near the farm, a hazard which will depend in part on the extent and character of the watershed above the farm.

Not all soils are easy to drain, a fact which sometimes makes the difference in value between two soil types in similar areas. In discerning whether a soil is easy or difficult to drain, the appraiser should examine the texture of both surface soil and subsoil. A clay surface soil allows the water to seep down slowly. A subsoil with a tight, impervious layer will of course prevent water from going down or coming up. Drain tile in this soil would have to be above the impervious layer in order to do any good. A sandy subsoil, on the other hand, usually allows the water to pass through too fast for the good of the plants.

A map of the tile drainage system should be obtained if available. If not, evidence of previous crops, drainage outlets, and information from the operator of the farm or neighbors may have to suffice. Sometimes it is possible to tell shortly after a rain where the tile drains are located, whether they are functioning properly, and whether they are adequate. Sometimes the drainage system on a farm may be part of a larger system, a record of which is available at the county courthouse or elsewhere. This record, which may include the original cost and amount spent on maintenance, adds to the completeness and accuracy of the appraisal.

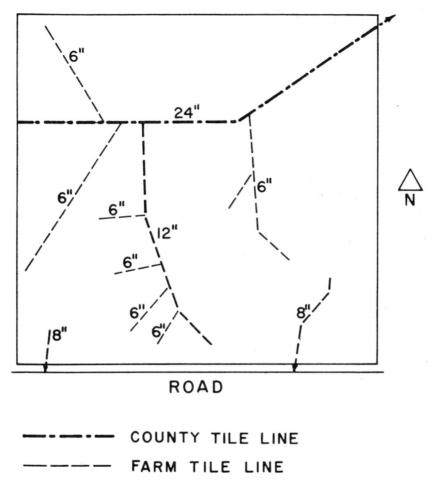

FIG. 20.5. *Drainage map for a level midwestern farm.*

The procedure to be followed in valuing the undrained area of a farm presents an appraisal problem. Whether or not the area can be drained profitably should be noted in the appraisal. If drainage is out of the question because of cost or physical obstacles, the area should be considered in its present condition. Although not always easy, it is often necessary to estimate the cost of drainage, either underground tile drains or surface ditches. Small undrained spots are often quickly recognized as unprofitable to reclaim, but the prospective buyer of a farm may be especially interested in the cost of draining a large area.

If the farm is in an irrigation district, there is probably available an excellent set of records on water costs, maintenance, map of the ditches, and other information of this character including a land rating. The land rating made in some of the federal projects has been used as a basis for levying charges against the land, the idea being to assess the land for the irrigation development according to its productivity.

It is difficult to generalize regarding appraisals within irrigation districts since these districts are relatively small, intensively farmed areas, each one with conditions peculiar to itself. It is possible, however, to caution the appraiser to make a thorough examination of the water right enjoyed by the farm, the risk from poor subsoil drainage and formation of alkali, and finally, as important as any, the financial setup of the district, including bonded indebtedness.

WEEDS AND OTHER DETRIMENTS

No attempt will be made here to take up all the additional influences on yield. Weeds, however, are such a common enemy of the farm that they deserve special comment. The farmer's battle against weeds has significance for the appraiser insofar as weeds may cut down the long-time average crop yield of the farm. This means that attention should be focused on those weeds which have the worst effect on yield and at the same time are the most difficult to eradicate.

The appraiser will want first of all to acquaint himself with the noxious weeds indigenous to a given locality. In the Midwest, for instance, European bindweed or creeping jenny, Canada thistle, and quack grass are regarded as particularly bad weeds. Knowing the weeds and being able to identify them in the field is not enough, however. The appraiser will also need to estimate what those weeds mean in terms of reduced yield and how much it will cost to eradicate them. This last calls for observation and information on the latest methods of control and eradication of the more noxious varieties. If an expense of $20 an acre will be necessary to eradicate a bad area of weeds, this amount multiplied by the acres infested would represent a proper deduction from the value of the farm.

An appraiser will add to the completeness of his farm picture by indicating on the farm map or describing in the body of the report the location, extent, and cost of eradication of noxious weeds found on the farm. This minor item, though often important, is just one of the many which the appraiser has to observe and evaluate as he works his way over the farm. Other unfavorable factors are stoniness, occurrence of boulders, brush, or trees in the midst of cropland, insect pests, and numerous other items, many of them common only in a limited area.

CLIMATE

Only those factors which might vary on an individual farm have been considered up to this point. Climate in its effect on yields is not likely to vary within a farm, but the effect between farms is often of consequence and, over a wide territory, may be decisive. "Climate" includes many items: precipitation, evaporation, sunshine, length of day, length of frost-free season, temperature, hail, and winds. Each of these items can be further subdivided. For example, precipitation includes the following: annual average, distribution throughout the year, variability from year to year, and frequency of "gully-washers." Only a small part of the world's surface has a climate which is exactly right for any one crop. The danger of unfavorable weather is one of the chief concerns of farmers everywhere.

Every farmer must adjust his farm operations to the climate, soil, and topography of his farm. Of these three natural factors, climate may be the easiest to overlook. Any person can see the topography any day of any year. A trained person can see the soil any day when it is not frozen. Anyone can see the weather the day he is on the farm. But climate is weather over a long period of time. The only way to see the climate is to study the records. There are two kinds of climatic records. One of these is the accumulated experience of the farmers of the region, and the appraiser should give serious consideration to this experience. Another kind of climatic record is that kept by the United States Weather Bureau or similar agencies in other countries, easily available and well worth thoughtful study by the appraiser.

Rainfall is probably the most important climatic factor influencing yield. A study of the precipitation map for the United States (Fig. 20.6) indicates, for example, the western boundaries of the corn area. Rainfall west of the Mississippi River declines gradually at first but more rapidly beyond the western boundary of Missouri, Iowa, and Minnesota until in the western part of Nebraska and South Dakota the annual rainfall amounts to less than 20 inches. Still farther west, territory with less than 15 inches of rainfall annually will be found in Montana, Wyoming, and Colorado. Here even wheat production becomes hazardous.

The amount of rainfall during the critical months of the crop season may sometimes be far more significant than the total precipitation figure. Fortunately weather records of the USDA are available to answer, in part at least, such questions.

Another effective climatic factor is length of growing season

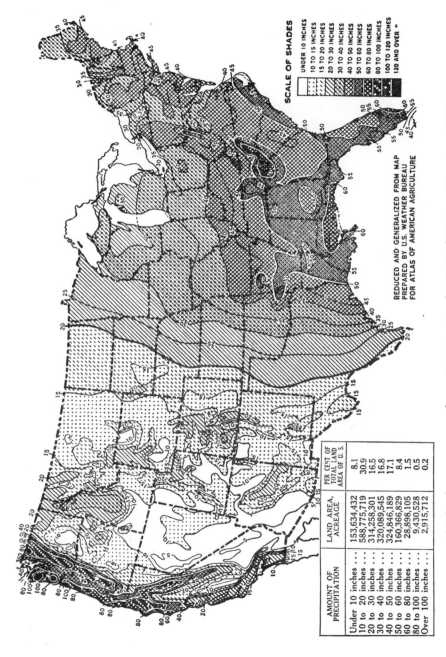

AMOUNT OF PRECIPITATION	LAND AREA, ACREAGE	PER CENT OF TOTAL LAND AREA OF U.S.
Under 10 inches	153,634,432	8.1
10 to 20 inches	588,775,719	30.9
20 to 30 inches	314,258,301	16.5
30 to 40 inches	320,089,545	16.8
40 to 50 inches	324,846,189	17.1
50 to 60 inches	160,366,829	8.4
60 to 80 inches	28,898,105	1.5
80 to 100 inches	9,430,528	0.5
Over 100 inches	2,915,712	0.2

SCALE OF SHADES

UNDER 10 INCHES
10 TO 15 INCHES
15 TO 20 INCHES
20 TO 30 INCHES
30 TO 40 INCHES
40 TO 50 INCHES
50 TO 60 INCHES
60 TO 80 INCHES
80 TO 100 INCHES
100 TO 120 INCHES
120 AND OVER "

REDUCED AND GENERALIZED FROM MAP
PREPARED BY U.S. WEATHER BUREAU
FOR ATLAS OF AMERICAN AGRICULTURE

FIG. 20.6. *Average annual precipitation in the United States. (Courtesy USDA.)*

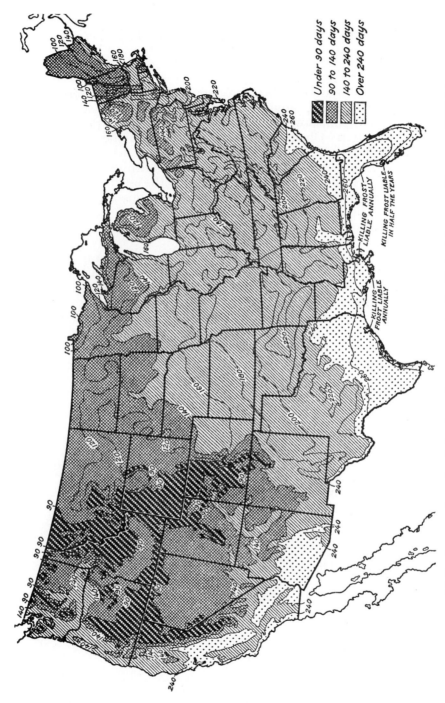

Under 90 days

90 to 140 days

140 to 240 days

Over 240 days

KILLING FROST LIABLE ANNUALLY

KILLING FROST LIABLE IN HALF THE YEARS

KILLING FROST LIABLE ANNUALLY

FIG. 20.7. *Average length of the frost-free season in the United States.* (*Courtesy USDA.*)

(Fig. 20.7). The number of frost-free days, for instance, definitely determines the northern limit to the cotton-producing territory and the northern boundary to the area where corn can be produced as grain. For special crops, length of growing season is particularly important.

When estimating yields, the appraiser should have in mind not only the rainfall, length of growing season, and such other factors as temperature, windstorms, and hail but also the year-to-year variation. Obviously a farm getting 30 inches of rainfall year after year with little variation is more desirable than one receiving the same average of 30 inches but with violent fluctuations, such as 10 inches one year and 50 the next. Likewise a farm where the growing season is relatively constant from year to year will be more desirable than a similar farm with a growing season that fluctuates greatly, with the first frost in August one year and October the next.

A striking example of climatic variation that appraisers may encounter in some districts is susceptibility to frequent hail storms. Hail insurance companies have these "hail belts" carefully mapped with higher premium rates in such areas to compensate for the extra risk. The appraiser should note this added risk and reflect it in his appraised value of the land in such an area, using the higher cost of hail or other crop insurance as an index of the added risk.

QUESTIONS

1. Describe each of the soil areas appearing on the appraisal maps prepared in connection with problems at the end of Chapter 19. In this description include texture, depth of surface soil, subsoil, fertility, and other soil factors affecting crop yield. Make preliminary yield estimates of major crops in the area for each soil condition shown.
2. After completing the problem outlined above, individuals may be asked to compare their preliminary yield estimates with each other and give reasons in support of their figures.

REFERENCES

Baver, L. D. *Soil Physics,* 3rd ed., John Wiley and Sons, Inc., New York, 1956.

Bear, F. E., *Soils and Fertilizers,* 4th ed., John Wiley and Sons, Inc., New York, 1953.

Buckman, H. O., and Brady, N. C., *The Nature and Properties of Soil,* 6th ed., Macmillan, New York, 1960.

Frevert, R. K., Schwab, G. O., Edminster, T. W., and Barnes, K. K., *Soil and Water Conservation Engineering,* John Wiley and Sons, Inc., New York, 1955.

Kohnke, H., and Bertrand, A. R., *Soil Conservation,* McGraw-Hill, New York, 1959.

Luthin, J. N., *Drainage of Agricultural Lands,* Amer. Soc. of Agron., Madison, Wis., 1957.

Millar, C. E., Turk, L. M., and Foth, H. D., *Fundamentals of Soil Science,* 3rd ed., John Wiley and Sons, Inc., New York, 1957.

Roe, H. B., and Ayres, Q. C., *Engineering for Agricultural Drainage,* McGraw-Hill, New York, 1954.

Schwab, G. O., Frevert, R. K., Barnes, K. K., and Edminster, T. W., *Elementary Soil and Water Engineering,* John Wiley and Sons, Inc., New York, 1957.

Stallings, J. H., *Soil Conservation,* Prentice-Hall, Englewood Cliffs, N.J., 1957.

Thompson, L. M., *Soils and Soil Fertility,* 2nd ed., McGraw-Hill, New York, 1957.

USDA, *Climate and Man: The 1941 Yearbook of Agriculture,* Washington, D.C.

———, *Soil: The 1957 Yearbook of Agriculture,* Washington, D.C.

———, *Soil Survey Manual,* Handbook No. 18, Washington, D.C., 1951.

———, *Water: The 1955 Yearbook of Agriculture,* Washington, D.C.

21

SOIL RATINGS–INDEXES, YIELDS, AND SCS CLASSIFICATION

ONE OF THE MOST promising developments in the field of soil science as far as the appraiser is concerned is the soil productivity rating, which designates in a single index figure or an estimated crop yield the combined effects of many soil characteristics. A productivity rating commonly summarizes for a given soil type the effect on yield of surface and subsoil texture, of depth of surface soil, of fertility, reaction, topography, and drainage—a summarizing which makes it possible to compare one soil type with another. It must be remembered, however, that soil productivity ratings are still in the developmental stage and are as yet only approximations with limited use. The appraiser should be familiar with the manner in which these ratings are constructed in order to use them correctly.

The soil productivity rating idea fostered by the Soil Survey Division of the USDA has taken hold in many states. As a result a number of state agricultural experiment stations have published ratings for soils in their respective states. In some cases these ratings are in the form of indexes, in others in the form of estimated yields. Several bulletins giving the ratings for individual states have been published. In general, the method of preparing the index or yield estimate has been for members of

343

the soil department in the state agricultural experiment station to pool their judgment of the productive ability of the soils within their state. Each member of the department contributes to this combined judgment his observation of experiments and of farming results over the state on different types of soils. Although experimental results have thrown some light on variation between types, for the most part the ratings are subjective estimates and not objective data.

In the following discussion several different methods of rating will be presented, either because they represent important variations or because they illustrate the use of ratings in different sections of the country. Consideration will be given first to examples of index ratings, then to examples of yield ratings. At the end of the chapter a special type of soil rating called "land capability classification" will be presented.

INDEX RATINGS

CORN SUITABILITY RATING

A common practice followed in evaluating Corn Belt soils is use of the corn suitability rating. Each soil phase or condition is given a rating between 1 and 10, with 1 the best and 10 the poorest. For example, Nicollet loam is given a corn suitability rating of 1, Webster silty clay loam a rating of 1.5, Clarion loam a rating of 2, and Storden loam a rating of 5. To figure the suitability rating for a farm with 60 acres of Nicollet, 46 acres of Webster, 85 acres of Clarion, and 9 acres of Storden, you would use the following procedure, assuming the ratings given applied to the soils on this farm:

Soil Type	Acres	Rating	Total
Nicollet loam	60	1	60
Webster silty clay loam	46	1.5	69
Clarion loam	85	2.	170
Storden loam	9	5	45
Total	200		344

$$\text{Corn suitability rating} = \frac{344}{200} = 1.7$$

Each soil type or phase is weighted by its area in acres. The overall rating is a means of converting the individual ratings for different soil conditions, both on cropland and pasture, into a single index, thus making it easy to compare one farm with another in soil productivity.

One danger in the use of the corn suitability system, a danger which applies to all rating systems, is a mechanical application of the indexes. If the appraiser applies a rating of 1 to all Nicollet loams, a rating of 2 to all Clarion loams, and so on, he is in for trouble. There is a wide variation within each soil type and often even within each soil phase. One Webster silty clay loam, for instance, may be better drained than another and consequently may yield more than the other. One area of a given soil type may have a deep layer of surface soil, another area of the same soil type may not. This second area should be identified as a shallow phase if there is a marked difference, but if there is not enough difference to warrant using different phases, then it is necessary to show the difference in the index rating. The ratings, therefore, for any soil type or phase are only an average. When the appraiser has identified the soil type or phase, he has a second task—that of determining whether the soil is average for the type or phase. If it is average, this justifies application of the average rating; but if it is better or poorer, it should have a better or poorer rating than the average.

Another farm of similar soils might be appraised, for example, where approximately the same acreages of the four soil conditions were found as in the farm mentioned earlier. But in this instance the Webster was not as well drained, the Clarion loam was slightly below average, and the Storden was more eroded. The appraiser in this case might assign ratings of 1, 4, 3, and 6, with a resulting corn suitability rating of 2.8. This would mean that this second farm was not as productive overall as the first farm even though it contained the same acreages of the same general soil conditions.

Use of soil ratings obviously is not a mechanical procedure; on the contrary it requires a high order of knowledge and skill. It is this kind of skilled evaluation which makes appraisal a judgment process performed by a well-trained professional person rather than a routine technical procedure.

This corn suitability rating system can be applied readily to other crops. It works well of course when one crop such as corn, cotton, or wheat predominates, but it could also be used with a composite crop rating plan. In such a composite rating, each soil would be rated from 1 to 10 in terms of its general suitability for crop production.

ORCHARD SOIL RATINGS

A special soil rating for apple orchards has been published for Orleans County in New York State, a county located along the south shore of Lake Ontario.[1] The various soil types, 69 in all, were classified into four groups—good, medium, medium to poor, and poor orchard soils. In this list 9 soils were rated as good, 12 as medium, 15 as medium to poor, and 33 as poor. Four colors—yellow, green, blue, and red—were used on the county soil map accompanying the bulletin to indicate the four grades of orchard soil. Soil of course was not the only factor used in establishing the four grades. Distance from the lake was also included because soils in protected areas close to the lake were better orchard soils than the same kinds of soils in the unprotected areas in the southern part of the county. The advantage of such a map as this one for Orleans County is the ease with which one not familiar with the technical description of soil types can distinguish the different grades of soils when used for a specific purpose. It is to be hoped that work of this kind may be expanded with soils ranked for other crops as well as for orchards.

STORIE INDEX OF CALIFORNIA

R. Earl Storie of the California Agricultural Experiment Station has devised a unique method of rating the productivity of soil. Although this method has been applied specifically to California soils, Storie considers it applicable to soils generally. The chief characteristic of this method is multiplying four factors, $A \times B \times C \times X$, to obtain a final index of productivity.

> This method of soil rating, known as the Storie Index, is based on soil characteristics that govern the land's potential utilization and productive capacity. It is independent of other physical or economic factors that might determine the desirability of growing certain plants in a given location.
> Percentage values are assigned to the characteristics of the soil itself, including the soil profile (factor A); the texture of the surface soil (factor B); the slope (factor C); and conditions of the soil exclusive of profile, surface texture, and slope — for example, drainage, alkali content, nutrient level, erosion, and microrelief (factor X). The most favorable or ideal conditions with respect to each factor are rated at 100 per cent. The percentage values or ratings for the four factors are then multiplied, the result being the Storie Index rating of the soil.[2]

Table 21.1 shows how the different factors, A, B, C, and X, are broken down according to detailed classification, factor B

---

1. A. T. Sweet, *Soils of Orleans County, New York, in Their Relation to Orchard Planting*, Cornell Univ. Agr. Exp. Sta., Bull. 637, 1935.
2. R. Earl Storie, *Revision of the Soil-Rating Chart*, Calif. Agr. Exp. Sta., Berkeley, 1959. This leaflet is a revision of the soil-rating chart published originally by the same author in Bulletin 556, *An Index for Rating the Agricultural Value of Soils*, 1933.

representing the texture ratings discussed in Chapter 20. An example of how Storie uses this factor system in rating California soils follows:

> Index for Acl-CC (Altamont clay loam, rolling): This is a brown upland soil from shale parent material; bedrock at a depth of 3 feet. Rolling topography, moderate sheet erosion, with occasional gullies.

Rating in Percentage

Factor A: Altamont series, profile group VIII........ 70
Factor B: clay loam texture 85
Factor C: rolling topography 90
Factor X: moderate sheet erosion with shallow gullies 70
Index rating $= 70\% \times 85\% \times 90\% \times 70\% = 37\%$

After the percentage rating is obtained, the soil is classified into one of six grades, each grade representing a certain range in the percentage scale. Soils having a percentage rating between 80 and 100 are in Grade 1 (excellent); 60 to 79 in Grade 2 (good); 40 to 59 in Grade 3 (fair); 20 to 39 in Grade 4 (poor); 10 to 19 in Grade 5 (very poor); less than 10 in Grade 6 (nonagricultural).

The rating index developed by Storie, although a definite improvement over the simple score-card system, retains some of the disadvantages attached to the score card. The rigidity present in the score card carries over to the Storie index when the four factors are given equal weight, assuming that they are equally important in their contributions to value. One soil with an

TABLE 21.1. Soil-Rating Chart (Storie Soil Index Rating $=$ Factor $A \times$ Factor $B \times$ Factor $C \times$ Factor X)

FACTOR *A*—Rating on Character of Physical Profile

Soil Group	Rating in Percentage
I. Soils on recent alluvial fans, flood plains, or other secondary deposits having undeveloped profiles	100
x—shallow phases (on consolidated material), 2 feet deep	50– 60
x—shallow phases (on consolidated material), 3 feet deep	70
g—extremely gravelly subsoils	80– 95
s—stratified clay subsoils	80– 95
II. Soils on young alluvial fans, flood plains, or other secondary deposits having slightly developed profiles	95–100
x—shallow phases (on consolidated material), 2 feet deep	50– 60
x—shallow phases (on consolidated material), 3 feet deep	70
g—extremely gravelly subsoils	80– 95
s—stratified clay subsoils	80– 95
III. Soils on older alluvial fans, alluvial plains, or terraces having moderately developed profiles (moderately dense subsoils)	80– 95
x—shallow phases (on consolidated material), 2 feet deep	10– 60
x—shallow phases (on consolidated material), 3 feet deep	60– 70
g—extremely gravelly subsoils	60– 90

TABLE 21.1. *(continued)*

Soil Group	Rating in Percentage
IV. Soils on older plains or terraces having strongly developed profiles (dense clay subsoils)...	40– 80
V. Soils on older plains or terraces having hardpan subsoil layers	
at less than 1 foot...	5– 20
at 1 to 2 feet..	20– 30
at 2 to 3 feet..	30– 40
at 3 to 4 feet..	40– 50
at 4 to 6 feet..	50– 80
VI. Soils on older terraces and upland areas having dense clay subsoils resting on moderately consolidated or consolidated material.......	40– 80
VII. Soils on upland areas underlain by hard igneous bedrock	
at less than 1 foot...	10– 30
at 1 to 2 feet..	30– 50
at 2 to 3 feet..	50– 70
at 3 to 4 feet..	70– 80
at 4 to 6 feet..	80–100
at more than 6 feet..	100
VIII. Soils on upland areas underlain by consolidated sedimentary rocks	
at less than 1 foot...	10– 30
at 1 to 2 feet..	30– 50
at 2 to 3 feet..	50– 70
at 3 to 4 feet..	70– 80
at 4 to 6 feet..	80–100
at more than 6 feet..	100
IX. Soils on upland areas underlain by softly consolidated material	
at less than 1 foot...	20– 40
at 1 to 2 feet..	40– 60
at 2 to 3 feet..	60– 80
at 3 to 4 feet..	80– 90
at 4 to 6 feet..	90–100
at more than 6 feet..	100

FACTOR *B*—Rating on Basis of Surface Texture

Texture	Rating in Percentage	Texture	Rating in Percentage
Medium-textured:		Gravelly:	
very fine sandy loam.........	100	gravelly fine sandy loam......	70–80
fine sandy loam..............	100	gravelly loam................	60–80
loam.......................	100	gravelly silt loam............	60–80
silt loam....................	100	gravelly sandy loam..........	50–70
sandy loam..................	95	gravelly clay loam............	60–80
loamy fine sand.............	90	gravelly clay................	40–70
silty clay loam..............	90	gravelly sand................	20–30
clay loam...................	85	Stony:	
Heavy-textured:		stony fine sandy loam.........	70–80
silty clay...................	60–70	stony loam..................	60–80
clay.......................	50–60	stony silt loam..............	60–80
Light- or coarse-textured:		stony sandy loam............	50–70
coarse sandy loam...........	70–90	stony clay loam..............	50–80
loamy sand.................	80	stony clay..................	40–70
very fine sand..............	80	stony sand.................	10–40
fine sand...................	65		
sand.......................	60		
coarse sand.................	30–60		

TABLE 21.1. *(continued)*

FACTOR *C*—Rating on Basis of Slope

Slope	Rating in Percentage
A—Nearly level (0 to 2%)	100
AA—Gently undulating (0 to 2%)	95–100
B—Gently sloping (3 to 8%)	95–100
BB—Undulating (3 to 8%)	85–100
C—Moderately sloping (9 to 15%)	80– 95
CC—Rolling (9 to 15%)	80– 95
D—Strongly sloping (16 to 30%)	70– 80
DD—Hilly (16 to 30%)	70– 80
E—Steep (30 to 45%)	30– 50
F—Very steep (45% and over)	5– 30

FACTOR *X*—Rating of Conditions Other Than Those in Factors *A*, *B*, and *C*

Conditions	Rating in Percentage
Drainage:	
well drained	100
fairly well drained	80– 90
moderately waterlogged	40– 80
badly waterlogged	10– 40
subject to overflow	variable
Alkali:	
alkali free	100
slightly affected	60– 95
moderately affected	30– 60
moderately to strongly affected	15– 30
strongly affected	5– 15
Nutrient (fertility) level:	
high	100
fair	95–100
poor	80– 95
very poor	60– 80
Acidity:	
according to degree	80– 95
Erosion:	
none to slight	100
detrimental deposition	75– 95
moderate sheet erosion	80– 95
occasional shallow gullies	70– 90
moderate sheet erosion with shallow gullies	60– 80
deep gullies	10– 70
moderate sheet erosion with deep gullies	10– 60
severe sheet erosion	50– 80
severe sheet erosion with shallow gullies	40– 50
severe sheet erosion with deep gullies	10– 40
very severe erosion	10– 40
moderate wind erosion	80– 95
severe wind erosion	30– 80
Microrelief:	
smooth	100
channels	60– 95
hogwallows	60– 95
low hummocks	80– 95
high hummocks	20– 60
dunes	10– 40

$A \times B \times C \times X$ of $100 \times 90 \times 100 \times 100$ may or may not equal an $A \times B \times C \times X$ of $90 \times 100 \times 100 \times 100$ even though according to the Storie index they come out exactly the same. It is true that adjustments might be made in the percentages given to A or B or C or X in any given case so that the correct answer could be obtained, but this would presuppose the answer.

Another difficulty with the Storie method is the determination of the percentages which are the backbone of the system and yet are only estimates based on personal judgment. A better estimate of productivity might be obtained by listing the various factors affecting the productivity of a given soil, weighting them in each individual case, and estimating the productivity in terms of crop yield or a general productivity index. The Storie index, designed for California where a wide variety of crops is grown, has an advantage in that it is not based on any one crop. Moreover, those using the Storie index point to its relative simplicity and to the advantage it has over those productivity measures which do not show any breakdown into constituent parts such as A, B, C, and X factors.

LAND RATING BY LE VEE AND DREGNE

A variation of the Storie system, as proposed by LeVee and Dregne of New Mexico,[3] was briefly explained in Chapter 19. A code system is used and each of four factors is estimated and then the four factors are multiplied, just as in the Storie system, to give the final rating.

The four factors in the LeVee-Dregne method are different from those in the Storie method in some important respects. The first of the four factors is the soil profile which is made up of surface texture, permeability of subsoil and of substratum, effective soil depth, and presence of geologic factors that limit plant growth. This factor in an example is assumed at 96.

The second factor is slope which in this example is rated 89. The third factor is erosion, rated 88. Finally, there is a fourth factor, a special one, which includes all other soil productivity influences. In this example a high water table which affects crops adversely is assumed. The rating for this fourth factor will be 75. This sample soil then has symbols and factor percentages as follows:

		Rating in Percentage
1.	3231 soil profile	96
2.	2% slope	89
3.	17 V erosion	88
4.	W2 water table	75

---

3. W. M. LeVee and H. E. Dregne, *A Method for Rating Land*, N. Mex. Agr. Exp. Sta. in cooperation with U.S. Soil Conservation Service, Bull. 364.

The final rating for this soil would be 96% × 89% × 88% × 75% which equals 56%. In this instance the effective limiting factor is the high water table which brings the overall rating down from 75% to 56%.

A difference between this method and the Storie method is assigning definite percentages to all factors rather than allowing a range. In the Storie system a 2% slope has a rating of 95 to 100%, and in the LeVee-Dregne system a rating of 89%, no more and no less.

In most respects, however, the two systems are similar with the advantages and disadvantages of the Storie method applying in much the same manner to the LeVee-Dregne method. For example, to say that a high water table should have a percentage of 75, as in the LeVee-Dregne system, or even a range of 40–80, as in the Storie system, assumes that the person making the rating knows the specific effect which a high water table has on yields and thus on the land rating. Nevertheless, the existence of these two rating systems and of other variations may well be an indication of progress toward a more systematic and objective method of evaluating soils.

YIELD RATINGS

Yield estimates for different kinds of soil are being made by soil scientists in the USDA and state agricultural experiment stations. The estimates have been appearing in county soil survey reports and in some instances in reports covering the major soils in a state.[4]

More than one management level is usually assumed in making the yield estimates in order to give the range of probable yields likely to be obtained by farmers. For example, in the soil survey of Newton County, Mississippi, located in the east central region of the state, the two levels of management are spelled out as follows:

A.   Yields to be expected under the management ordinarily practiced in the county.
B.   Yields to be expected under improved management, which includes use of planned crop rotations; selection of crops suitable for the soil; adequate fertilization of all crops; return of organic matter and crop residues to the soil; and, where needed, terracing, farming on the contour, and providing adequate drainage.[5]

Yield estimates for these two management levels are shown in Table 21.2. In these comparisons it is evident that there is a

4. *Principal Soils of Iowa*, Spec. Rept. No. 42, Iowa Agr. Exp. Sta., Ames, 1965.
5. *Soil Survey, Newton County, Mississippi*, USDA in cooperation with Miss. Agr. Exp. Sta., 1960, p. 44.

TABLE 21.2.   Several Newton County, Mississippi, Soils under Two Management Levels, *A* and *B*

Soil Type	Cotton		Corn		Permanent Pasture	
	A	B	A	B	A	B
	(*lb.*)	(*lb.*)	(*bu.*)	(*bu.*)	a	a
Ochlockonee						
fine sandy loam, local alluvium phase	450	700	45	85	2	1
Cahaba						
very fine sandy loam, level phase	400	625	45	75	3	2
eroded gently sloping phase	350	600	35	60	3	2
Ora and Dulac soils						
eroded very gently sloping phase	300	500	25	45	4	3
eroded gently sloping phase	250	400	20	45	4	3
severely eroded gently sloping phase	150	300			5	3

Source: *Soil Survey, Newton County, Mississippi,* USDA in cooperation with Miss. Agr. Exp. Sta., p. 45.

a Acres per animal unit. Average number of acres required to furnish adequate grazing, without injury to the pasture, for 1 animal unit for a grazing season of 221 days. An animal unit is equivalent to 1 cow, steer, or mule; 5 hogs; or 7 sheep.

wide variation between the Ochlockonee fine sandy loam and the Ora and Dulac eroded soil, the former yielding roughly three times the latter in both cotton and permanent pasture.

A question may be raised as to how the yields in this instance were estimated. The answer, which comes from the report, follows:

> The yield estimates are based on observations made during the course of the survey and on interviews with farmers and other agricultural workers. Some research data were available, and they were considered in making the estimates. The yields . . . are estimated averages for the county, not for any particular farm or tract. They indicate, however, the response the different soils will make when management is improved.[6]

Another example, this one from the soil survey of Geary County, Kansas, shows variations in wheat yields and range carrying capacity for different soils. Geary County is located in the Flint Hills bluestem area in northeastern Kansas.

A typical group or toposequence of soils in the southeastern portion of Geary County is pictured in Figure 21.1. Yields were estimated for the soils in this group that are usually in crops. The two levels of management used in producing wheat are explained as follows:

> A.   Yields are those obtained when grain crops are grown most of the time in no systematic rotation; legumes are grown less often than once in 8 years; and little or no lime or fertilizer is used.

6. Ibid.

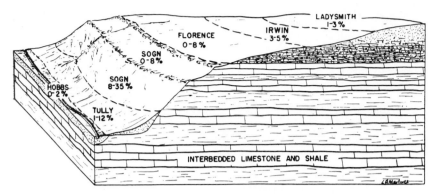

FIG. 21.1. *Typical grouping or toposequence of soils found in Geary County, northeastern Kansas. (Courtesy Kans. Agr. Exp. Sta.)*

B. Yields are those obtained when crops are rotated and erosion is controlled according to the needs of the soil; lime and fertilizers are applied in kinds and amounts indicated by soil tests; legumes are grown at least once every 5 years; suitable crops are selected; seedbeds are properly prepared; and weeds, insects, and diseases are controlled.[7]

Yield estimates for wheat under the two management levels taken from the report are compiled in Table 21.3.

Two striking comparisons in these wheat yield estimates are evident: the wide gap between the Hobbs and the severely eroded soils under either the *A* or *B* management level, and the lack of difference between the *A* and *B* management levels for

TABLE 21.3. Yield Estimates for Wheat Grown in Geary County, Kansas, under Two Management Levels, *A* and *B*

	Wheat Yield Estimates	
Soil Type	*A*	*B*
	(*bu.*)	(*bu.*)
Ladysmith silty clay loam		
0 to 1 percent slopes	20	30
1 to 4 percent slopes	18	30
Irwin silty clay loam		
0 to 4 percent slopes	15	25
4 to 8 percent slopes	15	20
severely eroded	12	18
Tully silty clay loam		
1 to 4 percent slopes	23	25
4 to 8 percent slopes	20	20
severely eroded	15	18
Hobbs silt loam	27	27

Source: *Soil Survey, Geary County, Kansas,* USDA in cooperation with Kans. Agr. Exp. Sta., p. 9.

7. *Soil Survey, Geary County, Kansas,* USDA in cooperation with Kans. Agr. Exp. Sta., 1960, p. 9.

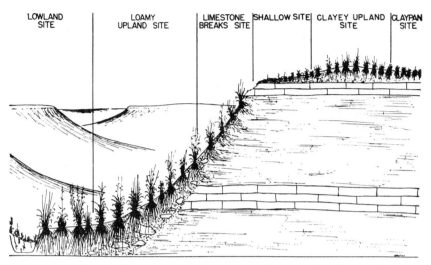

FIG. 21.2. *Typical grouping of range soil sites in Geary County, northeastern Kansas. (Courtesy Kans. Agr. Exp. Sta.)*

the Hobbs soil and to some extent for the Tully. These discrepancies indicate the importance of the appraiser's familiarity with the soils in the area he is appraising. He should be able to judge not only the present yielding ability of a soil but also the likely response from improved management.

Range carrying capacity was also estimated in the Geary County soil survey. A diagram showing the grouping of soil conditions is presented in Figure 21.2. Instead of soil conditions we have range sites, a somewhat broader term including climate and topography as well as soil. The range carrying capacity for

TABLE 21.4.   Estimated Range Carrying for Different Range Sites in Geary County, Kansas

Major Range Sites	Suggested Stocking Rate (acres required per head)	
	Excellent[a]	Good[a]
Loamy lowland site	3½– 5	5 – 7
Loamy upland site	4 – 5½	5½– 8½
Limestone breaks site	5 – 7	7 – 10
Shallow site	5 – 7	7 – 10
Clayey upland site	5 – 7	7 – 10
Clay pan site	6½– 8	8 – 12½

Source: Geary County study, pp. 10–13.
[a] A range is described as in "excellent" condition if 76 to 100 percent of the stand is of the same composition as the original stand. It is in "good" condition if the percentage is between 51 and 75. The figures given refer to acres needed per head for a grazing season of 5 months.

the different range sites shown in Figure 21.2 is estimated in Table 21.4.

Marked differences in pasture yield are evident in the figures presented. Almost twice as many animals can be pastured on a tract in the loamy lowland site as on a similar sized tract on clay pan. These range yields tie in closely with the yields in crops. This is evident when the soils in Figure 21.1 are checked for their range productivity. For example, Hobbs silt loam is placed in the loamy lowland site, Florence and Tully soils in the loamy upland site, Sogn in the limestone breaks and shallow sites, Irwin and Ladysmith in the clayey upland site, and severely eroded Irwin in the clay pan site.

A final example is from the area of Iowa in which Farm Z, pictured in Chapter 19, is located. The principal soils on this farm are Lindley and Weller. The typical toposequence or catena of soils in this area is shown in Figure 21.3. The Lindley and Weller soils, it may be noted, are located in the wooded area on the left side of the figure.

The two levels of management used in estimating yields for soils in this area are described as follows:

> The *A,* or average, level of management is based on the following assumptions: (1) The nearly level and gently sloping soils are used intensively for grain crops; (2) no definite rotations are used, but corn and oats are the main crops grown; fields are seeded to a meadow of red clover and timothy no oftener than once in 5 years; (3) pastures are unimproved and consist mainly of bluegrass; (4) adequate amounts of lime and some manure are used, but little commercial fertilizer is applied; and (5) no terracing or contour cultivation is practiced.
>
> The *B,* or superior, level of management is based on the following assumptions: (1) Suitable crop rotations are used along with practices to control erosion . . . and (2) lime and fertilizer are applied according to needs indicated by soil tests.[8]

Further comment on the assumptions underlying the yield estimates under the two management levels is provided in the following:

> In preparing the estimated yields for both levels of management, the technicians assumed that: (1) The level of management had been applied long enough, approximately 10 years, so that the yield figures would reflect the full effect of the practices applied; (2) the estimated figures represented the average yields expected over a 10-year period; (3) cultural operations were timely; (4) weeds, diseases, and insects were controlled according to the best known methods; and (5) suggested varieties of

---

8. *Soil Survey, Lucas County, Iowa,* USDA in cooperation with Iowa Agr. Exp. Sta., 1960, p. 8.

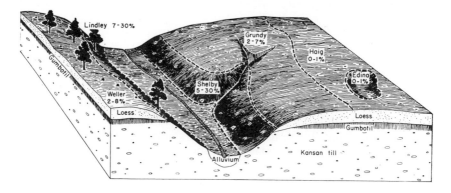

FIG. 21.3. *Typical grouping or toposequence of soils found in Lucas County, southern Iowa. (Courtesy Iowa Agr. Exp. Sta.)*

crops were grown, and recommended rates of planting were followed.

The estimated yield figures are considered to be fairly reliable estimates of the yields that can be expected at present. Improved farming practices, including better methods of fertilization and the use of improved varieties of plants, may result in greater yields in the future.[9]

SCS CLASSIFICATION

The classification system used by the Soil Conservation Service (SCS) is essentially a rating system except that the basis of rating is not value or yield but limiting factors. As the following explanation will indicate, the SCS classification, although not designed for appraisers, can be extremely useful to them.

The capability classification shows the relative suitability of soils for crops, grazing, and other purposes. It is based on the needs and limitations of the soils, the risk of damage to them, and their response to management. In the capability classification soils are grouped into classes, subclasses, and units (sometimes called management groups).

The broadest grouping is the *capability class* of which there are eight. All the soils in one class have limitations and management problems of about the same degree but may be of different kinds. Productivity may vary widely among soils within one capability class.

The next grouping, the *subclass*, is used to indicate the dominant kind of limitation within a class such as susceptibility to erosion or wetness.

The *capability unit*, which is the lowest level in capability

9. Ibid., pp. 8–9.

grouping, is made up of soils that need about the same kind of management and are similar in risk of damage, in suitability for use, and in productivity. Below the capability unit are the soil types and phases which were discussed in Chapters 19 and 20; these are the smallest soil mapping areas, the ones which are grouped to form the capability units, subclasses, and classes.

In explaining the capability system the map of an actual farm, Figure 21.4, will be used. This is the usual soil map discussed in Chapter 19. Here are identified two major items—soil types and capability classes. All capability classes except V and VIII are present on this one farm and are designated in white. The other white numbers and letters designate three factors— soil types, slope, and erosion. In some cases the soil types and soil capability classes have the same boundaries. For example along the east border of the farm there is a small area of 1AO; this is a soil type—Arenzville silt loam, 0–2 percent slope, uneroded—and this same area is also designated Land Class I. This particular area of Land Class I is the only Class I soil on the farm.

In the upper left corner three soil types are shown:

1. 31B2—Dubuque silt loam, deep, 2–5 percent slope, moderately eroded
2. 31C2—Dubuque silt loam, deep, 5–10 percent slope, moderately eroded
3. 31D27—Dubuque silt loam, 10–15 percent slope, moderately eroded

The first two soils, 31B2 and 31C2, were placed in Land Class II and the third soil, 31D27, was placed in Land Class III. The separation between Classes II and III was made on the basis of slope. This illustrates just one of the differences between Class II and Class III soil. The list of major differences between the eight capability classes provides a key to the successful use of capability maps in farm valuation.

Eight Capability Classes

Limitations which restrict the use of a soil are the major guides in setting up capability classes. Class I soils at one extreme have virtually no limitations. Class VIII soils at the other extreme have such severe limitations that they have little if any use. The first four classes are suited to cultivation, the second four are not.

CLASSES I–IV: Land suited to cultivation

Class I: Soils with few limitations that restrict their use[10]

---

10. This part based on USDA explanation of capability classification.

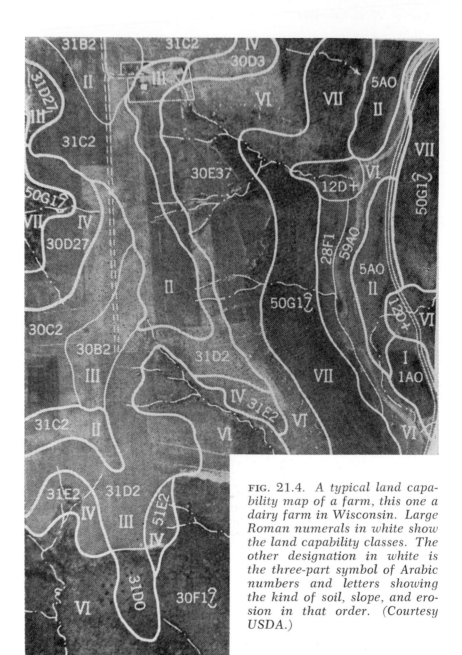

FIG. 21.4. *A typical land capability map of a farm, this one a dairy farm in Wisconsin. Large Roman numerals in white show the land capability classes. The other designation in white is the three-part symbol of Arabic numbers and letters showing the kind of soil, slope, and erosion in that order. (Courtesy USDA.)*

These soils are:
- — nearly level, with no erosion hazard
- — deep, generally well drained, easily cultivated
- — well adapted to holding water, well supplied with plant nutrients or highly responsive to fertilizer
- — not subject to damaging overflow
- — productive and suited to intensive cropping

In irrigated areas these soils are:
- — protected by relatively permanent irrigation works
- — nearly level, with deep rooting zones
- — easily worked with favorable water-holding capacity
- — not subject to salts, overflow, erosion, or water table problems

Class II: Soils with some limitations that reduce the choice of plants or require moderate conservation practices

These soils will have the following limitations either singly or in combination:
- — gentle slopes, moderate susceptibility to wind or water erosion
- — soil depth less than ideal
- — somewhat unfavorable soil structure and workability
- — slight to moderate salinity or alkalinity
- — occasional damaging overflow
- — wetness that can be corrected by drainage
- — slight climatic limitations

Class III: Soils with severe limitations that reduce the choice of plants, require special conservation practices, or both

These soils will have limitations either singly or in combination of the following types:
- — moderately steep slopes with high susceptibility to erosion
- — frequent overflow causing crop damage
- — poor permeability of the subsoil
- — wetness even after drainage
- — shallow depth of surface soil
- — low moisture-holding capacity
- — low fertility not easily corrected
- — moderate salinity or alkalinity
- — moderate climatic limitations

Class IV: Soils with very severe limitations that restrict the choice of plants, require very careful management, or both

The limitations of these Class IV soils either singly or in combination are:
- — steep slopes with severe susceptibility to erosion
- — severe effects of past erosion
- — shallow soils and low moisture-holding capacity

— frequent overflows with severe crop damage
— excessive wetness, severe salinity or alkalinity
— moderately adverse climate

CLASSES V–VIII: Land not suited to cultivation

Class V: Little or no erosion hazard but have other limitations that are impractical to remove and make soils unsuitable for cultivation

Examples of conditions on Class V soils:
— on bottomlands subject to frequent overflow
— a short growing season that prevents normal production
— stony or rocky soils
— ponded areas where drainage is not feasible

Class VI: Soils with severe limitations that make them unsuited for cultivation

Examples with limitations which cannot be corrected are:
— steep slopes, severe erosion hazards, effects of past erosion
— stoniness, shallow rooting zone
— excessive wetness or overflow
— salinity or alkalinity, severe climate

Class VII: Soils with very severe limitations that make them unsuited for cultivation

Examples of limitations more severe than in Class VI are:
— very steep slopes, erosion, shallow soil
— stones, wet soil, salts, alkali, and unfavorable climate

Class VIII: Soils with limitations that prevent their use for commercial plant production. Such uses as wildlife preserves, watershed protection, and recreation are possible

Examples of limitations on Class VIII soils which prevent their use for crop production are:
— erosion, stones, wet soil, low moisture-holding capacity
— salinity or alkalinity, severe climate
— badlands, rock outcrops, sandy beaches, river wash, mine tailings, and other nonarable factors

A graphic representation of the eight classes is presented in Figure 21.5. The contrast between the barren rock on the hillside labeled Class VIII and the level cropland identified as Class I is evident.

Four Subclasses

In some areas capability classes are divided into subclasses to show important limitations. For instance if part of a Class

LAND CAPABILITY CLASSES			
SUITABLE FOR CULTIVATION		NO CULTIVATION-PASTURE, HAY, WOODLAND AND WILDLIFE	
I	REQUIRES GOOD SOIL MANAGEMENT PRACTICES ONLY	V	NO RESTRICTIONS IN USE
II	MODERATE CONSERVATION PRACTICES NECESSARY	VI	MODERATE RESTRICTIONS IN USE
III	INTENSIVE CONSERVATION PRACTICES NECESSARY	VII	SEVERE RESTRICTIONS IN USE
IV	PERENNIAL VEGETATION-INFREQUENT CULTIVATION	VIII	BEST SUITED FOR WILDLIFE AND RECREATION

FIG. 21.5. *An area of land showing all eight SCS capability classes, from the level Class I soil in the middle area to the barren hillside of Class VIII land in the background.* (*Courtesy USDA.*)

III soil area were especially wet it would be separated from the rest and given a designation of IIIw; if another part had a severe erosion hazard it would be separated and marked IIIe. The four subclasses used by the SCS are:

Subclass e (*Erosion*): That portion of any major class in which the dominant limitation is susceptibility to erosion or past erosion damage.

Subclass w (*Wetness, excessive*): That portion of any major class in which the dominant limitation is excess water. The common limitations are poor soil drainage, wetness, high water table, and overflow.

Subclass s (*Shallow, stony, etc.*): That portion of any major class in which the dominant limitation is in the root zone — shallow soils, stones, low moisture-holding capacity, low fertility difficult to correct, and salinity or alkalinity.

Subclass c (*Climatic limitation*): That portion of any major class in which the dominant limitation is climatic, such as high temperature or lack of moisture.

Units

The capability unit is a group of soil types or soil type phases having similar productivity, similar limitations, and requiring similar management. For example, if a wet area of Class III soil were situated so that one part was subject to more flooding than another, the two parts might be separated into units labeled IIIw-1 and IIIw-2. The reason for this separation is that these two parts require different kinds of soil treatment.

CAPABILITY MAP

There are two main steps in making a soil capability map: (1) preparation of a regular soil map of a farm usually with an aerial map as the base; (2) grouping of the individual soil areas into land classes, subclasses, and units. In Figure 21.4 the first step was delineating the soils with slope and erosion notations as shown in white numbers and letters for the soil types. The second step was drawing the white lines which represent the grouping of soils into major land classes. These maps, including the office file as well as the one held by the farm operator, are colored with each land class having a separate and distinct color so the reader can note the location and extent of the different land classes at a glance.

Important for the appraiser is the method followed by the SCS personnel in drawing the capability class boundaries. Specific instructions are used in making the groupings into classes, subclasses, and units. Special emphasis is given to limitations based on climate, wetness, toxic salts, slope and hazard of erosion, previous erosion, moisture-holding capacity, and soil depth. As an example, soil depth is explained as follows in the instructions:

> Effective depth includes the total depth of the soil profile favorable for root development. . . . Where the effect of depth is the limiting factor the following ranges are commonly used: Class I, 36 inches or more; Class II, 20–36 inches; Class III, 10–20 inches; and Class IV, less than 10 inches. These ranges in soil depth between classes will vary from one section of the country to another depending on the climate. In arid and semiarid areas, irrigated soils in Class I are 60 or more inches in depth. Where other unfavorable factors occur in combination with depth, the capability decreases. For example, in one locali-

ty, Quinlan silt loam (10–20″ deep) with 0–3 percent slope is Class III, but the same kind of soil with 3–5 percent slope is Class IV.[11]

It is evident from this soil depth example and also from the emphasis on climatic limitations that capability maps are adapted to local areas. Consequently the appraiser should acquaint himself with the standards, procedures, and practices used by the SCS in the district in which he is appraising. This can be done at the local SCS office where maps and other information are filed for farms in the SCS program.

USE OF CAPABILITY MAPS BY THE APPRAISER

With an understanding of the capability classification and of the capability map-making procedure, the appraiser is in a position to make excellent use of these maps. The soil types and phases shown as a first step in the making of a capability map provide the basic material for an appraisal map. The land classes, subclasses, and units established in the second step provide additional data useful in making his yield estimate. The appraiser should keep in mind that the capability unit provides the most nearly homogeneous group of soils other than the soil type or soil phase. The capability subclass and class require increasing levels of generalization about the productivity and management requirements of the soils in the group.

There is one major difference between the capability map and the appraisal map which needs to be stressed—the difference in management assumed in each case. A soil conservation plan under *improved* management is the objective in a capability map, while estimated yields under *typical* management are the objective in an appraisal map. The SCS technician is concerned with soils and land classes as they indicate the kinds of practices that will be most desirable in developing a long-term conservation program under improved management. The appraiser is primarily concerned with soils and land classes as aids in estimating the productivity of a farm under typical operation.

To acquaint himself with the methods used by the SCS, the appraiser may find it desirable to visit the local SCS office. In these offices are duplicate copies of the capability maps, farm plans, and other information placed in the hands of individual farm operators. Besides, a visit with the soil technician and other personnel at the local office can provide the appraiser with

11. USDA, *Soils Memorandum,* SCS-22, Washington, D.C., May 1958, pp. 20–21.

a wealth of background briefing on the soils in the area including important points that might be easily overlooked, thus saving hours of frustration in the field and adding immeasurably to the appraiser's confidence in his yield estimates.

A capability map for the farm being valued can be traced for an appraisal map. In doing this, soil types and phases as well as the units, subclasses, and classes should be traced. This will give a more detailed and authoritative picture of soils on the farm than the appraiser would have had if he had mapped the farm himself.

With the tracing the appraiser should walk over the farm to check the different kinds of soils and capability classes, to fix their character in his mind, and to judge their yielding ability under typical management. Judging and estimating yields under typical management is a vital step. The appraiser should not take estimated yields from someone else without checking to make them his own. In evaluating farm productivity under typical management, he can use to excellent advantage the SCS data, but these materials do not replace his own check on soil productivity nor his own estimated yields. These estimated yields he will establish after he has walked over the farm, checking his observations with the available data prepared by others.

CONCLUSION

The numerical completeness of the soil productivity ratings which are being published may suggest to some that appraisal will soon be reduced to the simple task of identifying the soil types on a given farm, affixing the proper ratings, and converting these into value. This is far from true. Difficulty has been encountered in some areas in applying ratings to soils on individual farms because of the wide variation in productivity existing within a given soil type. Ratings are not made to be applied to individual farms but to a soil type on the average as it occurs throughout the state or wide area. The appraiser faces no easy task when he applies the ratings to the soil on individual farms. Soil ratings should be considered for what they are intended to be—aids in appraising productivity, not absolute figures to be applied wherever a given soil type is found.

The same general conclusion which applies to ratings applies also to the SCS classification. It is not a value rating system; rather it is a method of grouping soils according to their limiting factors. An appraiser needs to understand how the SCS classes, subclasses, and capability units are defined and mapped in order to use them successfully in establishing a productivity grade for a farm.

QUESTIONS

1. Obtain, if possible, bulletins or other material giving soil productivity ratings for your state or region. (Yield estimates are being included in the more recent county soil survey reports.) Compare these yield estimates with the yield estimates you have made in answer to previous problems. Discuss use of soil productivity ratings, including indexes, such as the corn suitability rating and Storie system; and yield estimates, such as those published in the county soil survey reports. How do they apply to soils in your area?
2. A trip may be arranged to the nearest SCS local office where an explanation can be obtained of the procedures followed in making a capability classification map and in formulating a conservation plan to go with it. If possible a farm should be selected that will be appraised in connection with the class work.
3. A tracing can be made of an actual capability map of an individual farm and a trip made over the farm to check the capability classes and to estimate crop yields under typical management.
4. What are the differences between capability classes, subclasses, and units?
5. Explain the difference between the basic objectives behind the making of a capability map and an appraisal map.

REFERENCES

Current bulletins, county soil survey reports, soil judging, and similar information explaining or showing soil or land capability classification in your area. Materials of this kind are published by the Soil Conservation Service in cooperation with state agricultural experiment stations and state agricultural extension services.

USDA, *Soil: The 1957 Yearbook of Agriculture*, Washington, D.C., pp. 400–411.

———, *Soil Survey Manual*, Handbook No. 18, Washington, D.C., 1951.

22

CROP YIELDS—STATES, COUNTIES, AND TOWNSHIPS

Crop yields reported by the USDA, by state and federal census authorities, and by other agencies are a valuable aid to the appraiser. They provide a system of bench marks that enable the appraiser to compare his productivity estimates with reported crop yields. One of the chief advantages of the crop yield reports by the different state and federal agencies is that they are provided on a continuous basis without charge. All the appraiser has to do is to ask for them and use them.

PURPOSE AND USE OF CROP YIELD MATERIALS

How this crop yield material can be used as bench marks in an appraisal can be illustrated by examples showing yield averages as they vary over time and from one area to another. From the time angle the averages will start with the 1870s, with the main emphasis on the situation since 1930. From the area angle the United States as a whole will be used first, followed by states, counties, townships and finally by individual farms.

The purpose of this crop yield analysis should not be overlooked. It gives the appraiser a check on the yield estimates he makes for the soil types on an individual farm. For instance, yield estimates (based on soil type differences) made for Farms

A, X, Y, and *Z* will be checked against published yield figures available for the areas in which these four farms are located. This procedure shows how an appraiser can tie together crop yields published officially for states, counties, and townships with yield estimates for different soil types as discussed in Chapter 21. With these crop yield averages in mind the appraiser is able to make his crop yield estimates for an individual farm with more confidence than if he had only the official crop yield figures or only soil type yield estimates.

If the territory of an appraiser covers only a few counties or even as much as an average-sized state it should not be difficult for him to build up a file of data on yields that will prove highly useful in making individual appraisals. The nature of this file can be seen in the examples that follow.

CROP YIELDS FOR THE UNITED STATES AS A WHOLE

A first requirement in the appraiser's crop yield file is an overall view of the United States yield situation—a long-range record of crop yield averages for several important crops. If we assume the appraiser is located in the Midwest, his choice of crops should include corn of course but also one or more crops from outside the area, like cotton and wheat, to show what is happening in the country as a whole.

A long-range view of crop yields for the United States is provided in Table 22.1. Twenty-year averages are used in this table, starting with the 1870–89 period and continuing through 1929. Beginning in 1930, ten-year averages are used because of the marked changes taking place in these recent years. The twenty- and ten-year averages iron out the year-to-year fluctuations and, starting with 1870, give a long view of the situation for each of the crops included. The long period is important because an appraiser's judgment on crop yields is forti-

TABLE 22.1.   United States Yields for Corn, Cotton, and Wheat

Years	Corn	Cotton	Wheat
	(bu.)	(lbs.)	(bu.)
1870–1889	26	173	13
1890–1909	27	188	14
1910–29	26	173	14
1930–39	24	206	13
1940–49	34	266	17
1950–59	44	363	20
1960–67	67	479	26
1968	78	511	28

Source: USDA. Yields are on basis of harvested acres.

fied by an understanding of what has happened, not just in the immediate past but over a long span of years.

Table 22.1 and the two tables which follow have an open-end form; that is, individual years can be added beyond the date of the table so that the yields as they are reported can be inserted. The appraiser may find it helpful to follow this practice with his yield tables, as this focuses attention on changes in yields that are occurring.

There is a definite break in Table 22.1 which started with the 1930–39 period. Up to this time there was practically no change in any of the three yield averages. Actually only cotton broke loose in the 1930–39 period, making a substantial rise above the average of the preceding sixty years. Corn was lower in the 1930–39 period than in any of the preceding periods but this was due to the severe drouth years—1930, 1934, and 1936—which occurred during this period.

The 1940–49 decade brings into sharp focus the revolutionary increase in crop yields which had started in the thirties and was still under way in the sixties. Table 22.1 gives quantitative evidence of this crop yield revolution for the country as a whole for the three crops of corn, cotton, and wheat; the rise which gathered speed in the forties showed no signs of slackening in the fifties and was continuing in the sixties. This means that the appraiser has to keep his eye fixed on crop yields as they come out each year to judge whether the upward trend has stopped.

In Table 22.1 it is evident that the yields in the fifties were up 80 percent over the level of the thirties and up 30 percent over the forties. If allowance is made for the unusually dry years in the thirties it can be said that corn yields in the fifties were up 60 percent over what would have been normal in the thirties. In the sixties yields for all three crops took another big jump, an unusually large one for corn but a substantial one for both cotton and wheat. Will research and development in the seventies give another boost for this decade? Since the trend is still upward, with the highest yields on record at the end of the sixties, it is reasonable to assume that at least a substantial increase over the sixties can be expected.

STATE YIELDS

Crop yields in the states will be concentrated on corn, since the Midwest is being used as an example. Although there was little change in the United States as a whole until after 1930, there were some changes in the states before 1930 that merit attention. For area comparison the three states of Illinois, Iowa,

TABLE 22.2. Corn Yields for Illinois, Iowa, and Nebraska

Years	Illinois	Iowa	Nebraska	U.S.
	(*bu.*)	(*bu.*)	(*bu.*)	(*bu.*)
1910–19	36	39	24	26
1920–29	36	40	26	27
1930–39	36	37	15	24
1940–49	50	51	28	34
1950–59	59	56	35	44
1960–67	84	79	56	67
1968	89	93	73	78

Source: USDA. Figures are for all corn per harvested acre.

and Nebraska will be used in order to show the effects of declining precipitation moving westward from Illinois.

The effect of rainfall variations is shown in striking fashion in Table 22.2. Nebraska on the western edge of the Corn Belt has felt the effect of rainfall shortages with lower yields on the average and much wider fluctuations than either Iowa or Illinois. The devastating effects of the drouths in the thirties pulled the average corn yield for Nebraska down to 15 bushels per acre, well below the national average. On the other hand, good years in Nebraska, such as in the 1920–29 period, brought the corn yield up near the U.S. average. When valuing land in Iowa and Nebraska the appraiser will find rainfall fluctuations in the two states go far in explaining differences that exist between the two states in the value of land devoted to corn as a principal crop. The explanation is based not only on the lower rainfall average and lower average yields in Nebraska but also on the wider fluctuation in rainfall and yields from one year to another.

Another interesting fact in Table 22.2 is the increase in yield in Illinois compared to Iowa. Back in the 1910–29 period, Iowa yields were higher on the average than Illinois yields; but in the 1950–59 period, corn yields in Illinois were definitely higher than in Iowa. And in the 1960–67 period the gap widened. This may be an important reason behind the much faster rise in farmland values in Illinois compared to Iowa during this period. The USDA reported Illinois farm real estate values at $174 an acre on March 1, 1950, and at $446 on March 1, 1960, a rise of 156 percent.[1] During this same period Iowa farm real estate rose from $161 an acre to $350, a rise of 119 percent. There were other factors involved besides yields, but yields were an important single factor.

_____

1. USDA, *Current Developments in the Farm Real Estate Market*, Washington, D.C.

TABLE 22.3. Corn Yields for Selected Counties in Iowa

Years	Cedar (high county)	Story (Farms A, X)	O'Brien (Farm Y)	Monroe (Farm Z)	Wayne (low county)	State
	(*bu.*)	(*bu.*)	(*bu.*)	(*bu.*)	(*bu.*)	(*bu.*)
1910–19	43	41	40	32	30	39
1920–29	45	43	40	36	34	40
1930–39	50	44	42	27	24	37
1940–49	59	55	51	39	35	51
1950–59	69	59	55	42	41	56
1960–67	92	85	76	61	65	79
1968	108	102	70	80	92	93

Source: USDA and the Iowa Annual Farm Census. Figures are for all corn harvested per acre. Highest and lowest counties were selected on basis of 1925–54 yield averages.

COUNTY YIELDS

State yields obviously cover up many important variations between counties. This is true of all three of these midwestern states—Illinois, Iowa, and Nebraska—for which corn yields were given in Table 22.2. For the purpose of showing yield variations between counties, Iowa was chosen in order to check corn yield estimates made for Farms A, X, Y, and Z against county corn yields and later against township yields. Counties selected and shown in Table 22.3 include the three in which these four farms are located and, in addition, the county with the highest and the county with the lowest corn yield per acre.

Corn yields in the counties, as in the states and in the country as a whole, started an impressive rise in the thirties that carried through the forties, fifties, and sixties. With the exception of Monroe and Wayne counties and the state average, all other yield averages increased in the thirties. Cedar County, which made the most substantial rise, is in the eastern part of Iowa where the drouths of the thirties were not as severe as in other parts of the state. It is also interesting to note that Cedar County is relatively close to Illinois, and both enjoyed a big spurt in the fifties and sixties.

Farms A and X are in Story County whose average corn yield is above the state average. Farm Y is in O'Brien County which is near the state average, and Farm Z is in Monroe County whose corn yield average is close to the lowest in the state. It is evident, therefore, that Farms A, X, Y, and Z are located in high, average, and low corn-yielding counties respectively.

By keeping an open-end table with the ten-year averages at the top, the appraiser can refresh his memory at a glance as to what has been happening in yield variations over the years in each county and in the state as a whole. Comparison of one county with another, both in the earlier years and in the present, readily discloses any major shifts like the relatively rapid gain of Cedar County in the fifties and the gain of Wayne County over Monroe in the sixties.

TOWNSHIP YIELDS

County yields, because of soil and other factors, naturally cover up variations within a county. For this reason it is useful to use township yield averages when available. Since they are available for Iowa in the Annual State Farm Census conducted by assessors, they will be used, as shown in Table 22.4, for the townships in which Farms A, X, Y, and Z are located and for the average of the three townships in the state with the highest and the three with the lowest average corn yields. Three townships are included in the high and low township averages to remove the influence of any abnormal situations.

In most periods the average of the three high townships is at least double the average of the three low townships. A yield difference between the two townships in Story County shown in Table 22.4 is noticeable. If the Story County average for the years 1925–54 is given a percentage of 100, Franklin Township rates a percentage of 98 and Lafayette a percentage of 106 for these same years. Townships change their relative position also. Hartley, for example, was below the county and state averages through the forties, but in the fifties it shot up above both the county and state averages. Guilford Township was consistently below both the county and state averages but also consistently above the average of the three low townships except in 1960–67.

The wide variation in corn yields in Iowa is shown in Figure 22.1, a map prepared from township yield data. The three high townships are located in Cedar and Scott counties in the east central portion of Iowa. Story County with its two high-yielding townships is located in the center of the state, Hartley Township in O'Brien County is located in northwest Iowa, Guilford Township in Monroe County is located in the second tier of counties from the southern border about midway across the state, and the three low-yielding townships are in Wayne and Decatur counties on the southern boundary not far from Monroe County. The counties and townships represented in Table 22.4 are distributed among all four of the major corn-yielding areas of the state portrayed in Figure 22.1.

A map like Figure 22.1 gives the appraiser a ready guide for checking yields. If the appraiser is in high corn-yielding

TABLE 22.4. Corn Yields for Selected Townships in Iowa

Years	Three high twps.	Townships Franklin (Story Co., Farm A)	Lafayette (Story Co., Farm X)	Hartley (O'Brien Co., Farm Y)	Guilford (Monroe Co., Farm Z)	Three low twps.	State ave.
	(bu.)	(bu.)	(bu.)	(bu.)	(bu.)	(bu.)	(bu.)
1930–39	39	34	34	34	24	19	37
1940–49	66	54	57	48	38	33	51
1950–59	72	56	60	58	40	36	56
1960–67	99	86	88	75	57	60	79

Source: USDA and the Iowa Annual Farm Census. Figures are for all corn harvested per acre. The three highest yielding townships are Springdale and Fairfield in Cedar County, and Lincoln in Scott County. The three lowest yielding townships are Clinton and Monroe in Wayne County, and Morgan in Decatur County. These high- and low-yielding townships were selected on the basis of the 1925–54 corn yield averages.

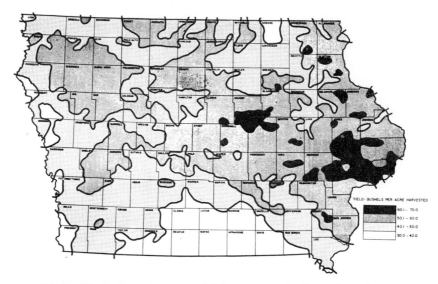

FIG. 22.1. *Variations in corn yield per acre in Iowa based on township yield data for the period 1945–54. The three high townships are in black area in eastern Iowa, the three low townships in white area in south central part of the state. (Source: USDA and the Iowa Annual Farm Census.)*

territory he can use this as a check against the yield estimates he makes for the farm he is appraising. Such a map enables an appraiser to know at all times the yielding ability of the area in which he is working. His special task is to determine how the farm he is appraising rates in comparison to the average of the area in which the farm is located.

The farms appraised and used as examples in this book are shown in Table 22.5, each compared with its township, county, and state average. Cropland on Farm *A* is above the yielding

TABLE 22.5. Estimated Corn Yields for Individual Farms and Reported Yields for Townships, Counties, and Iowa

	Corn Yield Estimate by Appraiser	1960–67 Average		
Farm		Township	County	State
	(*bu.*)	(*bu.*)	(*bu.*)	(*bu.*)
Farm *A*	103	86	85	79
Farm *X*	110	88	85	79
Farm *Y*	80	75	76	79
Farm *Z*	60	57	61	79

Source: USDA and the Iowa Annual Farm Census. Figures are for all corn harvested per acre.

ability of the township, county, and state averages of 1960–67. Farm X is all tillable with an estimated yield well above the township, county, and state averages. Farm Y, which is subject to overflow from the river running through it, still is higher than the township, county, or state averages. Farm Z, with only a small amount of cropland, has a yield estimate above the average for the township, just a little below the county, and well below the state average.

These individual farm yield comparisons, which complete the crop yield analysis, provide the appraiser with final bench marks on yields. He will find it worthwhile to make this check before reaching a conclusion on the productivity of the farm he is appraising. As an illustration, assume the appraiser has arrived at a preliminary corn yield estimate of 100 bushels an acre for corn on Farm A after he has completed his field inspection of the cropland and has given each of the soil types a corn yield rating. On checking this 100-bushel corn yield against the township, county, and state averages, however, he realizes that he is slightly low. In explanation he recognizes that probably he has been adversely and to some extent unconsciously affected by the pasture and waste along the open ditch adjoining the cropland on this farm.

He compares his 100-bushel estimated yield with the 110-bushel average he had made previously on Farm X, and compares it also with the yield on other farms he has appraised in the general vicinity. After this detailed review of all the evidence he decides his estimated yield is too low. He then goes back over the soil type yields, adjusting them in line with his new overall judgment. This brings the estimated yield up to 103 bushels an acre which he concludes is a satisfactory answer from all angles.

On the other hand, the appraiser may come up with a corn yield estimate, like the one of 110 bushels an acre for Farm X, which needs no adjustment when it is compared with the township, county, and state averages and with the estimates made for other farms in the vicinity. The point is that the final productivity estimate made by the appraiser will be more reliable if the appraiser checks it against published crop yields for the township if available, and for the county and state averages in all cases.

INDIVIDUAL FARM YIELD DATA

Crop yields for individual farms are useful to the appraiser, especially when they cover a period of years and are compared with reported yields for township, county, and state. The appraiser should be on the alert for yield records kept by farmers who measure their crop production, and from landlords who keep an accurate record of the grain or other crops that are sold from their farms. Professional farm managers are an excellent source for this type of information since one of their tasks is to maintain records of crop production on the farms they manage. In checking individual farm yields an important point is to get the total production figure and divide it by the acreage figure. Another desirable procedure is to obtain the crop sales record from landlords who sell their entire crop share, or from farmers who sell their entire crop for cash, as in the case of farmers who raise and sell wheat, soybeans, or cotton. When these sales are divided by the acres in the crop, a more reliable yield estimate is obtained than when a rough guess of the yield is obtained without regard to total production and acres.

HIGHLY VARIABLE YIELDS

Determination of yield averages in areas where yields fluctuate widely presents a special problem. Nebraska's corn yields are good examples of this problem. Where crop yields within a region are highly variable, a record covering a long period of years is especially desirable. The Great Plains, where extreme weather variations have a large effect on wheat yields, is such an area. Specific data of the type desired have been selected from the reports on wheat yields by counties in North Dakota. Four counties from different parts of North Dakota were selected, and the yields for planted acres, not harvested acres, are shown in Table 22.6. Acres abandoned during the year are not figured in harvested acres, but they should be included, as they are in seeded acres, in order to figure the yields from all the acres planted to wheat.

Short-period averages are dangerous in areas where yields fluctuate widely. The ten-year wheat averages in Table 22.6, like the ten-year corn averages for Nebraska in Table 22.2, do not remove the highs and lows due to periods of drouth or extremely good years. If the appraiser using Table 22.6 had taken the 1960–67 wheat average for Williams County (19.8 bushels per seeded acre) he would have been misled because this period was much better than any other in the 48-year period. On the other hand, if he had taken the 1930–39 average of 4 bushels he would have been misled in the other direction. Even 20-year averages, as evident in Table 22.6, were not stable.

The amount of fluctuation as well as the average should be given proper weight by the appraiser. It is one thing to have a farm which yields an average of 11 bushels of wheat with little variation from year to year, and something entirely different to have a farm with the same average but with variations from nothing one year to 20 bushels the next year. The figures for

TABLE 22.6. Average Wheat Yields on the Basis of Planted Acres for Four Representative North Dakota Counties

Year	Williams (northwestern)	Hettinger (southwestern)	Wells (central)	Grand Forks (eastern)
	(bu.)	(bu.)	(bu.)	(bu.)
1920	9.7	6.4	9.3	10.8
1921	12.5	1.9	11.4	8.7
1922	16.6	14.7	14.1	14.7
1923	9.8	6.2	5.0	8.9
1924	16.6	13.6	15.6	19.0
1925	8.4	11.6	14.6	14.5
1926	8.2	5.6	6.3	13.2
1927	12.7	13.3	14.2	12.0
1928	16.5	17.1	14.7	16.8
1929	10.4	8.8	8.4	11.3
1930	7.7	7.7	10.5	15.3
1931	.1	4.2	2.9	6.5
1932	8.3	11.7	9.2	11.7
1933	5.7	3.5	7.4	14.4
1934	.2	.5	1.3	7.2
1935	4.2	4.6	7.4	7.6
1936	.4	.0	.1	6.3
1937	.0	2.0	4.1	16.6
1938	6.2	1.5	7.3	15.0
1939	7.0	9.3	12.1	12.6
1940	8.8	9.7	10.2	11.8
1941	21.2	17.2	15.4	18.5
1942	23.2	21.1	17.1	22.4
1943	22.8	17.0	16.8	19.2
1944	19.6	16.6	15.0	17.5
1945	16.2	13.9	16.2	18.9
1946	13.8	13.5	13.3	18.9
1947	14.9	13.8	11.4	18.2
1948	13.8	14.8	13.1	17.5
1949	10.1	6.1	8.6	13.6
1950	17.3	10.6	9.5	15.4
1951	9.8	13.9	12.5	16.4
1952	8.4	9.1	10.4	14.3
1953	15.7	8.0	8.0	14.7
1954	9.4	8.3	5.6	14.6
1955	18.3	14.0	15.4	17.6
1956	12.8	7.8	18.3	23.4
1957	14.3	21.7	16.1	25.9
1958	13.8	14.4	24.9	35.1
1959	10.4	10.8	12.5	28.3
1960	17.5	15.9	22.9	26.4
1961	6.4	8.4	11.2	23.0
1962	28.0	22.9	30.0	30.2
1963	26.1	18.9	22.2	25.6
1964	22.8	18.1	28.4	30.4
1965	21.0	22.7	23.8	30.6
1966	20.0	23.4	20.3	23.8
1967[a]	17.0	25.5	21.4	29.7
	10-year averages			
1920–29	12.1	9.9	11.4	13.0
1930–39	4.0	4.5	6.2	11.3
1940–49	16.4	14.4	14.0	17.6
1950–59	13.0	12.8	13.4	20.6
1960–67	19.8	19.5	22.5	27.5
	20-year averages			
1920–39	8.0	7.2	8.8	12.1
1940–59	14.7	13.6	13.7	19.1
	48-year average			
1920–67	12.8	11.9	13.1	17.6

Source: North Dakota Statistical Reporting Service, USDA.
[a] Preliminary.

Williams and Grand Forks counties, especially for the years 1937 and 1962, provide a good contrast in this respect.

PASTURE YIELDS

Pasture yields have to be estimated by the appraiser as well as yields of tillable crops. It happens, however, that less attention has been given to pasture yields, including the carrying capacity of range land. Consequently the appraiser has little reported material to rely on in this area. Nonetheless, as noted in Chapter 21, soil scientists have rated soils on the basis of their carrying capacity in animal units. Studies have also been made of pasture yields. One of these showed "the amount of feed necessary to maintain the animal and produce the gain in weight or the livestock products produced by the animals while on pasture."[2] This feed requirement was then figured in terms of an equivalent in corn.

On the basis of this method, L'Hote obtained an average pasture yield of 11.2 bushels of corn equivalent per acre for the area sampled in Missouri. The sample included 9,837 acres of permanent pasture in a total of 364 fields situated in nine different counties well distributed over the state. Of importance to the land appraiser is not only this average of 11.2 bushels but also a classification of yields by soil types which is presented in Table 22.7. Soil types differ widely in productivity according to this table. This evidence, however, covers only one year.

Different methods of figuring pasture returns give widely different answers according to a study made in Kentucky.[3] The different methods and the annual value per acre produced by the pasture, using 1949 Kentucky prices and basic data from a research study at the University of Illinois,[4] were:

Value of feeds replaced by pasture forage $68
Value of milk produced from pasture forage 96
Clippings sold as hay from the field 85
Rental value from TDN's replaced by pasture 18

Insufficient account is taken in the first three methods of the costs involved, as pointed out by Young and Nesius. There is a big difference between pasturage and corn stored in a crib which can be fed when desired. Pasturage may or may not be

2. Homer J. L'Hote, *Measuring the Productive Value of Pastures*, Mo. Agr. Exp. Sta., Bull. 443.

3. H. M. Young, Jr., and E. J. Nesius, *Planning for Economic Production of Pasture*, Ky. Agr. Ext. Ser., Bull. 498.

4. W. B. Nevens, R. W. Touchberry, and J. A. Prescott, Jr., "A Method for Estimating the Feed-Replacement Value of Pasture Forage," *J. Dairy Sci.*, 32:894–900.

TABLE 22.7. Yields in Corn Equivalent of Permanent Pastures by Soil Types, Missouri

Major Soil Types	Yield per Acre in Bushels of Corn Equivalent
Marshall, Wabash	18.1
Baxter, Avilla	13.6
Summit	13.4
Putnam, Lindley, Memphis	11.7
Putnam, Lindley	10.2
Shelby, Lindley	9.8
Clarksville	8.8
Cherokee, Bates	8.0

Source: Homer J. L'Hote, Mo. Agr. Exp. Sta., Bull. 443, p. 33.

available, depending on the weather, and it has to be used when it is available or else it is lost for the most part. If it is clipped, there are heavy harvesting costs to be figured. If the value of milk produced from pasture is figured, there are large labor, housing, and other costs in addition to feed that have to be considered. Accordingly, the $18 rental figure is the most reasonable one to use in valuing the production of pasture, assuming that this figure represented a good competitive market reflecting normal supply of and demand for pasture.

In all cases it is well for the appraiser to observe not only the carrying capacity of the pasture and the rental rates charged for the pasture but also the relative supply and demand conditions which apply to the pasture in question. In some areas it is difficult to find a farmer who wants to rent pasture, while in others pasturage is scarce and can be rented easily at the going rate.

QUESTIONS

Prepare table and graph showing average yield for state or county for one or more crops commonly grown in your region. Include at least 15 years if possible. Compute averages for 5, 10, and 15 years. How would you use this material in estimating crop yields for appraisal purposes?

REFERENCES

USDA, *Agricultural Statistics*. See the annual report for most recent year; also reports prepared by State Statistician.

23

TYPICAL MANAGEMENT AND HIGHEST AND BEST USE

TYPICAL MANAGEMENT, an essential part of a farm valuation, often presents difficulties because the actual management on the farm may not be typical. The operator may be so lacking in "know-how" that he is not getting results that an average operator could be expected to get; at the other extreme the operator may be following practices which give him much higher yields than an average operator would get. This was illustrated by an unwarranted high value placed on a poor sandy farm by an appraiser who was misled by an excellent operator who was getting unsusually high yields. In this instance too much emphasis was placed on the man and not enough on the land.

Highest and best use, also called most profitable farm type, is another essential which presents difficulties because the type being followed on a farm may not be the highest and best use for the farm. This was brought out in an appraisal of a level, all-tillable farm with one-third of the land in bluegrass pasture because the owner-operator wanted it that way. Another example was a hilly, eroded farm planted entirely in corn by an owner-operator who wanted all he could get from the farm as soon as possible.

Since these two principles—typical management and highest and best use—are fundamental in farm valuation, and since they present difficult and complicated situations, this chapter is devoted solely to their explanation.

TYPICAL MANAGEMENT

Since a commercial farm has value because it yields a net return, it is necessary to assume some kind of management producing this net return. The kind of management recommended for appraisal is typical management.

Typical management means most likely or average management. In a technical sense the word "typical" means the modal average, the common group, or the class with the most items. For example, the net incomes on 20 farms of average size varied in a recent year from a high of $12,200 to a low of minus $220. The average was $7,200, and more than half had incomes between $5,800 and $8,800. The modal or typical group was in this $5,800 to $8,800 range. An appraiser valuing the $12,200 net income farm would be looking at an abnormal management situation—an unusual one out of line with what might be expected. And the same could be said for the minus $220 net income farm. In short, typical management is the average or most likely operation for a given farm, not one of the extremes.

Actual management is emphasized by some appraisers in preference to typical. The argument is that with the actual you are sticking close to reality and not indulging in estimates or guesses. Unfortunately, however, this management reality can be misleading, especially when an excellent manager causes an appraiser to judge a farm better than it is, or when a poor manager causes an appraiser to judge a farm at a lower value than it deserves.

The recommended procedure is to note actual management while making the appraisal, then determine what the most likely management would be and compare it with the actual. If the typical and actual are the same there is no argument. But if the operator is considerably above or below average in the kind of management one would expect, the situation is unstable; in that case typical rather than actual management should be used.

There are still other reasons for not emphasizing actual management. Every year some farmers have to quit farming because of injuries or sickness, some farmers die from accidents or disease, and some farmers decide to sell or rent out their farms and engage in other business. Also the operator may be nearing retirement so that his management obviously would not continue much longer. Changes like these plus the normal turn-over of tenants make it clear that the appraiser should not accept the management of the present operator as a base for income estimates of typical management.

Finally, typical management rather than actual is preferred because it is used by buyers and sellers of farms when they are arriving at a price at which they are willing to buy or sell. What income would a prospective owner-operator or investor-landlord expect? What he would expect, of course, would depend on what he or his tenant would be able to obtain from this particular farm. The buyer planning to operate the farm would assume his own management in the first place, but in making his bid he would not base it on his ability to produce extra high yields. If he based his bid on prospective high yields and high income, he would be bidding and paying more for the farm than other prospective purchasers would be able to pay, more than he would need to pay to get the farm. Furthermore, by paying for a farm as high a price as his management might justify, he would be losing all the advantage of his superior management, for it would go into land value instead of coming to the purchaser in the form of annual high net income. In short, there would be no point in the superior manager paying more for a farm than he has to pay, and he most certainly does not need to pay a price based on his excellent operating ability.

Typical management might be seen toward the end of a farm real estate auction where competition had been reduced to only three bidders, one a landlord-investor and the other two farmers wanting to operate the farm themselves. The landlord-investor bids on the basis of the net income he thinks he can obtain with the type of tenant the farm will attract, but he also thinks of the asking price for other comparable farms because he does not want to pay more than the going market price. The two farmers, one an excellent and the other an average manager, bid on the basis of what they think the farm will produce under their management, but they too keep in mind the going market price as a top bidding price. Before the sale started they probably would have sized up this farm in value as compared to other similar farms that have sold recently or are on the market at present. However, in the excitement of bidding, they may go above this maximum.

If the excellent operator has the funds and the desire he may get the farm with the high bid. But his high bid will not be based on using up all the potential net income he estimates he can make on the farm—only as much as is necessary to outbid the other two and not to exceed to any large extent the going prices for comparable farms. The landlord-investor may figure the most likely tenant could produce on this farm a net income of $4,000 annually, and the average operator may figure he can net $4,000, while the excellent operator may figure a net of $5,000. If all buyers are thinking in terms of a 5 percent return, $80,000 would be the approximate maximum for the first two buyers. The excellent manager might be able to buy the farm with a bid of $81,000, since there are no other top managers competing with him. His net income would support a price as high as $100,000, but this is far above the going price of farms of this quality, and he undoubtedly would not have to go that high.

BETTER OPERATORS ON BETTER FARMS

Better managers, it is evident from this discussion, tend to get on the better farms. Farm management specialists have long observed this tendency. However, it is only a tendency; the variations in earnings among farmers in most areas indicate the imperfect working of this principle. So it can be said in general but not in every case that the successful operator, because of his superior management, is able to outbid his less capable competitors either in purchasing or in leasing a farm. That the superior operator does not always exercise this right explains why the best farmers are not always found on the best land. There is no denying the fact, however, that the landlord with the outstanding farm is in a position to attract the outstanding tenant because both will prosper by getting together. The most likely operator of an above-average farm would be a superior manager.

In summary, typical management for a given farm is determined by the kind of manager this farm is most likely to attract. A farm that is above the community average would attract an operator above the community average, a farm below the community average would most likely have an operator that was not up to the community average, and an average farm would most likely have an average operator. Although this principle does not always work perfectly in practice, we do find that it is operating and that buyers and sellers are consciously or unconsciously following it in a general way in the land market.

HIGHEST AND BEST USE

Highest and best use means placing the land in the use which will pay the highest net return—choosing the farming type which is likely to pay the highest net return. In the Corn Belt, for example, the highest and best use is likely to be a crop-livestock combination involving corn, other feed grains, soybeans, and some form of livestock to be fattened from the feed grains. In the dairy area some combination of crops and dairy cows will probably yield the highest net return. And so it goes over the country with wheat, tobacco, citrus fruit, cotton, peanuts, range livestock, and other types emerging as the highest and best income producers in certain areas. Soil, climate, and other natural conditions along with prices of the products and government programs determine in each case the combination which is most likely to give the highest net return over the long pull.

Highest and best use of farmland can be seen in the map showing types of farming (Fig. 23.1). In this map it is evident that climate is an especially potent factor in determining where certain products are grown, as for example in fixing the northern boundary of the cotton area. Rainfall, temperature, and growing season are the specific factors which have a great deal to do with the distribution of crops throughout the country.

Within the Corn Belt there are many variations, from state to state and within each state. Variations in farm type within a state are shown for Illinois in Figure 23.2. This shows nine separate regions in Illinois whereas the United States map in Figure 23.1 showed only three general regions. A study of limiting factors such as pasture and woodland, brought out in Figure

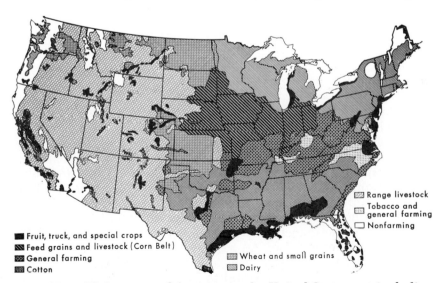

Range livestock

Tobacco and general farming

Nonfarming

Fruit, truck, and special crops
Feed grains and livestock (Corn Belt)
General farming
Cotton

Wheat and small grains
Dairy

FIG. 23.1. *Major types of farming in the United States not including Alaska and Hawaii. Illinois, which is referred to later, has three of the nine types shown. (Courtesy Bureau of Agricultural Economics.)*

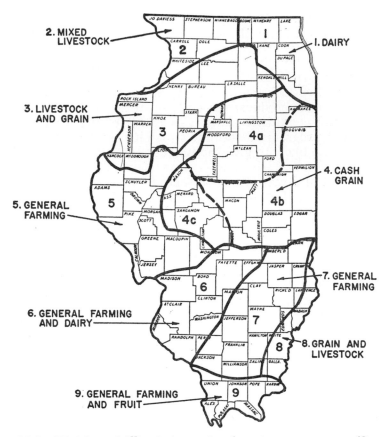

FIG. 23.2. *Division of Illinois into nine farming-type areas. (Source: Ill. Agr. Exp. Sta., Bull. 601.)*

23.3, indicate basic reasons for the division between livestock and cash-grain farming in Illinois. The emphasis on dairying in northeast Illinois near Chicago, as indicated in Figure 23.2, is based on the short distance to a large market for dairy products.

Finally, two farms side by side may vary enough to make it most profitable for one to be in dairying and the other in livestock fattening. Here again natural factors are the determinant. One farm may be near a river and have steep hillsides and an abundance of permanent pasture which can be used most profitably in a dairy enterprise. An adjoining farm farther away from the river and with mostly level and fertile land would be best suited to corn and soybeans and to fattening of livestock.

The mixture of farm types in an area is illustrated by the distribution of cash-grain farms in Illinois as shown in Figure

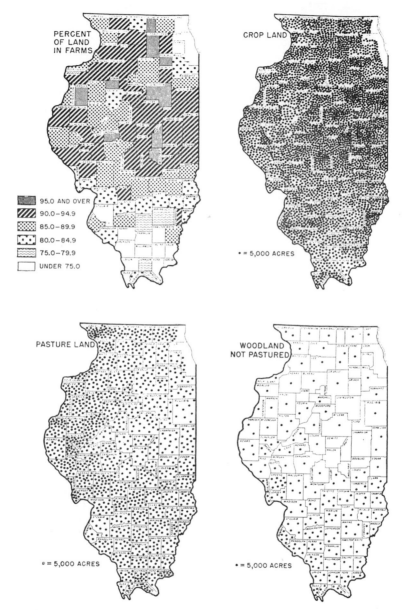

FIG. 23.3. *Type of farming factors in Illinois: Percentage of land in farms, cropland, pasture land, and woodland not pastured. (Source: Ill. Agr. Exp. Sta., Bull. 601.)*

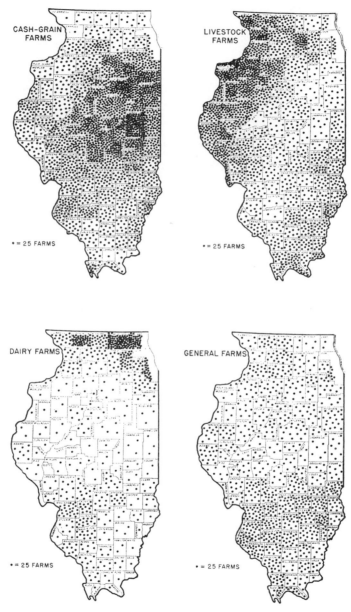

FIG. 23.4. *Location of major farm types in Illinois. A farm was classified as cash-grain, livestock, dairy, or general if the product or group of products designated accounted for 50 percent or more of the value of all products sold. These four types included 96 percent of the commercial farms in Illinois. (Source: Ill. Agr. Exp. Sta., Bull. 601.)*

23.4. Cash-grain farms are scattered over most of the state although there are more in east central Illinois than in any other part. The two areas with smallest numbers are in northern and southern Illinois. With the exception of the northeast district near Chicago, these north and south areas have extensive amounts of permanent pasture (Fig. 23.3) which make cash grain relatively unprofitable. In the northeast district the pressure for dairy products in Chicago creates the comparative advantage shift from cash grain to dairying.

A similar analysis of the farm type situation in other states will reveal (1) the reasons why the predominant type has become established and why it is holding its position in certain areas of the state and (2) the reasons why variations exist side by side within any area. When an appraiser goes into a dairy area to appraise a farm he may expect to find a farm adapted to dairying, but instead, because of certain limiting factors affecting this particular farm, he may find that the highest and best use of this farm is for fattening hogs and cattle.

TYPE OF FARMING TEST

In determining the highest and best use of a farm, a type of farming test is recommended. The chief purpose of this test is to find the most likely combination of enterprises for a given farm. The test concerns not crops alone but the whole farm business, including the mutual relationships between crops and livestock. The percentage of receipts from different enterprises may be used as a measure of farm type. The main enterprise may be cash grain, dairying, poultry, beef raising or feeding, hogs, sheep raising or feeding, truck crops, fruit, or any combination of these or others. On any given farm careful analysis may show one or two or even three enterprises that stand out as distinctly more profitable and more likely than the rest.

A type of farming test includes three steps as follows:

1. To establish the nature and extent of any limiting factors. This includes all natural and economic factors. A large amount of permanent pasture, for instance, is a controlling factor in selecting the most profitable farm type; that is, types are limited to the few that can utilize efficiently this large amount of pasture. In Illinois, as the figures illustrated, permanent pasture separated livestock and cash-grain farming areas. Because of the importance of such limiting factors as permanent pasture, the appraiser must be certain that he has carefully noted and evaluated all of them, and that he has the lines correctly drawn between tillable and nontillable land. This is not always easy because land which might be tilled is sometimes left in

pasture, while some land so steep or otherwise limited that it should be in permanent pasture is in crops.

2. To decide on a high profit combination of crops for the tillable land. The appraiser must take into consideration soil types and capabilities, topography, erosion, climate, equipment and machinery, distribution of labor on different crops, and relative prices of crops. His aim is to choose those crops which the farm appears best fitted to produce.

3. To select the livestock enterprises, if any, that may be dictated by permanent pasture, profitable crop combinations, and such economic factors as market outlets and relative prices. Nearness to a large city, as shown in the Chicago example, might make production of market milk the most profitable type to follow even though the farm seems better suited to livestock fattening.

Following the choice of livestock in the third step, it may be necessary to make some minor adjustments in the crop plan to make the two—crops and livestock—fit together. If dairying is selected as the highest and best use because of a good nearby market, it may be desirable to increase alfalfa hay and cut down on corn.

In selecting farm type in terms of highest and best use, the ability and aptitude of the operator have not been stressed because the operator usually seeks a location where the farm type he wants is profitable. The operator can move, the farm cannot. A farmer interested in operating a market milk farm would naturally seek a farm where this type is profitable. Similarly a farmer wanting to produce fruit would seek areas where fruit production is carried on efficiently and extensively. This is true of all types except for those farms where any one of four or five different types could be carried on with approximately equal returns. Such a situation often occurs where land is all tillable and the grain raised can be either sold as cash grain or fed to different kinds of livestock which do not require extensive pasture. Another such situation holds near large cities where soil may not be an important factor but where purchase of feed may make possible a choice between such types as eggs, live poultry, and market milk.

DEPRECIATING OR APPRECIATING CROP PLANS

Once the most profitable crop-livestock combination has been determined it should be checked against the combination in actual use. When the two are the same, appraising is simplified because no adjustments are necessary. When they are different, such as the examples of level land in bluegrass or eroded land in corn, an abnormal situation exists which requires analysis.

Two abnormal situations which require special treatment are: (1) a crop plan that depreciates the land and (2) a crop plan that builds up the land. If the crop plan is so heavy that it causes serious erosion and would, if continued, result in eventual loss of the land for crops, the annual income at present will not be a correct estimate of what will be received in the future, because a declining series of incomes will result. On the other hand, a crop plan that appreciates (builds up) the land will produce an income at present which will be much less than that expected in the future. A young orchard that will soon begin bearing fruit or a badly gullied farm that is being reclaimed or a wet area that is being drained are examples.

The easiest plan to follow is to assume a crop program which neither increases nor decreases the producing ability of the land. If the land is depreciating, a lower level than the present one is assumed in order to allow for the decline in the future. If the land is appreciating, a higher level than the present one is assumed to allow for the increase in the future. The difficulty in this procedure is in estimating the proper level to use. Even so, it is a realistic approach to an estimate of a farm's productivity.

Another more detailed approach is to estimate the likely decrease or increase in production by years. Appreciation can be handled by assuming the level of production achieved at the end of the appreciation period and by making appropriate deductions for the years between the present and the time at which the new higher level is reached. Similarly, the reverse situation of depletion can be handled by assuming the eventual lower level of productivity and by making additions for the years between the present and the time at which the new lower level is reached. This depleting situation can be treated the same as a resource being used up, such as a gravel pit from which the gravel is being sold or an orchard where the yield is declining and the trees are not being replaced. An explanation of how to compute the additions and deductions in these procedures was given in Chapter 13 on capitalization.

EXAMPLES

The principles presented in this chapter are partly of an abstract nature because management and type of farming are factors not easy to pin down. The examples which follow have been added to show the reader how these principles can be used in two concrete appraisal situations.

1.   Farm X. (See Chapter 19 for appraisal map.)

Typical management. The location of this farm five miles from a city of 30,000 population in a well-improved, highly productive land area will attract an outstanding farmer. The disadvantage of this farm is its small size of 120 acres. However, there is some possibility of renting or buying nearby land. An attractive modern home, a well-landscaped farmstead, and good farm buildings are added features that would draw a good tenant or buyer.

Highest and best use. Since the land is all tillable, the only limiting factor is the small size. The most profitable crop combination would be corn-soybeans-corn. Livestock fattening of either hogs or cattle or both would be the logical addition to the crop plan. There is virtually no risk from erosion so the estimated crop plan, with fertilizer assumed, would maintain the fertility at its present level.

2.   Farm Z. (See Chapter 19 for appraisal map.)

Typical management. This farm is not an economic unit because of its small amount of cropland. Adjoining land is also mostly pasture so it would not be possible to add much cropland. But this farm has an advantage in its excellent location on a paved highway in an area with few paved roads. The typical manager for this farm would be a part-time farmer who might combine livestock dealing or trucking with a small amount of farming. Since the house is good and the location is very good, a part-time farmer can be assumed.

Highest and best use. The small amount of cropland and the large amount of pasture are severe limiting factors. Crops would be negligible except for hay. Not enough good cropland is available to justify a large investment in machinery. Hiring custom machinery service would be logical. The pasture and buildings favor raising cattle or pasturing livestock. The good location favors a part-time farmer in the livestock trucking business. The crop plan set up for this farm would maintain the present fertility level of the farm. An alternative would be sale of farm, except for farmstead and small area of adjoining land, to a neighbor for expansion of neighbor's farm.

QUESTIONS

1.   Obtain and study bulletins and other reference materials from your agricultural experiment station or from the USDA relating

to types of farming in your area. Farm management studies are also helpful on this subject.

2. Analyze the most likely or typical management situation on the farm or farms you have appraised. Apply the type of farming test. Discuss crop plans from standpoint of increasing or decreasing farm productivity.

REFERENCES

Appraisal of Real Estate, 5th ed., American Institute of Real Estate Appraisers, Chicago, 1967. See especially section on highest and best use.

See also discussions on management in current farm management text books.

24

RATING WHOLE FARMS AND LARGE AREAS

It is not enough to rate individual features of a farm in preparing an appraisal. A judgment or rating of the farm as a whole is essential because the sum of the individual parts does not necessarily give the value of the farm as a unit. Consequently, in the first part of the chapter the emphasis is on the overall evaluation of the farm.

One step beyond farm rating is area rating. These area ratings are a different type of appraisal, described later in the chapter, which are made or contracted for by tax assessment authorities, loan agencies, or other groups. The emphasis in this case is on classification of the land in a given area, with the variation that exists being indicated by quality grades. This is a relatively new and promising field, especially in tax reassessment studies.

COMPARISON JUDGMENTS

The major objective in whole-farm and area appraising is comparison judgment. It is mainly a shift from comparing and rating individual features in a small area—such as soil on an individual farm—to comparing and rating all the features of a farm rolled into one, or of rating whole farms one with another over a large area. In doing this the appraiser needs special skill. It is not a special skill in using mathematical formulas or in using score card procedures because neither of these by itself will enable the appraiser to accomplish the task. *It takes judgment.* This judgment is whetted to a sharp edge by practice in making comparisons. These comparisons in turn lead to decisions which, whether they cover individual features on a single farm, whole farms, or a whole area, constitute the heart of appraisal. Appraisal, in essence, is the art of making comparison judgments or decisions.

RATING THE FARM AS A WHOLE

The overall rating of a farm serves as a summary or check on the detailed inventory an appraiser makes of a farm. In the process the appraiser brings into focus not only soil and yield but also buildings, farmstead arrangement, road type, location, and farm size. He thinks of the farm as a whole and not as the sum of its individual parts. By taking this broad view he is able to correct omissions or to adjust emphasis of individual details. In fact the whole may give a different impression than the sum of the parts; the individual features when added may indicate an "A" rating, but the overall impression may be that it is only a "B" farm. The usefulness of this "whole-farm" comparison and rating technique will become more apparent to a beginning appraiser after he has made a number of appraisals which he can then compare one with another.

If the rating system is made up of five classes with A the best and E the poorest, the farm being appraised will rate as an A, B, C, D, or E farm. A variety of rating or classification systems may be used, and many different kinds are used, but the most common is this 5-class type. Variations include the 10-class system of AA, A, BB, B, etc., or the 15-class system of A+, A, A—, etc. Whatever the system, the task of the appraiser is the same—to weigh, to evaluate, and to reach a decision regarding the quality of the farm or tract being appraised.

The success of the rating system depends first on how well the classes are defined and second on how well the appraiser performs his task of identifying and placing the individual farms in their proper classes. In the case of definition it is not just a question of having it clear cut; it must provide enough difference between classes—for example between an A and a B farm—so that appraisers and those who use and interpret appraisals can recognize the differences.

A farm rating system can be divided readily into four parts as follows: (1) estimating productivity (in production units), (2) rating noncrop factors, (3) assigning weights to the productivity and noncrop portions, and (4) calculating the overall rating. To show how this can be done two Midwest farms are used—Farms X and Y pictured earlier in Figures 19.2 and 19.4 in Chapter 19.

PRODUCTION UNITS

The first step in rating a farm is converting all the production of the farm into units of the same amount; in the case of Farms X and Y, into bushels of corn. Unless some common unit like this is used, it is next to impossible to add together such unequal quantities as bushels of soybeans, oats, or corn, and tons of hay and silage. The computations involved in converting the productivity of Farms X and Y into production units are shown in Table 24.1. The factors used in this table to convert the different products into the equivalent of bushels of corn follow:

1 bu. of corn: 1 unit	1 ton legume hay: 18 units
1 bu. of oats: .5 unit	1 ton mixed hay: 15 units
1 bu. of soybeans: 2 units	1 ton silage: 7 units
1 bu. of barley: .75 unit	1 acre pasture (av.): 15 units
1 bu. of wheat: 2 units	1 acre rotation pasture: 25 units

These conversion factors are based either on average price relationships (such as 1 bu. of soybeans selling for twice the price of 1 bu. of corn) or on feed equivalents (such as one ton of legume hay roughly equal to the digestible nutrients in 18 bushels of corn). An appraiser would not necessarily follow the system outlined here but would construct a conversion system adapted to the situation in his own area.

From the unit computations in Table 24.1 and the rating scale in Table 24.2 it is evident that Farm X with an average of 90 production units per acre would be classed as an A farm on productivity, while Farm Y with a 51-unit production per acre would fall into the C class. These ratings are based on production without reference to the noncrop factors.

TABLE 24.1. Production Unit Chart for Farms X and Y

	Farm X				Farm Y			
Crop	Acres	Yield	Factor	Units[a]	Acres	Yield	Factor	Units[a]
Corn	70	110	1	7,700	70	70	1	4,900
Soybeans	40	38	2	3,040	30	30	2	1,800
Oats					10	50	.5	250
Hay					10	3	18	540
Pasture	5	...	25	125	35	...	20	700
Farmstead	5	...	...	...	5	...	...	...
Total	120			10,865	160			8,190
Average	...	...	...	90	...	...	...	51

[a] Units computed as follows: acres × yield × factor. Example: soybeans for Farm X are computed: 40 × 38 × 2 = 3,040 units.

TABLE 24.2   Midwest Productivity Rating Scale

Range in Production Units[a]	Percentage Scale[b]	Farm Rating
120 +	100	AAA
100–119	90	AA
80– 99	80	A
60– 79	70	B
40– 59	60	C
20– 39	50	D
0– 19	40	E

[a] One unit equals one bushel of corn or equivalent.
[b] Scale is an arbitrary conversion of production units into percentage points.

NONCROP FACTORS

Among the noncrop factors the most important are buildings, farmstead arrangement, location, road type, and farm size. Location includes such factors as distance to town, schools, and churches. The best way to handle the noncrop items, since they are much more intangible than productive factors, is to set up an ideal or 100 percent condition and measure each farm in terms of the extent to which it comes up to this ideal.

An ideal or 100 percent situation for noncrop factors would be a farm of optimum size for profitable operation with a relatively new, well-adapted set of buildings. This would include well-built service buildings that would store the grain, hay, silage, and other crops likely to be raised, house the machinery, and provide suitable quarters for the livestock the operator would be likely to have. The ideal farm would also include a well-built, well-designed modern house appropriate for the farm. The farmstead would be attractively landscaped with windbreak and plantings; the buildings would be efficiently located. Telephone and electricity would be available. Mail service would be provided at the road in front of the farmstead. The road would be paved, providing an all-weather access to the nearby town or city, and would not have any objectionable traffic features. The nearby town or city would provide market outlets, farm supplies, shopping facilities, and cultural opportunities with good schools, library, hospital, churches, and other desirable community organizations and services.

An ideal farm may seem unreal; however, there are farms that come close to this 100 percent rating on noncrop items, and Farm X is one of them. Farm X has to be graded down because the buildings are not new and in some cases are partially obsolete. In addition the traffic on the paved road in front of the farmstead is too heavy for comfort. But outside of these handicaps, Farm X has practically everything; and on the basis of 100 percent for ideal, comparison judgment would indicate it deserves an 85 percent noncrop rating. Farm Y, with good buildings but on a gravel road five miles from a small town which offers few advantages for the farm family, is rated 70 on noncrop items.

In contrast to the 100 percent rating, a zero rating would indicate a farm with dilapidated buildings, an unattractive farmstead, located far from town and schools, on a dead-end, almost impassable road.

WEIGHTING AND CALCULATING

How much weight should be given to the production units and how much to the noncrop factors depends on the importance of each in the value. In some cases such as with dairy farms near large cities, buildings and location may far outweigh crop production. In other cases where location and buildings are relatively unimportant, the weights may be as high as 4 or 5 for production to only 1 for noncrop factors. In this process the comparison judgment has to play a large part. Within an area, however, once the proper weights have been worked out for different situations, a consistent pattern can be expected. In the case of Farms X and Y the weights and ratings came out as shown in Table 24.3. For Farm X the production percentage of 80 was multiplied by 4 and the noncrop percentage of 85 by 1, which when totaled was 405. This total was then divided by 5, the sum of the weights, which gave the final percentage rating of 81.

Farms will be appraised which have a low production rating and a high noncrop rating, and vice versa. Situations like this give average whole-farm ratings which would be misleading if separate production and noncrop ratings were not provided.

Unusual factors have to be given special attention in rating. A farm with a roadside produce stand, a farm with a lake front-

TABLE 24.3. Rating Chart with Weights

| | Production | | | Noncrop Factors | | Whole Farm[a] | |
	Units	Per-cent	Weight	Per-cent	Weight	Per-cent	Rating
Farm X	90	80	4	85	1	81	A
Farm Y	51	60	4	70	1	62	C

[a] Whole farm ratings are obtained by multiplying the percentages by the weights and dividing this total by the sum of the weights. Example for Farm X: $80 \times 4 = 320$, plus $85 \times 1 = 405$; then $405 \div 5 = 81$.

age, a farm close to town, or an unimproved tract are the types which require this special treatment. Judgment based on a knowledge of the local situation has to be used in figuring the influences and applying the proper weights in the rating process. For instance, if farms are being expanded because the average size is too small for profitable operation, small-sized units which can be added to existing units will be rated high because they command a premium. In such situations small unimproved tracts without buildings may sell as high as tracts of the same size with buildings because the purchasers want additional land and not more buildings.

This system of rating farms on production and noncrop factors was used by the author in preparing review appraisals of 95 Midwest farms for a lending agency. Both production and noncrop ratings were made. The lending agency was chiefly concerned with a production rating to determine the income potential behind the loans, but it wanted a noncrop rating also because this had a bearing on the farm's sale value. A tabulation of the 95 farms on a composite production and noncrop rating showed the following:

Rating	No. of Farms
A	10
B	18
C	42
D	23
E	2

The wide spread in the ratings indicated to the lending agency the existence of more variation among these farms than they had realized. The advantage of this rating procedure is that the comparisons, each farm with all the others, reveal mistakes which otherwise might slip through unnoticed.

RATING LARGE AREAS

Rating a large area such as a whole county has been so successful in numerous instances that farm loan agencies, tax authorities, and research groups have been active in conducting projects of this general type. The appraiser has two direct interests in these area projects: (1) He may be called on to take charge of a project and supervise the work. (2) He may find the completed rating map a highly useful guide in any future individual valuations he makes in the area.

These area appraisals have been called by different names but are often referred to as economic land classification studies.

The objective in these studies is to combine and translate the physical features of the land into an economic rating. The main purpose of the rating is to express differences in quality throughout an area. However, the emphasis varies depending on the use to be made of the ratings. To give the reader a bird's-eye view of the different kinds of area appraisal, a number of examples will be cited from among those made for farm loan agencies, tax authorities, and the general public.

AREA STUDIES FOR LOAN AGENCIES

An agency making farm loans may want an area classified as to quality for home office evaluation of loan applications and appraisals which come from this area. Numerous studies of this type have been made, especially by federal land banks and insurance companies, the two major institutional farm mortgage lenders. An example of a Corn Belt county taken from the area appraisal studies of the Federal Land Bank of Omaha, is shown in Figure 24.1.[1]

Three classes of land are indicated in Figure 24.1. In making this classification, the guiding principle was desirability of the area as a place to live, to farm, and to make money. The more desirable the area, the higher was the rating. Specific factors used in measuring desirability were stability and level of production and income, type and desirability of soils relative to use made of them, transportation facilities, community standards, and relationship of taxes to community services and conditions.

Not all farms in the No. 1 areas in Figure 24.1 were top farms. The classification was not designed to give individual farm ratings. However, the dominant class of farms in the Class 1 areas were A farms; in the Class 2 areas, B farms; and in the Class 3 areas, C farms. When loan agency officials receive applications or appraisals from such a county, they can locate the applications and appraisals on the county classification map and obtain a general idea as to the quality of the land on the farm and in the neighborhood.

AREA ASSESSMENT STUDIES

Assessors whose duty it is to value all of the tracts in a district for property tax levies find area appraisal studies partic-

---

1. Aaron G. Nelson, "Experience of the Federal Land Bank with Loans in Four North Central Iowa Counties," Ph.D. thesis, Iowa State University, Ames, pp. 20, 21.

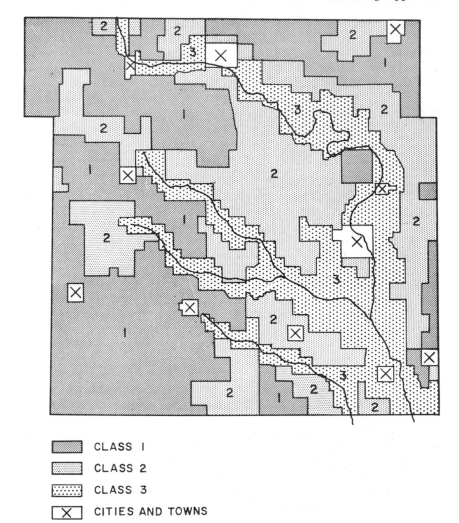

CLASS I

CLASS 2

CLASS 3

☒ CITIES AND TOWNS

FIG. 24.1. *Economic classification of land in a Midwest county for use in overall rating of loan territory. (Courtesy Aaron G. Nelson.)*

ularly useful in their work of comparing one tract with another in making equitable assessments. Montana has had an extensive program of this kind under way for more than fifteen years.[2]

Montana Study

The example presented in Figure 24.2 shows the classification of 1½ sections (960 acres) of a combination of dry farming

---

2. H. G. Halcrow and H. R. Stucky, *Procedure for Land Reclassification in Montana*, Mont. Agr. Exp. Sta., Bull. 459.

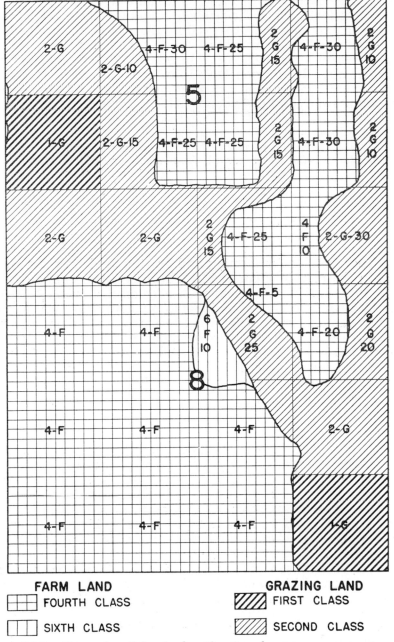

FIG. 24.2. *Example of land classification for tax assessment purposes in Montana. The area shown comprises 960 acres. Number in symbol indicates class of land, the letter F or G indicates farming or grazing land. (Courtesy Mont. Agr. Exp. Sta.)*

and grazing land. The nature of the land, whether farming or grazing, is indicated by the symbols F and G. The class of land is indicated by the number preceding the F or G, the number of acres indicated by the number following the letter. The symbol "4-F-30" means fourth-class farming land, 30 acres. The grading system for dry or nonirrigated farmland in bushels of wheat per acre and the system for grazing land in acres for 10 months grazing season per 1,000-pound steer or equivalent are as follows:

Grade	Farmland (bu. wheat per a.)	Grazing Land (a. needed per 1,000-lb. steer)
1	22 and over	18 and under
2	16 – 21	19 – 27
3	12 – 15	28 – 37
4	8 – 11	38 – 55
5	under 8	56 – 99
6	none	100 and over

These grades for farming and grazing land were the ones used in the map in Figure 24.2. The procedure recommended in starting one of these studies is a detailed drawing of land classification boundaries in three to six townships which are used as sample areas. Soils, yield histories, and other relevant data were used in fixing these boundaries. The results in the sample townships were checked with the citizens, the county commissioners, and the State Board of Equalization. Then the procedure was extended from the sample townships to the rest of the county.

Soil Appraisal In Germany

An elaborate system of soil comparison for tax assessment was developed in Germany prior to World War II and carried forward after the war. The system included the selection of the best farm in the country (a farm in Magdeburg) which was rated 100, use of bench mark or comparison farms on which net incomes were figured, and use of an appraisal guide to rate every farm or parcel in relation to the comparison farms and the Magdeburg farm.

The comparison process started with district comparison or guide farms which were selected in every district of the country. Net return was figured on each of these district comparison farms to give an index of productivity which could be used in rating this farm in relation to 100 for the Magdeburg farm. The next step was selection of subcomparison farms in local areas and the determination of soil value indexes for these local bench

Ackerschätzungsrahmen

Boden-art	Ent-ste-hung	Zustandsstufe						
		1	2	3	4	5	6	7
S	D		41 — 34	33 — 27	26 — 21	20 — 16	15 — 12	11 — 7
	Al		44 — 37	36 — 30	29 — 24	23 — 19	18 — 14	13 — 9
Sl (S/lS)	D		51 — 43	42 — 35	34 — 28	27 — 22	21 — 17	16 — 11
	Al		53 — 46	45 — 38	37 — 31	30 — 24	23 — 19	18 — 13
	V		49 — 43	42 — 36	35 — 29	28 — 23	22 — 18	17 — 12
lS	D	68 — 60	59 — 51	50 — 44	43 — 37	36 — 30	29 — 23	22 — 16
	Lö	71 — 63	62 — 54	53 — 46	45 — 39	38 — 32	31 — 25	24 — 18
	Al	71 — 63	62 — 54	53 — 46	45 — 39	38 — 32	31 — 25	24 — 18
	V		57 — 51	50 — 44	43 — 37	36 — 30	29 — 24	23 — 17
	Vg			47 — 41	40 — 34	33 — 27	26 — 20	19 — 12
SL (lS/SL)	D	75 — 68	67 — 60	59 — 52	51 — 45	44 — 38	37 — 31	30 — 23
	Lö	81 — 73	72 — 64	63 — 55	54 — 47	46 — 40	39 — 33	32 — 25
	Al	80 — 72	71 — 63	62 — 55	54 — 47	46 — 40	39 — 33	32 — 25
	V	75 — 68	67 — 60	59 — 52	51 — 44	43 — 37	36 — 30	29 — 22
	Vg			55 — 48	47 — 40	39 — 32	31 — 24	23 — 16
sL	D	84 — 76	75 — 68	67 — 60	59 — 53	52 — 46	45 — 39	38 — 30
	Lö	92 — 83	82 — 74	73 — 65	64 — 56	55 — 48	47 — 41	40 — 32
	Al	90 — 81	80 — 72	71 — 64	63 — 56	55 — 48	47 — 41	40 — 32
	V	85 — 77	76 — 68	67 — 59	58 — 51	50 — 44	43 — 36	35 — 27
	Vg			64 — 55	54 — 45	44 — 36	35 — 27	26 — 18
L	D	90 — 82	81 — 74	73 — 66	65 — 58	57 — 50	49 — 43	42 — 34
	Lö	100 — 92	91 — 83	82 — 74	73 — 65	64 — 56	55 — 46	45 — 36
	Al	100 — 90	89 — 80	79 — 71	70 — 62	61 — 54	53 — 45	44 — 35
	V	91 — 83	82 — 74	73 — 65	64 — 56	55 — 47	46 — 39	38 — 30
	Vg			70 — 61	60 — 51	50 — 41	40 — 30	29 — 19
LT	D	87 — 79	78 — 70	69 — 62	61 — 54	53 — 46	45 — 38	37 — 28
	Al	91 — 83	82 — 74	73 — 65	64 — 57	56 — 49	48 — 40	39 — 29
	V	87 — 79	78 — 70	69 — 61	60 — 52	51 — 43	42 — 34	33 — 24
	Vg			67 — 58	57 — 48	47 — 38	37 — 28	27 — 17
T	D		71 — 64	63 — 56	55 — 48	47 — 40	39 — 30	29 — 18
	Al		74 — 66	65 — 58	57 — 50	49 — 41	40 — 31	30 — 18
	V		71 — 63	62 — 54	53 — 45	44 — 36	35 — 26	25 — 14
	Vg			59 — 51	50 — 42	41 — 33	32 — 24	23 — 14
Mo				45 — 37	36 — 29	28 — 22	21 — 16	15 — 10

FIG. 24.3. *Arable land appraisal guide used in German soil system. Figures indicate relative soil value ratings for tax assessment purposes with 100 considered as the best soil in the country. (Source: Publications of the Office of Military Government for Germany (U.S.), Economics Division, Food and Agriculture Branch. Prepared by Philip M. Raup.)*

mark farms which were related to the district comparison farm, which in turn was related to the Magdeburg farm.

With the district and local comparison farms rated, the next step was the individual farm or parcel appraisal which was conducted with the use of an appraisal guide form. A copy of this guide as developed for cropland is shown in Figure 24.3. Since the guide is in German the following translation of the principal terms is presented:

> Ackerschätzungsrahmen: Arable land appraisal guide
> Bodenart: Soil type
> Entstehung: Origin
> Zustandsstufe: Stage of development

S	Sand	D	Diluvial
s	Sandy	Al	Alluvial
L	Loam	Lö	Loess
l	Loamy	V	Weathered
T	Clay	Vg	Weathered (Stony)
Mo	Moor		

Most of the terms will be readily understood; LT means heavy loam, and Diluvial means transported by glaciers, Alluvial transported by water, and Loess transported by winds. One of the terms which needs more explanation is "Stage of development." As used it refers to the present condition of the soil in relation to its process from original rock to its optimum condition for plant production and on to its old stage where it is worn out and poor. In the guide chart the first stage is the optimum and the seventh is the old poor soil. On the soil guide (Fig. 24.3) the best land comparable to the Magdeburg farm is shown as an alluvial or loessial loam in the first stage of development.

In appraising individual farms throughout a local area a team consisting of a leader (appraiser) and five or six others do the field work. A team which the author saw in operation near Frankfurt included a leader, two soil surveyors who determined the nature of the soil, two helpers who did the soil boring, and a local farmer who checked the results. This farmer was an important member of the team because he represented the local farmers whose lands were being valued for tax purposes by this procedure.

There are two qualifications that should be pointed out as weaknesses in this "best" farm comparison system. One is that the relationships change with time. Dairying, for example, may become more profitable than the raising of sugar beets or vice versa, with the soils that support these two farm types changing in value relationship to each other. The second point is that there are technical difficulties in comparing farms that are entirely different, such as vegetable raising versus wheat raising, or irrigation farming versus cattle raising on dry land. But within a similar type of farming there is much to commend the Magdeburg system with its emphasis on a standard of *one best farm,* with all other farms compared to it through a series of district and local bench mark or key farms.

GENERAL PURPOSE AREA STUDIES

Land classification studies in some regions have been made by experiment station personnel for multipurpose use. These purposes include use by farmers buying a farm, by lenders making farm loans, and by assessors making assessments. Two of these general purpose studies are included, one from New York State which has pioneered in land classification work and the other from the wheat areas of Saskatchewan in western Canada.

New York State Classification

The first example, shown in Figure 24.4, is from Cayuga County, New York. Here five major land classes are based on appraisals and analyses of the income potentialities of the individual farms in the area.[3] The five land classes run in reverse order from Class I (the poorest) to Class V (the best). The major purpose of these New York land class maps is to help farmers in buying farms, in using their land resources, and in determining how much debt they can carry and how much they can invest profitably in their farms. The land classes indicate the general degree of farming success likely in the various land class areas. The likelihood of success increases from the lowest to the highest class.

In the process of developing the land class boundaries, individual farms were studied in detail by means of soil and topographic maps, climatic information, sale prices of farms, and other pertinent materials. A survey record was obtained from the occupant of every eighth dwelling in the farming areas; this provided information on the farm business and any nonfarm employment.

A brief explanation of the land classes in Figure 24.4 follows:[4]

Class   I — Chances for success in full-time commercial farming extremely small

Class   II — Chances for success in full-time farming very small

Class   III — Chances for financial success moderate with the price relationships that can be expected

Class IIIx — Areas which could support a more prosperous agriculture; development will require imagination, skill, and capital

3. H. E. Conklin and J. J. Nolan, *An Economic Classification of Farms by Areas, Cayuga County, New York,* Leaflet 6, N.Y. State College of Agriculture, Ithaca, 1957.
4. Ibid.

Class IV — Chances for financial success good
Class   V — Major proportion of the farms have high to very high
            income expectancies

Although maps of the kind indicated in Figure 24.4 are likely to
go out of date in time, they are especially useful to farmers buying farms, to assessors, and to loan agencies.

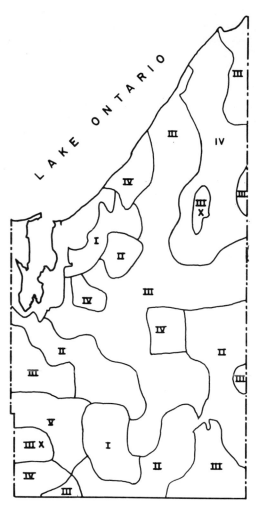

FIG. 24.4. *Economic land classes for northern Cayuga County, N.Y.
Roads, farmsteads, and other buildings not shown. (Based on map in
Leaflet 6, N.Y. State College of Agriculture.)*

Saskatchewan Classification

The final example is from an economic classification of land suitability for wheat production in Saskatchewan, Canada. A small sample showing the classification in one township—Township 9 north, Range 16 west (northern of Weyburn)—is shown in Figure 24.5.[5] These land class maps which cover most of the farming areas of the province have been used to separate the land suited to wheat production from that suited to grazing, to indicate the areas where grain growing did not pay, and to provide data for equitable real estate taxation.

In setting the land class boundaries, the specialists took into consideration wheat yields and physical factors such as soil, topography, and stoniness. In addition, they studied costs and returns to determine the income expectancy for different kinds of land on various sized units. When the land classes were first developed in the late thirties, a wheat farm of 480 acres was the average size. Since then the average size has increased and cultural practices have been improved, but the differences between the land classes remain substantially the same. The five land classes were established in terms of wheat yields per quarter section as follows:

Land Class	Average Wheat Yields per Quarter Section
I	350 bu. or under
II	351–475 bu.
III	476–720 bu.
IV	721–900 bu.
V	Over 900 bu.

Additional data are given for each class; for example, Class V soil is generally of a heavy, clay texture with level to undulating topography, good drainage, and no stones.

Land Classes I and II were singled out as the problem areas when the classification was first worked out. Although some large-scale wheat operations have developed on a profitable basis on Classes I and II land in recent years, these two low-yielding classes are still considered the critical ones, especially with an operation as low as 480 acres in size.

In concluding this area rating discussion it is evident that a wide variety of procedures is being used. No attempt was made

5. The original map from which Figure 24.5 was taken was prepared by the Economics Division, Marketing Service, Canada Dept. of Agriculture and the Farm Management Dept., University of Saskatchewan, 1954. A supplementary statement by R. A. Stutt was issued by the Canada Dept. of Agriculture, Ottawa, Canada, in 1959.

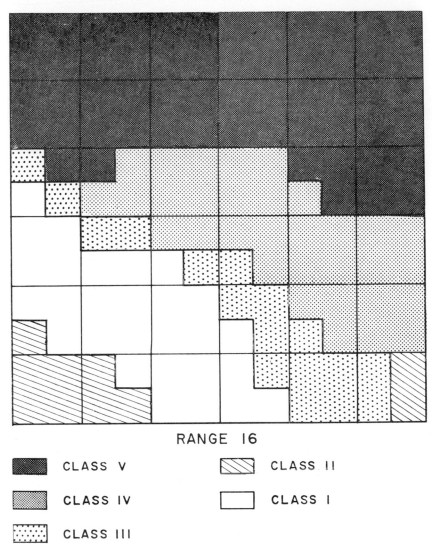

RANGE 16

▓ CLASS V		▨ CLASS II	
▒ CLASS IV		☐ CLASS I	
⋯ CLASS III			

FIG. 24.5. *Economic classification for wheat farming of a township in eastern Saskatchewan. Five different land classes are shown.*

to include all types but to give examples from widely separated geographic districts—from Germany to New York State to Montana and Saskatchewan—to illustrate solutions developed to meet pressing appraisal problems. All of the examples cited have strong points and weaknesses as well; each has certain adaptations to its area and to the use for which it was designed. In time new types will undoubtedly emerge, and those now in use will be changed. This is as it should be because area appraisals are still in their infancy.

QUESTIONS

1. Prepare a "whole-farm" rating system for the area in which you are appraising. Include rating symbols, production unit conversion factors, method of handling noncrop features, and weights for items considered in the rating.
2. Plan and carry out on a pilot basis an area appraisal. This sample study might include assessments for tax purposes in a small area such as a 4-section square. Assessments which are public records could be obtained, plotted, and checked in the field. Another sample study might include use of soil maps, other physical data, and economic information in making a field determination of land classes in a small farming area in which considerble variation in productivity exists.
3. What is the difference between a comparison judgment and a comparison between two score card results?
4. Of what use is the whole-farm rating?
5. Explain the difference between economic land class rating and physical productivity ratings.

REFERENCES

USDA, *Land: The 1958 Yearbook of Agriculture*, Washington, D.C., pp. 362–70.

8

APPRAISAL TYPES

THREE well-known types of appraisal, each with special characteristics that set it apart from the others, are considered in the following three chapters. Each of the types—assessment, condemnation, and loan—calls for an appraisal value, but the emphasis in each case is distinctly different.

Some agricultural enterprises have characteristics which set them apart and present unusual appraisal problems. Livestock ranches, irrigation farming, fruit and nut growing, and timber culture are several of the more important enterprises that fall in this group. In the final chapter the particular features of these enterprises and the valuation of allotments are discussed.

25

TAX ASSESSMENT APPRAISALS

Tax assessments differ from other appraisals in two important particulars: (1) They are based on legal statutes specifying the basis of the appraisal and rights of the property owner. (2) They depend for their success far more on uniformity—relative value of one tract to another—than on their absolute value. As a consequence, one of the important objectives in real estate assessment is the equalization of values among individual tracts, districts, and counties.

THREE CHARACTERISTIC TENDENCIES

A study of tax assessments in a typical district over a period of years will usually reveal three characteristic tendencies:

1. A tendency for any set of assessments to remain fixed or to change slowly.
2. Any change that does occur is likely to be a downward change, with some properties being reduced more than others.
3. A tendency for assessments to show an excessive concentration about the average value in the area.

The tendency of assessment valuation to remain fixed offers no difficulties in explanation. Any upward change in the appraisal will no doubt bring forth a complaint or at least a query from the owner as to the reason for the increase. The assessor, being elected directly by the people or appointed by an elected officer, feels no incentive to make upward revisions in valuations. The line of least resistance is to lower valuations and, if not, then to leave them just as they are.

The practice of lowering values may be a relatively harmless method of letting the property owner down easily. If the taxing district plans to raise $100,000 in property taxes, it makes little difference, provided there is no limit to the millage, whether the property in the area is valued at $10,000,000 which would call for a tax levy of 10 mills on every dollar of property value, or is reduced to $5,000,000 which would call for a 20-mill tax. There would also be a temptation to lower assessed values if the state used the local tax valuations to levy a state property tax, because the local area would benefit by putting its values as low as possible.

The tax assessor undoubtedly finds necessary adjustments easier to make by reducing certain valuations and leaving others as they are than by slightly reducing some and raising others. A property owner is much less likely to complain if his value is left where it is even if his neighbor's is reduced than if his value is raised and his neighbor's reduced. Of course if actual building construction has taken place, the valuations may be raised without any cause for complaint.

In view of these conditions, it is no wonder that valuations for tax purposes tend to decline. For instance, tax values may at first approximate sale values on this base. As time goes on, the changes may gradually reduce the average until the assessed values actually represent only 75 or even 50 percent of sale value, assuming sale values have not changed. When taxing valuations get down to 50 percent, taxing authorities may double the valuations, only to have these valuations gradually recede again through the process described above.

The third characteristic, excessive concentration about the average, can be explained by the peculiar situation faced by the assessor. He finds his greatest difficulty in explaining differences, especially wide differences. Establishing these differences is not an easy task, and providing convincing evidence may be even more difficult. Careful examination by the assessor may show him that a given tract of land is worth more than an adjoining piece, but he will find it easier to give each about the same value than to explain the difference. The man owning the better land will not complain, and the man owning the poor land does not like to say too much about how poor his land is because he may want to sell it some day or renew a loan on it.

LEGAL BASE

Assessment of real estate is performed in the process of levying taxes and accordingly falls under the jurisdiction of local and state governments. Since property taxes are levied in all 50 states, each has laws governing the procedure of real estate assessment for tax purposes. These laws include the definition of value which serves as the basis of assessment, the procedure to be followed by the assessor in arriving at assessment value, and the process by which an assessment is reviewed and equalized.

State legislatures have attempted to define in great detail the value standard to be used in assessment. The following definitions of property value from a representative list of states indicates the special effort exerted to make value for assessment clear and unequivocal:

Arkansas—"True market value"
Florida—"Full cash value"
Louisiana—"Actual cash value"
Maine—"Just value"
Missouri—"True value in money"
Pennsylvania—"Fair market value"

Practically all states qualify their definition of value with such words as "true," "full," "actual," or "fair" in an attempt to bring assessment values into line with market values. This is a result of the tendency for assessments to decline progressively below market value levels, one of the characteristics frequently associated with assessments.

ASSESSMENT UNIFORMITY

Above all the assessor has to make sure that he has valued the properties in his district in proper relation to each other. This is the first and cardinal principle of assessment. If A's property is about equal in value to B's, then A's assessment should be about equal to B's. Of course A is not likely to complain if his assessment is lower than B's, but in that case B is likely to complain and be fully justified in doing so.

It is far more important to have assessments in line with each other than to have assessments exactly equal to market value or some fixed percentage of market value. Theoretically assessments could be twice as high as market value and still work successfully if the law allowed this and all assessments were uniformly up to this level. In such a case the millage or rate of taxation would be just one-half of what it would be with assessments at market value. If, on the other hand, assessments were only one-half (or some other proportion) of market value—a more likely case—this situation should be an entirely satisfactory one provided all assessments were uniformly the same proportion below market value. In short, the objective of assessment is to provide a uniform base, relative to market value, for the equitable taxation of property.

Comparisons are important in achieving uniformity. In making comparisons, three tools are available—income, cost, and sales. Income is especially well adapted for comparisons between individual tracts of farmland in many areas, while cost is adapted to farm building comparisons. Sales are particularly useful in comparison of farms as a whole and as a check on income valuations of land and cost valuations of buildings. A reasonable procedure to follow in making assessments, therefore, is first to use the income of farmland and cost of farm buildings less depreciation and obsolescence, and second to check the results against sales.

BENCH MARKS AND STATISTICAL ANALYSIS

The first step in assessment is establishment of bench mark or key valuations. These are representative detailed appraisals to which the fieldmen and the assessor can tie or use as comparison standards. Bench mark properties should be scattered throughout the area if possible and should include examples of high, medium, and low value units. Income, cost, and sale value tests can be applied to these key units to make certain that the assessments of these units are satisfactory in every way. With these bench marks established, all other properties in the district can be assessed in line with them.

Various methods can be used in bench mark appraisals and in the follow-up valuation of all properties. Some of the detailed procedures being used successfully are briefly described in the section that follows.

In addition to bench marks an effort should be made to conduct a statistical study using regression analysis to determine the major land value factors and their effect on value. Assistance in a project of this type may be available from the state assessment office or from economists and statisticians at the state university. The studies by Ahmed and Parcher in Oklahoma and by Penn, Bolton, and Woolf in the Mississippi Delta (Chapter 18) are the type that could be especially helpful in a reassessment. In fact Parcher reports that this type of cooperation is operating in Oklahoma, with a regression formula provided by the university being used in reassessment work in several areas.

ASSESSMENT PROCEDURES

One of the best means of explaining the newer techniques in farm assessment is a description of an actual reassessment project. A good example in which income, costs, and sale values were used was a rural reassessment of Polk County, Iowa, carried out by Hugh l. Harter while he was county assessor. To start the project the county entered into a cooperative agreement with Iowa State University and the USDA to contribute to a detailed soil survey of the entire county. Later informal arrangements were made with the university for assistance on the assessment-sale ratio phase of the project.

Soil technicians and staff from the assessor's office worked closely in the field on the soil survey, thus achieving a mutual understanding of each other's problems and objectives. The assessor and his staff learned how the soil boundaries were drawn on the air photos which were used as base maps, and the soil technicians learned how important these soil boundaries could be when they meant differences in productivity and different assessments for neighboring farmers.

How these soil boundaries looked on the individual assessment cards can be seen in Figure 25.1, which shows the soil map for the SE¼ of Section 25, T81N R25W of the 5th p.m., less 2 acres for road, containing 158 acres.

Building data were entered on the back of the soil map card, as shown in Figure 25.2. Each building was measured, described, and valued as indicated. Since the dwelling is usually the most valuable structure on the farmstead, special notes, a sketch, and a building cost computation were provided on the bottom part of the card. After the replacement cost of the dwelling was figured at $3,336, a 50 percent physical depreciation was applied which brought the value down to $1,668. Then a functional depreciation or obsolescence factor of 20 percent was applied which brought the value down to $1,334, the final dwelling figure.

Next, soil information based on the soil map in Figure 25.1 was entered on assessment card 2 shown in Figure 25.3. To understand the method used it is first necessary to translate the numbers in each soil area. For instance the top number is the soil name—such as "119" in the upper right hand corner of Figure 25.1 which indicates Muscatine silt loam. The number below the soil indicates slope, and the number under slope indicates erosion.

Since assessments in Iowa have to be made by individual 40-acre tracts, the space on the card in Figure 25.3 was divided into 40-acre units with four of them on the card. Thus the soil

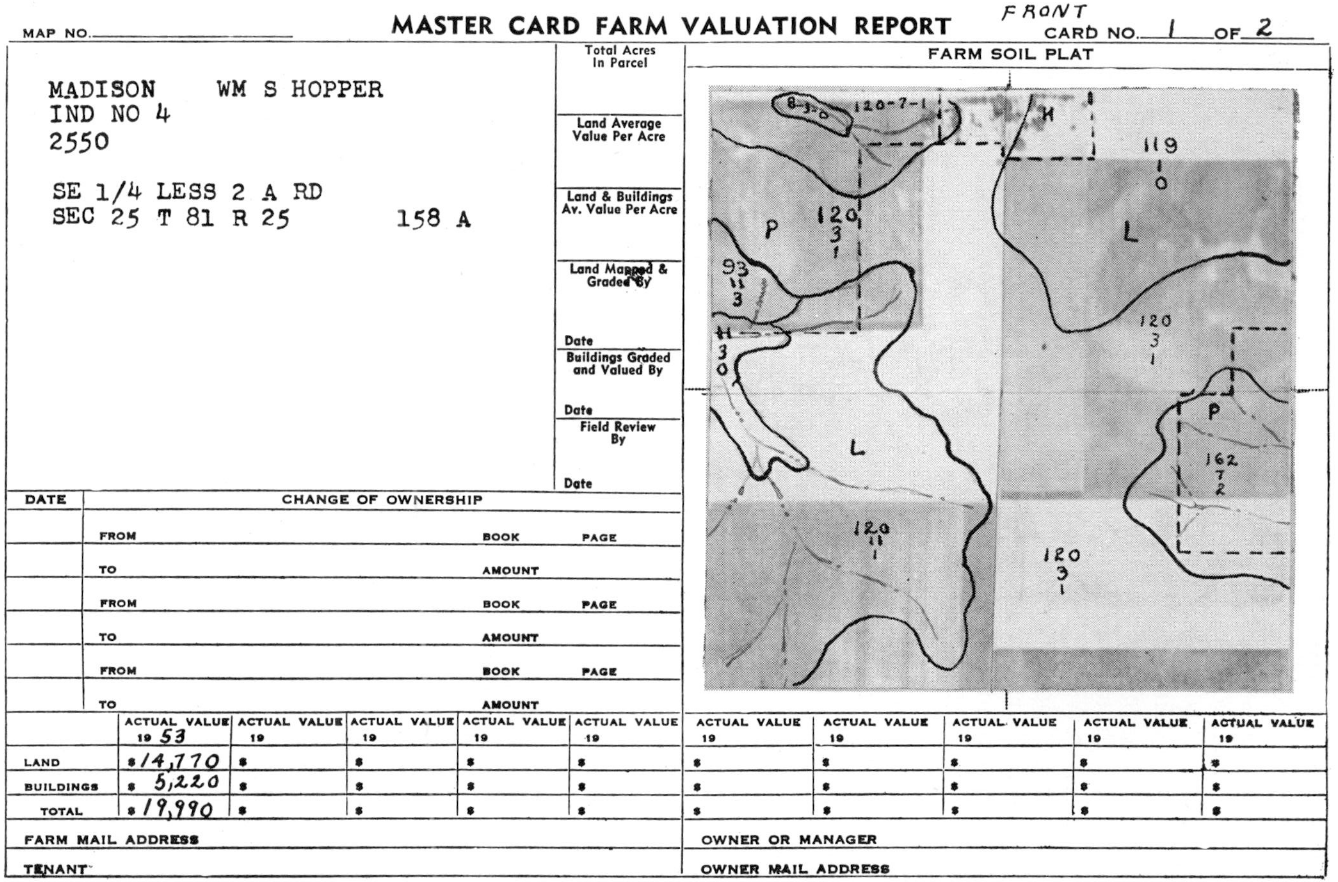

MAP NO.______

MASTER CARD FARM VALUATION REPORT

FRONT
CARD NO. 1 OF 2

MADISON WM S HOPPER
IND NO 4
2550

SE 1/4 LESS 2 A RD
SEC 25 T 81 R 25 158 A

Total Acres In Parcel

Land Average Value Per Acre

Land & Buildings Av. Value Per Acre

Land Mapped & Graded By

Date

Buildings Graded and Valued By

Date

Field Review By

Date

FARM SOIL PLAT

DATE | CHANGE OF OWNERSHIP

FROM BOOK PAGE
TO AMOUNT
FROM BOOK PAGE
TO AMOUNT
FROM BOOK PAGE
TO AMOUNT

	ACTUAL VALUE 19 53	ACTUAL VALUE 19	ACTUAL VALUE 19	ACTUAL VALUE 19	ACTUAL VALUE 19	ACTUAL VALUE 19	ACTUAL VALUE 19	ACTUAL VALUE 19	ACTUAL VALUE 19	ACTUAL VALUE 19
LAND	$14,770	$	$	$	$	$	$	$	$	$
BUILDINGS	$ 5,220	$	$	$	$	$	$	$	$	$
TOTAL	$19,990	$	$	$	$	$	$	$	$	$

FARM MAIL ADDRESS

OWNER OR MANAGER

TENANT

OWNER MAIL ADDRESS

FIG. 25.1. *Assessment card No. 1 used in the Polk County, Iowa, rural reassessment project, 1953. Soil boundaries are shown on an aerial map pasted on the card.*

BACK

BLDG NO.	OCCUPANCY	CONSTRUCTION	SIZE	GRADE	AGE	REMOD	COND	PHYS. DEPR.	FUNCT. DEPR.	REPL. VAL.	PHYS. VAL.	SOUND VAL.
1.	DWELL	1½ S & B FR	30x60x1.30	C+10	1908		F	50	20	3336	1668	1334
2.	BARN	1S & LOFT FR	36x36x2.56		1910	(1800)	G	40	20	2340	1404	1123
3.	DBL. CRIB	1S 8 BINS FR	12x24x1.80		1945	(1296)	G	18	20	3240	2657	2126
4.	HOG HSE	1S FR	10x30x1.90		1908	(288)	F	50	20	518	259	207
5.	CHIC HSE	1S FR	30x60x.80		1910	(300)	G	40	20	570	342	274
6.	CATTLE SHED	1S POLE FR	10x20x.70		1950	(1800)	G	8	20	144	132	106
7.	SHED	1S FR			1908	(200)	F	55	20	140	63	50
												5220

SKETCH AND COMMENTS   TOTAL

(6)   (3) ② ④ ⑦   1S ADD NO B
22   440   ⑫

DWELLING DESCRIPTION

FOUNDATION
- CONC. WALLS ✓
- CONC. BL. WALLS
- STONE WALLS
- PIERS

BSMT. AREA FULL ½ ¾

INTERIOR FINISH
- PINE DOORS ✓
- ENAMEL ✓
- HARDWOOD
- WALL BOARD

IMPROVEMENTS
- HOT AIR HEATING ✓
- PIPELESS FURNACE
- STEAM HEATING

WALLS
- SINGLE SIDING ✓
- LAP SDG. OR SHINGLES
- SOLID COMM. BRICK
- FACE BR. VENEER
- COM. BR. VENEER ✓

- HOT WATER HEATING
- CIRC. WARM AIR
- OIL BURNER
- COAL STOKER
- NO HEATING ✓
- BATHROOMS

ROOFING
- WOOD SHINGLE ✓
- ASPHALT SHINGLE
- SLATE
- METAL OR ROLL RFG.

- TOILET ROOMS 1
- ELEC. WATER SYSTEM ✓
- SEPTIC TANK ✓
- NO PLUMBING
- TILED BATH
- TILED TOILET

FLOORS
- DIRT FLOOR BSMT.
- CONC. FLOOR BSMT. ✓
- PINE FLOORS
- HARDWOOD FLOORS ✓

- ELEC. LIGHTING ✓
- FIREPLACE ✓

FIELD WORK BY   PRICED BY   REVIEWED BY

BUILDING COMPUTATION

440 s.f.	6.08	2675
80 s.f.	3.40	272
72 s.f.	1.20	86
s.f.		
s.f.		
s.f.		

BASE PRICE	6.70
BSMT.	-.21
ATTIC	+.14
FLOORS	-.14
HEATING SYST.	-.27
	6.08
AUTO. HTG. UNIT	
FIREPLACES	
PLUMBING	
TILING	
TOTAL	3033
COST FACTOR +10	303
REPLACEMENT VALUE	3336

FIG. 25.2. *Building information and valuation on back side of assessment card shown in Figure 25.1. Note space devoted to description, sketch, and cost valuation of dwelling.*

FARM VALUATION REPORT

N. E. QUARTER

MAP NO.	SLOPE	EROSION	NAME AND TEXTURE	ACRES	RATING
120	3	1		25.5	1.5
119	1	0		21.6	1.0
162	7	2		1.8	4.5
XX	XX	XX	TOTAL FOR QUARTER 48.90	39.00	1.3

N. W. QUARTER

MAP NO.	SLOPE	EROSION	NAME AND TEXTURE	ACRES	ACRES	RATING
120	3	1		33.00	22.00	1.5
93	11	2		21.00	2.8	7.5
11	3	0		3.60	1.2	3.0
120	11	1		24.00	6.0	4.0
120	7	1		18.60	6.2	3.0
8	3	0		1.60	.8	2.0
XX	XX	XX	TOTAL FOR QUARTER 101.80	39.0		2.6

S. W. QUARTER

MAP NO.	SLOPE	EROSION	NAME AND TEXTURE	ACRES	ACRES	RATING
120	11	1		21.60	30.40	4.0
11	3	0		3.60	1.20	3.0
120	3	1		12.60	8.40	1.5
XX	XX	XX	TOTAL FOR QUARTER 137.80	40.00		3.4

S. E. QUARTER

MAP NO.	SLOPE	EROSION	NAME AND TEXTURE	ACRES	ACRES	RATING
120	3	1		42.60	2.84	1.5
162	7	2		52.20	11.6	4.5
XX	XX	XX	TOTAL FOR QUARTER 94.80	40.00		2.4

ADJUSTMENTS

	N.E.Q.	N.W.Q.	S.W.Q.	S.E.Q.
1. NATURAL RESOURSES WORKING GRAVEL PIT COAL & ITEMS				
2. TOPOGRAPHY (A-GOOD, B-FAIR, C-POOR)				
3. EROSION (A-SLIGHT, B-MODERATE, C-SEVERE)				
4. DRAINAGE (A-GOOD, B-FAIR, C-POOR)				
5. LOCATION (A-GOOD, B-FAIR, C-POOR)	+3	+3	+3	+3
6. ROADS (A-PAVED, B-GRAVEL, C-DIRT)				
7. MARKET (A-GOOD, B-FAIR, C-POOR)				
8. WATER SUPPLY (A-GOOD, B-FAIR, C-POOR)				

ADJUSTMENTS

	N.E.Q.	N.W.Q.	S.W.Q.	S.E.Q.
9. NATURAL RESOURSES (A-PROVEN, B-PROBABLE, C-NONE)	-0	-3	-2	-1
10. PHYSICAL FEATURES				
11. CHURCH, SCHOOL, ETC. (A-GOOD, B-FAIR C-POOR)				
12. NUISANCES (A-SLIGHT, B-MODERATE, C-SEVERE)				
13. HAZARDS (A-SLIGHT, B-MODERATE, C-SEVERE)	-0	-1	-4	-1
14.				

FIG. 25.3. Soil productivity data showing calculation of corn suitability rating for each 40-acre tract. Adjustments for intangible and nonincome features are shown on bottom of card which is the front side of card No. 2 in rural reassessment project in Polk County, Iowa.

data for 40 acres in the upper right hand corner (NE¼ of the SE¼) was listed separately in the upper left hand part of Figure 25.3. This 40-acre tract contained three soils: the 119-1-0 area amounting to 21.6 acres with a corn suitability rating of 1; the 120-3-1 area amounting to 17 acres with a rating of 1.5; and the 162-7-2 amounting to .4 of an acre with a rating of 4.5. Each soil area was multiplied by the rating to give the figure just preceding "acres," i.e. 17 acres of "120" soil times 1.5 rating equaled 25.5. These weighted averages were then totaled and divided by total acres to give the average corn suitability rating for the tract. In this instance it was 48.9 divided by 39 acres which gave a rating of 1.3.

The corn suitability ratings for each tract were carried to the back of card 2, Figure 25.4, where base values were assigned. The values assigned came from a schedule in which value per acre had been worked out for corn ratings, using bench mark income appraisals in the determination.

A final step involved adjustments for intangibles, amenities, and nonincome features listed at the bottom of Figure 25.3. In this instance the only adjustments were those made for a favorable location, unfavorable physical features, and hazards. These adjustments were then applied to the base values in Figure 25.4 to give the final value of the land. The 100 percent value of the land was $14,770 which, added to the building value of $5,220, gave a total value for the 158-acre tract of $19,990, or $126.50 an acre. This was entered at the bottom of the first card, Figure 25.1. Since Iowa law at the time required assessment at 60 percent of value, the assessment total on this 158-acre tract was $11,994, or $76 an acre.

In developing the schedule of values according to corn suitability ratings, and in determining the dollar amount of adjustments, bench mark properties were used. One group of these bench mark properties was made up of appraisals on an income basis, the other group was made up of farms which had been sold recently in bona fide transactions. Income appraisals of land with a 1.5 rating came out around $275 an acre, while a 2.5 rating was around $220 an acre. Since the 100 percent assessment value desired by the authorities for land rating 1.5 was around $117 an acre, this was the figure established for 1.5 land, with $92 an acre for 2.5 land and so on. Sale values were estimated to be at 2.4 times the 100 percent level or 4 times the 60 percent level. Hence the $76 assessment value (60%) or $126.50 value (100%) of the sample farm in Figures 25.1 to 25.4 would have an estimated sale value of around $300 an acre. By checking income appraisals and actual sales of farms

BACK

CORN SUITABILITY RATING

CARD NO. 2 OF 2

TYPE FARMING	QUARTER	RATING	MINIMUM	MAXIMUM	BASE	ADJUST-MENTS	PER ACRE ACTUAL	ACRES	VALUE PER QUARTER	B OF R
LIVESTOCK DAIRY	N. E.	1.3			119	+3	122	39	$ 4758	
FIELD CROP POULTRY	N. W.	2.6			89	−1	88	39	$ 3432	
DIVERSIFIED	S. W.	3.4			70	−3	67	40	$ 2680	
	S. E.	2.4			94	+1	95	40	$ 3800	
NO. ACRES IN LEGAL NO. ACRES FARMED		TOTAL VALUE LAND						158	$ 14,770	

	YIELD TYPICAL	EVALUATION TABLE	
CROP SHARE			
CASH	CORN – BU. PER A.	1. CORN LAND TILLABLE ACRES	5. ORCHARD ACRES
LIVESTOCK	OATS – BU. PER A.		
OTHER	WHEAT – BU. PER A.	2. SMALL GRAIN TILLABLE ACRES	6. TIMBER & BRUSH ACRES
LANDLORD EXPENSES	SOYBEANS – BU. PER A.		
HIGH	HAY – TON PER A.	3. MEADOW TILLABLE ACRES	7. NON-AGRI-(WASTE) ACRES
MEDIUM			
LOW	PASTURE – RENT PER A.	4. PASTURE NONTILLABLE ACRES	8. BUILDING LOTS ACRES

FIG. 25.4. *Final values for each 40-acre tract on back side of card shown in Figure 25.3. Indicated values are based on corn ratings with adjustments added or subtracted.*

of different productivity and different intangible features, it was possible to obtain a range in assessed values in line with income and sale values. A farm which would produce only half as much as another farm, or would sell for half as much, was assessed for only half as much. If the income and sale value check did not appear reasonable—the sample farm value of $300 an acre appeared too high or low—then soil areas, productivity ratings, and adjustments were examined again to locate the errors.

Once the final assessments were determined, they were plotted on township maps and taken to township meetings to which all the farm owners in a township were invited. This provided the acid test of assessment uniformity, as the author who attended one of these meetings can testify, because each of the owners present spent the first part of the meeting studying the map, noting his own assessment and those of his neighbors, and spent the remainder of the meeting discussing the validity of the assessments. What made these meetings successful was the opportunity it gave the assessor to explain how he had arrived at the values, an explanation which almost always settled a potential complaint, especially from owners whose values were raised by the new assessment.

Assessments in other parts of the country and in other countries differ in detail from the Midwest example presented, but the principles that govern up-to-date assessments based on professional appraisals are much the same in all areas. In the Great Plains, where ranching and extensive wheat raising predominate, land classification has been found to be a successful base for tax assessment. The procedure followed is similar to that in Montana and western Canada as described in Chapter 24. In these areas the important objective is to get an overall evaluation of the land in terms of (1) its best use—such as wheat raising or grazing and (2) its productivity—such as bushels of wheat per acre or grazing acres required per animal unit. Once a good land classification has been made, the tax assessor has most of the information needed to place values on the land so as to satisfy the uniformity principle. Additional information needed includes location, schools, churches, roads, and the like, plus building costs and sales data.

Soil Conservation Service land-use maps provide the basis for assessments in certain areas. In one county top-grade tillable land was placed at $500 an acre by the assessors, second-grade land at $400, and lower grades proportionately. Land that was designated as first-class pasture on the land-use maps was placed at $100 an acre, while second-grade was listed at $60. Adjustments to these figures were made by the assessor to allow for distance to market. Buildings were treated separately.

In areas of California the Storie soil rating system, explained in Chapter 21, has been used in arriving at comparative income values for the assessment of farm tracts. In South Carolina sales have been used as bench marks to indicate the value of intangible and nonincome features. In Nebraska one of the most complete studies of tax assessment for farms was made, combining the use of soils and productivity data with economic ratings based on sales. Steps recommended by the authors of this study for a systematic assessment valuation were:

a. Make a soils map of the county.
b. Estimate the cropping system commonly used on each soil.
c. Estimate crop yields for the soil management system used on each soil.
d. Estimate a net income (economic) rating for each soil.
e. Measure the acreages of each soil on each tract.
f. Calculate a weighted average economic rating for each tract.
g. Estimate a first approximation of sale value for each tract on the basis of the economic rating. This first approximation would be the final estimate for land without buildings.
h. Adjust the first approximation for farm building value.
i. Adjust the first approximation for location of the farmstead.
j. Compute the final estimate of the sale value of tracts with buildings.[1]

On the basis of the experience in Saunders County, Nebraska, which was the basis of the study, the authors also recommended that a committee made up largely of local farmers be used at all stages of a reassessment project.

"RURBAN" OR RURAL-URBAN FRINGE AREA

The transition zone between city and country is increasingly becoming one of the primary headaches of the assessor, especially in areas where cities are expanding rapidly into the countryside. An acute problem faces the assessor in deciding when to raise assessments in rural-urban areas where values are rising. In some states, notably California, Florida, and Maryland, legislation has been enacted dealing with this problem.

The situation is illustrated by a chronological development in six stages by Martin D. Miller.[2] The six stages are shown in

1. Howard W. Ottoson, Andrew R. Aandahl, and L. Burbank Krisjanson, *Valuation of Farm Land for Tax Assessment,* Univ. of Nebr. Agr. Exp. Sta., Lincoln, Bull. 427, 1954, p. 4.
2. Martin D. Miller, "Appraisal of Urban-Rural Land for Assessment Purposes," *Assessment Administration.* Proceedings of the 24th International Conference on Assessment Administration, Cleveland, Ohio, 1958, pp. 6–10.

Figures 25.5 to 25.10. The area in the diagrams changes from farm values on an acre basis to site values for rural homes and businesses to front foot or square foot values as in urban areas. Along with the charts and explanation Miller suggests the following valuation approaches keyed to the six diagrams:

> *1939*—The area illustrated is rural, therefore rural land valuation approaches should be used.
>
> *1945*—While the area is still predominantly rural, a site value should be given to the highway corner of Farmer C. Farmer B's land value would not be increased since his son's house would be considered the same as a tenant's house.
>
> *1948*—Use site values for the highway-road intersection corner tracts which are developed. Use homesite values with corrections for the residential tracts developed from Farmer B's and Dentist E's farms.
>
> *1952*—Use site values for the highway-road intersection corner tracts which are developed. Suggest adding ½ of comparable corner site value to Farmer D's highway corner, even though not developed, as market for the corner tract is definitely established by other three developed corners. Continue use of homesite values for residential tracts originally a part of farms owned by Farmer B and Dentist E. Use unit front foot or preferably unit square foot land values for subdivision originally a part of Farm A. The square foot unit may be preferable to front foot because of subdivision layout.
>
> *1956*—Continue use of site values for road intersection corners. Suggest shift from unit homesite approach for tracts originally a part of Farms B and E to unit front foot with depth corrections. Suggest use of excess frontage correction to U.S. highway frontage owned by Farmer A and to state road frontage owned by Farmer B and Dentist E. Subdivisions should have unit front foot or unit square foot values applied.
>
> *1958*—The Village of Utopia is now incorporated but much of the land is still rurban in character. Continue use of approaches outlined for 1956 until development continues to point where urban land development predominates.

Any growing city or town provides examples of this expansion of the urban area into the farming area. Farm *A* presented at the end of Chapter 1, which increased from $100 an acre sale value in 1935 to a $730 an acre appraised value in 1968, is an example of this development. This type of expansion has also been noticeable along major highways which radiate from cities and towns with businesses of many kinds locating beyond the corporate limits where highway access is available. With the growing mobility of people this trend toward locating out in farming areas surrounding the urban centers is likely to continue.

Assessment policy in these rural-urban areas varies, but generally assessments are left on a farm productivity basis as

1939

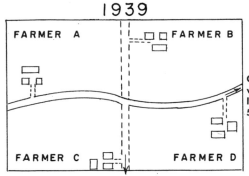

FARMER A

FARMER B

Center-
ville
limits
5 miles

FARMER C

FARMER D

Elementary School 6 Miles

FIG. 25.5—In the area known as "Four Corners," Farmers A, B, C, & D operate grain and hog farms. Their children go to a county elementary school located 6 miles to the south, and to high school in Centerville, located 5 miles east. Shopping is done in Centerville by all farmers. The north-south road is a state-maintained, graveled road. The east-west highway is designated to become a U.S. highway and has a blacktop surface.

1945

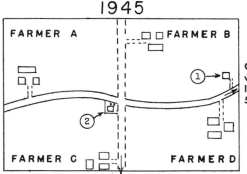

FARMER A

FARMER B

Center-
ville
limits
5 miles

FARMER C

FARMER D

Elementary School 6 Miles

FIG. 25.6—(1). Farmer B's son returns from military service and wants his own house for himself and his French war bride as a condition for remaining to work on farm. Farmer B agrees and builds son a home away from original farm dwelling.

(2). Farmer C sells ¼ acre on corner to an old friend, Joe, who builds a country store where he sells groceries, serves sandwiches, and has two gasoline pumps. Area shortly becomes known as "Joe's Corners."

1948

FARMER A

FARMER B

Center-
ville
limits
4 miles

DENTIST E
~~FARMER C~~

FARMER D

Elementary School 6 Miles

FIG. 25.7—(1). Farmer A sells ¼ acre on corner to oil company for service station.

(2). Farmer B's son moves to city as French bride (from Paris) can't stand farm. Farmer B sells son's house and ½ acre; gets good price for son's house and lot so he sells two other adjacent tracts fronting on the highway.

(3). East-west highway straightened and resurfaced.

(4). Farmer C bothered with arthritis, sells his farm to Centerville Dentist E who contracts with Farmer D to operate farm. Dentist needs cash to pay income tax so sells three tracts fronting on the highway.

428

FIG. 25.8—(1). North-south state road resurfaced with blacktop.

(2). Farmer A sells 5 acres to Centerville realtor for small subdivision.

(3). Gasoline station on corner of Farm A proves successful. Another oil company offers Farmer B a fabulous price for his corner; Farmer B sells.

(4). School Board tries to buy 5 acres from Farmer D for new consolidated school. Farmer D refuses to sell on grounds that "city folks are ruining our farms."

(5). Dentist E's wife needs new fur coat so he sells off three more tracts.

(6). Another oil company pays Joe large sum for his store. Joe retires to Florida.

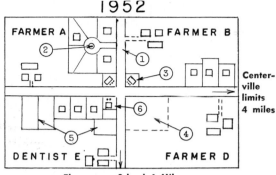

1952

Elementary School 6 Miles

FIG. 25.9—(1). East-west highway widened.

(2). Stop light erected at intersection.

(3). Farmer B sells remaining frontage on highway; buyer holds for development.

(4). School Board acquires 5 acres from Farmer D through condemnation proceedings. Area becomes known as "New School Corners."

(5). Farmer D disgusted so sells entire farm to developer. Developer subdivides farm and sells half of the lots during first year.

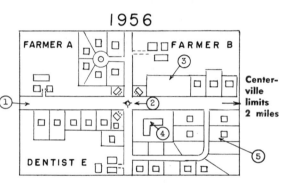

1956

FIG. 25.10—(1). School adds wing to take care of increased enrollment.

(2). Developer buys additional land from Farmer B to add to tract purchased 2 years ago. Developer constructs small shopping center and provides parking facilities.

(3). Centerville city council discusses annexation of growing area to the west. Group from New School PTA circulates petition stating evils of annexation and urges incorporation. After heated election "New School Corners" is finally incorporated as "Village of Utopia."

(Courtesy Martin D. Miller, see footnote bottom page 426.)

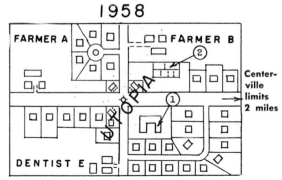

1958

429

long as the property is used solely for farming purposes. As soon as lots are sold for purposes other than farming, this establishes a new higher level of assessment for the tracts that are no longer used in farming.

AN ASSESSMENT IMPROVEMENT PROGRAM

Improvement in tax assessment comes from a concerted attack on the problem. Legislation, although helpful, cannot do the trick by itself. Five different points of attack are suggested for a tax assessment improvement program:

1. Improvements in administrative machinery
2. Improving internal before external equalization
3. Use of bench mark assessments
4. Use of assessment-sale ratios
5. Clinics or schools to provide assessors an opportunity to compare notes and improve their appraisals

Administrative machinery needed to provide the basis for good valuations includes a county in place of a township setup, appointed in place of elected assessors, advisory committees of citizens to work with the assessor, and specialist assistance from the state tax administration and the state college or university. The township, an acceptable unit for tax purposes in the horse-and-buggy days, is now too small a unit because it does not provide the basis for a full-time assessor. Elected assessors are too likely to be sensitive to the pressures of voters; in order to provide the necessary independence and also to provide the professional competence which goes with the task of valuation, an appointed official is desirable. The appointment can be made by a board representing the combined taxing bodies—schools, cities, and county government. Advisory committees of citizens enable the assessor to keep his value levels and his general administration in line with enlightened civic opinion. If the assessor allows himself to get too far from the taxpayers, he may lose their confidence even though he is doing a good job but a job they do not understand. Lastly, the state tax commission and the research and extension units of the state educational institutions have a responsibility to provide assessors with assistance in the form of information on value levels in use in other districts, and various other aids that they are not able to provide readily by themselves.

Local or internal equalization should come ahead of external equalization between taxing districts. That is to say, an intensive effort should be directed at getting individual property tracts assessed in proper relationship to each other in the local taxing district before much success can be had in adjusting the districts as a whole to bring them in line with each other. In a program of this kind the state tax commission and the state educational institutions should assist the local assessor and his fieldmen wherever possible to bring about this desired internal equalization. The next step is action by the state tax equalization body to bring the different counties, or whatever the units are, into line with each other by raising or lowering all assessments in the unit by a fixed percentage.

Bench mark or key assessments, as discussed earlier in this chapter, are a significant feature of an up-to-date program to improve assessments. In order to keep assessments in line throughout a tax district, some objective levels need to be established to which the different fieldmen as well as the county assessor can tie. If the taxing district is a large county, it would be desirable to have as many as 15 farms—including dairy, cattle-feeding, grain, and other types—and to spread them around the country so the fieldmen would not have far to go to check one of these bench mark units.

ASSESSMENT-SALE RATIOS

Assessment-sale ratio information can be of great help to the local assessor in achieving internal equalization, and to the state tax commission in external equalization. The ratio is calculated by dividing the assessment by the sale price of the property. The local assessor is especially concerned with the individual ratios. If these individual ratios cluster closely around the average, that is generally good evidence that the assessments are uniform as they should be; if the ratios show no tendency to concentrate about the average but vary through a wide range, this is evidence that improvement in the assessments is urgent.

Analysis of assessment-sale ratios gives an assessor a useful insight into the assessment situation in his district. In the first step in such an analysis, sales are tabulated and checked to exclude all but bona fide transactions. Second, assessments have to be matched against the sales to make sure the acreage and buildings are the same. Third, ratios are computed; and fourth, averages and coefficients of dispersion are figured to show the central tendency or average and the extent of variation around the average.

An example from two Illinois counties indicates the kind of analysis suggested and explained by the National Association

TABLE 25.1. Analysis of Assessment-Sale Ratios, Average and Dispersion from Rural and Urban Frequency Distribution of Ratios in Two Illinois Counties

| | Number of Ratios | | | |
| | County *A* | | County *B* | |
Assessment Ratio Groups	rural	urban	rural	urban
(*percent*)				
0 – 4	..	4	..	4
5 – 9	1	13	1	..
10 – 14	7	69	4	11
15 – 19	4	106	4	44
20 – 24	12	98	16	68
25 – 29	8	66	8	65
30 – 34	13	45	6	49
35 – 39	7	17	3	26
40 – 44	14	24	1	36
45 – 49	3	12	..	17
50 – 54	3	11	..	8
55 – 59	4	9	..	3
60 plus	21	31	..	8
Total	97	505	43	339
Median	38%	25%	24%	28%
1st Quartile	25%	17%	20%	22%
3rd Quartile	56%	33%	30%	38%
Coeff. Dispers.	41	35	21	28

Source: *Guide for Assessment-Sales Ratio Studies,* National Association of Tax Administrators, Chicago, Ill., 1954, p. 53.

of Tax Administrators.[3] Ratios for the rural and urban areas in two counties, based on bona fide sales which were checked and verified, are shown in Table 25.1. In this instance the median is used as the average because it gives equal weight to all ratios. The first and third quartiles are shown, these being the ratios which rank one-fourth from each end or one-fourth from the median. By relating the difference between the two quartiles to the median, as in the following formula, the coefficient of dispersion is obtained.

$$\text{Coefficient of dispersion} = \frac{\dfrac{\text{3rd quartile} - \text{1st quartile}}{2}}{\text{median}}$$

A low coefficient, such as in the rural area in County *B*, indicates assessment uniformity compared to the other areas shown. A high coefficient, such as in the rural area of County *A*, indicates a lack of uniformity, which can be readily seen in a

3. *Guide for Assessment-Sales Ratio Studies.* Report of the Committee on Sales Ratio Data of the National Association of Tax Administrators, Chicago, 1954, pp. 53–55.

glance at the distribution of the ratios. An objective of no dispersion (to have all ratios exactly the same) is not perfection, nor is it necessarily desirable, because it is well known that sales are not uniform. But a relatively low coefficient is desirable because it indicates that assessments are uniform. In a federal study about one-fifth of the areas studied had coefficients under 20 on nonfarm homes.[4]

Many assessors keep an up-to-date check on sale ratios as a guide. Annual comparisons provide additional insight to an assessor who is endeavoring to improve his assessments. Some of the assessors who have been computing ratios in their districts for years and making up annual averages say they would be lost without the information they obtain from this analysis.

Periodically, the federal government and utility companies make assessment ratio studies in which they provide county-by-county and state-by-state comparisons indicating the relationship between sales and assessments. Such a study by the federal government was conducted in the 1957 Census of Governments.[5] Average assessment ratios by states for all kinds of property were calculated from sales collected in sample counties. The average ratios by states for all property are shown in Figure 25.11. These ratios were obtained by totaling the sale amounts in all sales and dividing this figure into the total of all assessments for these properties.

Three aspects of the assessment-sale ratio warrant special consideration. In the first place, the principal advantage of these ratio studies is the information they give the local assessor which he can use in improving uniformity within his tax district. Consequently it is important that the assessor be made one of the central figures in any assessment-sale ratio studies. If a study is to be made, he should be brought in at the outset and have an opportunity to evaluate the results as they come in. An assessor is much more likely to use a study in which he has participated from the beginning than one which he knew nothing about until it was brought to him after it was finished.

In the second place, the sales which are used should be checked to determine those which are bona fide. In this connection the section in Chapter 5 on determining and verifying the sale price should be consulted. Many of the sales will be found to be unusable because of special conditions which were

---

4. *Taxable Property Values in the United States,* 1957 Census of Governments, U.S. Bureau of the Census, Washington, D.C., Vol. V, 1959, p. 85. A simple method of computing coefficient of dispersion by means of average deviation is illustrated on page 13 of this volume.
5. Ibid.; Lawrence A. Leonard, *Assessment of Farm Real Estate in United States,* USDA, ARS 43–117, 1960.

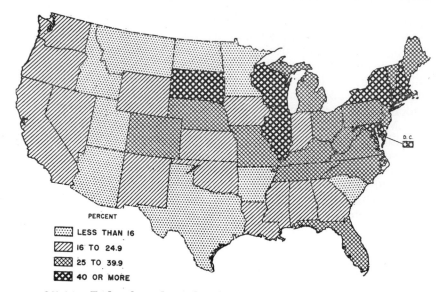

PERCENT
LESS THAN 16
16 TO 24.9
25 TO 39.9
40 OR MORE

FIG. 25.11. *Federal study of locally assessed real estate assessment-sale ratios by states during a 6-month period of 1956. (Courtesy Bureau of the Census.)*

part of the transaction. It is possible to eliminate most of the sales between relatives by omitting sales where the two parties have the same name, but it is not so easy to spot these sales where the names are different. Then there are the sales which involve other items besides real estate. In some instances it makes considerable difference in a farm sale during the summer months whether the growing crop is included in the sale. Then, too, during a rising land market, trouble can be experienced in getting accurate sales because the date of the deed frequently does not represent the date of sale. It may be a year or two before a contract to purchase eventuates into a deed.

In the third place, reliance should not be placed on individual sales. The assessment-sale ratio study usually is based on the assumption that sales are correct, and the correctness of the assessments can be determined by measuring the assessment against the sale. But there are numerous cases in the district with good assessments where the assessment is more nearly correct than the sale. For example at an auction a property sold for 25 percent above the market because of some personal competitive bidding. In this instance the ratio was low but the assessor did not worry about it because he correctly judged that the assessment was more accurate than the sale. Because of this variability in the accuracy of sales, a near-perfect frequency distribution of assessment-sale ratios should show some scatter on both sides of the average.

ASSESSMENT CLINICS

Clinics or schools where assessors have an opportunity to make assessments and compare their results with other assessors are an important part of any program for assessment improvement. At these meetings, bench mark properties can be appraised by those in attendance. Following the appraisal, the individual reports should be summarized and discussed. At one such clinic held at Iowa State University a total of 142 assessors turned in individual appraisals for three farm tracts, a set of farm buildings, four residential properties, and two commercial properties. One of the high points of the conference was the argument and discussion which developed after the reports were summarized and given to the group. In summarizing the information it is important to bring out the coefficient of variation in order to stress the amount of variability around the average. In some instances the variability may be slight while in other cases, where the assessors are not on sure ground, the variability may be wide. For example, at the Iowa clinic the coefficient of variability on the farm tracts was 14, 17, and 23 percent, while the coefficient on the two commercial properties was 28 and 30 percent—results which indicated lack of knowledge and training regarding commercial appraisals by many of the assessors. Of the three farm tracts, the assessors had little difficulty with the best one on which the variability was only 14 percent. On the third one, which was the poorest of the group, the assessors had some trouble in deciding just how poor it was, with 23 percent variability as a result.

In the conduct of these clinics and schools as well as in other educational endeavors, state tax commissions and state colleges and universities have an opportunity as well as a challenge. By working with assessors these agencies can be of great help in the improvement of real property valuation.

QUESTIONS

1. Explain the uniformity principle in tax assessments.
2. How are bench mark or key assessments made and used?
3. Show by example how soil productivity ratings can be converted into dollar assessments.
4. What else besides soil has to be included in an assessment?
5. Point out examples of rural-urban fringe areas where assessment

problems may exist. Discuss the situation from the farmer's point of view, from the public's point of view, and from the assessor's point of view.
6. Explain the difference between internal and external equilization of assessments.
7. Calculate several assessment-sale ratios using assumed figures. Obtain actual assessments for farms you have appraised, estimate sale prices, and calculate assessment-sale ratios based on these figures.

NOTE TO INSTRUCTOR

Plan a trip to local assessor's office to see forms used and to have the assessor explain, if he will, the methods he follows in making assessments. If data are available, compute assessment-sale ratios for several farms recently sold in the community.

REFERENCES

Appendix D, pages 511–19, containing excerpts on the following subjects from state assessment manuals:
    I. Rural Appraisal Procedure—Arkansas, Iowa, Kentucky
   II. Use of Aerial Photographs—Arkansas
  III. Appraisal of Farm Buildings—Iowa
  IV. Appraisal of Timber—California
   V. Market Values and Assessment-Sale Ratios—Missouri, Iowa
Appendix E, pages 521–22, containing information on publications issued by the International Association of Assessing Officers. Other references include those appearing as footnotes in this chapter.

26

CONDEMNATION
APPRAISALS

THE NATURE of the condemnation appraisal was discussed in Chapter 1. In this chapter legal concepts, principles, procedures, and examples are presented with emphasis on right-of-way takings which are particularly significant in view of interstate and other new highway developments. Other examples include power line and pipeline easements. A final consideration is the appearance of the appraiser in court as an expert witness to present and defend his appraisal.

LEGAL CONCEPTS

Condemnation or expropriation, as it is sometimes called, is the taking of land by government for public use through the exercise of the right of eminent domain. Strictly speaking, eminent domain means the superior right or power over property. This superior right is reserved in the United States and many other countries by government. In the United States the right of eminent domain is an inherent right of the state and federal governments against which specific safeguards have been established to protect the rights of property owners when governments exercise their rights of eminent domain by condemnation. The federal constitution contains the following provisions:

Fifth Amendment:
. . . nor (shall any person) be deprived of life, liberty, or property, without due process of law; nor shall private property

be taken for public use without just compensation.
Fourteenth Amendment:
          . . . nor shall any State deprive any person of life, liberty,
or property, without due process of law. . . .

The states have similar constitutional provisions and laws governing condemnation under the right of eminent domain. In Iowa, for example, the constitutional provisions are:

Article 1, sec. 9:
          . . . no person shall be deprived of life, liberty, or property, without due process of law.
Article 1, sec. 18:
          Private property shall not be taken for public use without just compensation first being made, or secured to be made to the owner thereof, as soon as the damages shall be assessed by a jury, who shall not take into consideration any advantages that may result to said owner on account of the improvement for which it is taken.

The words "just compensation" which appear in both the federal and Iowa constitutions and in the constitutions and laws of most states indicate the major function of the appraiser in this area of condemnation. His special task is to prepare for public bodies and private citizens appraisals which may be used in negotiation or in court. In either case, whether preparing an appraisal for a public body or a private citizen whose property is being taken, the appraiser is rendering a professional service in providing a valuation which represents "just compensation," a valuation which he stands ready to present and defend in court.

Since the critical test in most condemnation cases is the question of valuation—whether the amount offered is "just compensation"—it is essential that the condemnation appraisal be a well-reasoned statement which sets forth clearly and in detail the facts and judgments of the appraiser with respect to the property being condemned. If the appraiser eventually finds himself on the witness stand testifying under oath regarding his appraisal, he will no doubt find it to his advantage, especially during cross-examination, to have a thorough knowledge of every aspect of the property's value, and be able to give a logical and factual explanation of the process by which he arrived at his final valuation figure.

PRINCIPLES

Down through the years the courts have established principles which have been tested from every angle and found to be not perfect but the best guides to follow in determining "just compensation" in condemnation cases. There are two main principles or guides which concern the appraiser; the first is the kind of value accepted by the courts and the second is the process by which this value is reached. The first guiding principle is the use of market value as the basis for condemnation, and the second, which applies to partial takings, is that the total compensation is determined by the difference between the value before and the value after the taking.

MARKET VALUE

Market value has been chosen by the courts in preference to income value or cost value as the basis for compensating a person whose property is taken. This decision has been reached after many legal cases in which all kinds of appraisals have been considered. In general there have been two overriding reasons for the choice of market value by the courts. First, market value represents the yardstick which the owner who is losing his property must use in replacing the property taken. If a farmer sells his farm to the city for an airport, or a portion of it to a highway authority for a right-of-way, he is entitled to compensation that will enable him to replace this property; in short, sufficient compensation to make him "whole" again. In replacing the property condemned he will have to pay the going market price.

Second, market value has been a concept that could be tested by reference to actual sales or at least to a general knowledge of what properties like the one being taken would sell for if placed on the market or what it would be necessary to pay to buy a similar property. It is generally recognized that the real estate market is not a precise one, but it does exist and it is objective in the sense that there are cases of a willing buyer and a willing seller with approximately equal bargaining power making an agreement on the selling price of a given tract or farm. This market price provides the court's test of objective evidence on value.

Income value has been and is used in condemnation appraisals but it does not have the prestige or carry the weight of market value. In general, income data and income values are accepted and used by the courts as supporting evidence for market or sale value. The reason for this situation is clear on two counts. First, income does not represent all the aspects of property value. There are intangible and nonincome features, as brought out in Chapter 16, which may account for a sizable portion of a farm's value. Second, it is exceedingly difficult to pin down with objective tests the appraiser's estimates and choice of a capitalization rate. In sum, there is just too much leeway in figuring income value to satisfy the courts. This has come about partly because appraisers appearing for both sides in condemnation suits have come up with income values so far apart that at times it has been embarrassing to the appraisal profession.

Cost less depreciation as a valuation method has been recognized by the courts to some extent, but its use is limited. On farms there is little opportunity to use cost values except where buildings are included in a condemnation case. If a highway condemnation destroys a set of farm buildings the appraisal should include the replacement cost. If the buildings are old the owner will not be entitled to new ones but instead will be entitled to an amount which will compensate him in full for the depreciated value of the buildings taken.

BEFORE AND AFTER VALUE

Any condemnation that involves a partial taking of real estate introduces a special consideration, the damage to the remainder. If the highway authority purchases or condemns a whole tract there is no question or problem of damage, and compensation usually is limited to the full or fair market value of the farm. However, if one-half the farm is taken, compensation usually will involve more than one-half the market value of the whole farm because the remainder has been damaged by the part taken. In cases of this kind the guiding principle of the courts has been to figure the total compensation by valuing the property before and after the taking.

Another procedure which provides more detail is to value the land taken separately and the damages separately and add them together to obtain the total compensation. This method works well where a small percentage of the whole is taken, as for example a small strip for highway along the side of a farm. Also the appraiser can take the difference between the before and after valuations and allocate this sum into an amount for the land and for each of the individual items of damage that lower the value of the remaining farm.

HIGHWAY TAKING

Condemnation appraisals of various types have come into general use in the purchase by highway authorities of land for the interstate program. In recent years the cost of acquiring right-of-way has increased, which makes the appraisal part of the program even more significant than at first anticipated.

A major accomplishment in highway takings has been the negotiated purchase by highway authorities, thus avoiding the time and expense of condemnation proceedings in the courts. This has not been possible in some states because of the wording of state laws; but where direct purchases have been possible both parties, the highway authorities and the property owners, have found the practice a satisfactory one. To make this plan work satisfactorily, however, the appraisals prepared for the highway authority and for the property owners (where the owners have obtained them) must come close to "hitting the nail on the head"; the property owner is not going to accept an offer unless it is in line with what special court commissioners or a condemnation jury would award as "just compensation." If the offer is not acceptable the owner can turn it down and go to court. But the highway authority and public are also protected to some extent by the courts, because a property owner who turns down an offer and goes to court may not necessarily improve his position. One owner turned down an offer, went to court, and received a jury award which was $4,000 below what he could have received without going to court. In the area where this happened several condemnation cases awaiting court trial were quickly canceled and the original offers accepted. On the other hand, there are cases where owners have had professional appraisers prepare valuation reports which have been used as a basis for negotiating out-of-court settlements higher than the original offers. In Iowa over 90 percent of the interstate right-of-way acquisitions have been made by negotiated purchases based on prepared appraisals. The important point in this regard is that the courts, although not always consistent, are available as a last resort to protect the rights of the property owner in case he feels that the offer is not equal to "just compensation" for the taking.

HIGHWAY TAKING EXAMPLE

One of the most difficult condemnation appraisals is a partial taking. The problem is not so much the value of the whole farm or tract before the taking as it is the value of the remainder after the taking. This is illustrated by an actual case from southern Minnesota, the outline of which is shown in Figures 26.1 and 26.2. As these maps indicate, the projected highway passed through a 102-acre farm identified as Parcel No. 11. The highway cut the farm into two parts, taking 18 acres and leaving 25 acres with the farmstead on one side and 58 acres on the other side. Access to the other side from the farmstead was provided by an overpass and a frontage road.

A complete before and after condemnation appraisal of Parcel 11 made by Gerald F. Thorkelson appeared in the *Journal of the American Society of Farm Managers and Rural Apprais-*

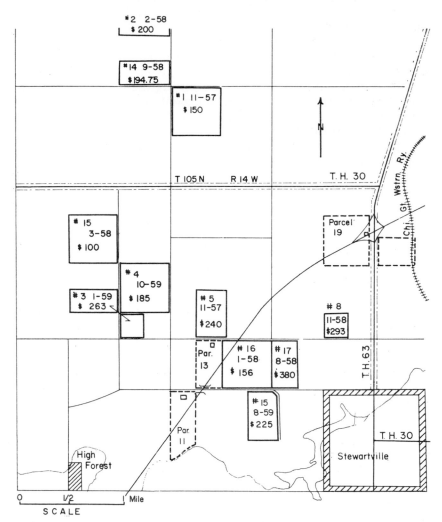

PART OF HIGH FOREST TOWNSHIP

FIG. 26.1. *Map showing projected interstate highway in southern Minnesota cutting across parcels 11, 13, 16, and 19. Parcel 11 is the 102-acre farm singled out for a special condemnation appraisal in this chapter. Sales of tracts used in establishing market value estimates are outlined with number, date (month and year), and price per acre. (Courtesy Minn. Dept. of Highways, Right of Way Section, E. R. Lorens, Chief Appraiser.)*

PARCEL II

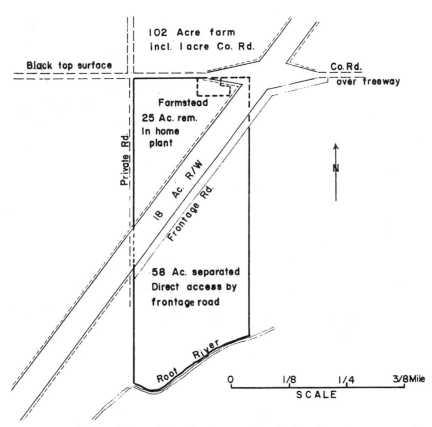

FIG. 26.2. *Map of Parcel 11 (102-acre farm) showing the area to be taken for projected interstate highway. (Courtesy Minn. Dept. of Highways, Right of Way Section, E. R. Lorens, Chief Appraiser.)*

ers.[1] The major conclusions are set forth in Table 26.1, in which figures have been rounded.

The evidence supporting the before value of $33,000 for the farm (Parcel 11) is much more convincing than that for the after value of $16,500. This situation illustrates the difficult problem an appraiser faces in reaching a closely reasoned estimate of value after the taking. In an article commenting on this

1. Gerald F. Thorkelson, "Rural Appraisals to Meet User Needs," *J. Amer. Soc. of Farm Mgrs. and Rural Appraisers*, Apr. 1961, pp. 63–76.

TABLE 26.1. Comparison of Before and After Valuations in Highway Taking Example

	Before Taking	After Taking
I. Income value estimate on landlord basis		
Gross income	$ 2,280	$ 1,592
Expenses	1,295	1,144
Net income to landlord	985	448
Capitalized value @ 3½%	$28,000	$13,000
Adjustments, plus	4,000	3,000
Market value based on income and adjustments	$32,000	$16,000
II. Cost value of buildings less depreciation and obsolescence		
All buildings	$15,000	$ 8,250
III. Sale value estimate		
Nine comparable sales analyzed, ranging from $156 to $412 an acre		
Estimate	$33,000	$16,500

Allocation of before and after difference of			$16,500
Land taken, 18 acres		$4,200	
Damages to remainder			
Reduction in building value	$6,750		
Separation of 58 acres	3,000		
Triangulation	2,000		
For fence	550	12,300	$16,500

Source: G. F. Thorkelson, "Rural Appraisals to Meet User Needs," *J. Amer. Soc. of Farm Mgrs. and Rural Appraisers,* Apr. 1961, pp. 70–76.

Thorkelson appraisal I brought out what I considered the strong and weak points.[2] The strong points were mainly in the "before" appraisal and the weak points in the "after" appraisal. In the after appraisal, gross income appeared low and expenses high, with a lower net income than seemed reasonable. In the allocation of damages farm size reduction could be included as a major factor accounting for part of the damage attributed to building loss.

The damage estimate of $12,300 is large compared to the value of the land taken. It is in this area of damages and their allocation that the appraiser needs more information; this is an area in which more research would be helpful. In reality the appraiser is operating in this damage area with poor tools and few objective tests.

It would be helpful, for example, if information and analysis were available on sales of cut-up and separate tracts after a taking. This would provide objective tests for part of the problem but not for the whole problem. The part not covered in-

---

2. "Comments by William G. Murray on Gerald F. Thorkelson's Appraisal," ibid., pp. 77–78.

cludes the damage represented by the reorganization which is often forced on the farm owner whose farm is split by a highway. For instance he may or may not be able to buy land to replace that which was taken, or he may have to change from dairying to some other major enterprise. These are intangibles difficult to evaluate; with some farmers about to retire and sell out they might present no problem at all, but with others (like the owner of Parcel 11) the taking could mean a complete readjustment of the farming business and a permanent curtailment of income.

HIGHWAY STUDY

A long step towards useful "after" information was made in a study by the University of Minnesota for the Minnesota Highway Department, a study which had the cooperation of the U.S. Department of Commerce and the Bureau of Public Roads.[3] The research covered an 8-mile stretch of interstate highway in southeastern Minnesota, 60 miles south of Minneapolis and St. Paul. The new highway passed through 28 farms taking 370 acres. Maps in Figures 26.3 and 26.4 show the situation in 1955 before construction of the highway was started in 1956, and again in 1959 after the highway had been completed in 1958.

This is what happened to the 28 farms:

10 farms: no change
 8 farms: farms sold and combined with other land in new
          shapes
 3 farms: enlarged by renting more land
 2 farms: combined into one unit
 2 farms: sold severed tracts
 2 farms: enlarged by buying severed tracts
 1 farm: reduced rented land

A first glance at the two maps (Figs. 26.3 and 26.4) would indicate that there were more farms in 1959 after completion of the highway than in 1955 before the project was started. This apparent increase is caused by the increased number of small tracts resulting from the new highway. A number of these small tracts were sold or rented to farmers some distance away which makes it difficult to compare directly the operating units in the before and after maps. Actually there were fewer operating units in 1959 than in 1955. According to the report:

---

3. Walter Gensurowsky and Everett G. Smith, Jr., *How Farmers Adjusted to an Interstate Highway in Minnesota*, Depts. of Agr. Econ. and Geog., Univ. of Minn., Sept. 1960.

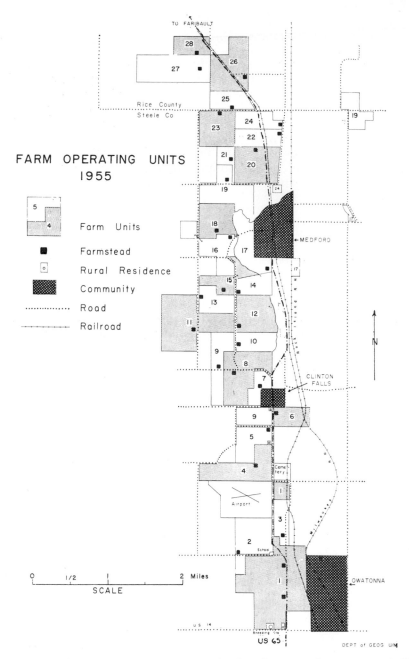

FIG. 26.3. *Farm operating units and other location features in 1955 before interstate highway was started in 1956. Difference between shaded and unshaded farm units used only for contrast.* (*Source:* How Farmers Adjusted to an Interstate Highway in Minnesota, *Depts. of Agr. Econ. and Geog., Univ. of Minn., Sept. 1960, p. 3.*)

446

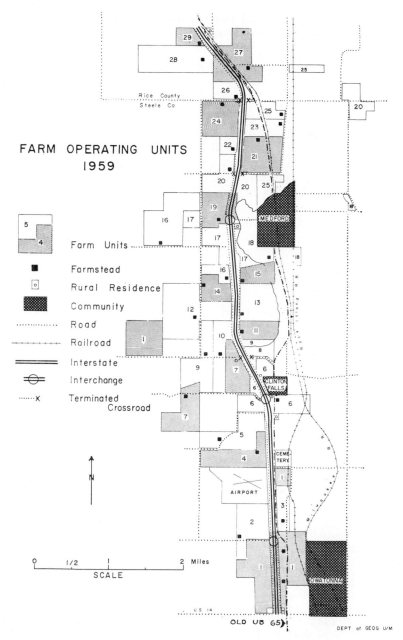

FARM OPERATING UNITS
1959

5
4 ——— Farm Units ....
■ ——— Farmstead
[o] ——— Rural Residence
——— Community
············· ——— Road
——————— ——— Railroad
══════════ ——— Interstate
⊖ ——— Interchange
·····x ——— Terminated
            Crossroad

0   1/2   1   2 Miles
         SCALE

DEPT of GEOG U/M

OLD US 65

FIG. 26.4. *Same area as in Figure 26.3 except date is 1959 after completion of the highway in 1958. The new highway trimmed acreage off 13 farms and split 15 farms. (Source: same as Fig. 26.3.)*

447

Between 1955 and 1959 in the study area, five of 28 farmers sold out and discontinued farming. Thus there was an 18 percent decrease in the number of farmers in the study area during this five-year period. Despite the loss of acres to the right-of-way, combination and consolidation of farms resulted in an increase in average farm size during the study period. A major determinant of this increase was the fact that five farmers left agriculture for retirement or other occupations.[4]

The awards or acquisition cost phase of the study indicated the same situation shown with Parcel 11—a heavy payment for damages compared to the payment for land. The market value of farmland and buildings in the area was estimated at $150 an acre. E. R. Lorens of the Minnesota Highway Department estimated sale values of the 28 farms at $250 an acre.[5] Market value for the 28 farms at $250 an acre would have given a total of $92,500 for the 370 acres taken. But payments for the 370 acres, including buildings on five of the farms, amounted to $317,000, or an average of $856 an acre. Thus the taking could be summarized as follows:

Actual payment, total	$317,000
per acre	856
Sale value estimate, per acre	250
Damages paid, per acre	606

Following is an interesting breakdown of the average of $856 an acre that was paid:

13 farms, acreage trimmed, award average ..	$ 442 an acre
10 farms, split but no buildings, award average	779 an acre
5 farms, split plus buildings, award average	1,409 an acre

Three conclusions from the study provide the basis for further study of the damage problem which is the most significant issue in the condemnation area. These conclusions are:[6]

Buildings:

Those five farms from which the State acquired buildings and land comprised 18 per cent of the farms in the study area in 1955, yet they received 33 per cent of the total award money. The average per acre value when right-of-way acquisition included both land and building complex was twice the value when land only was taken. It is apparent that if right-of-way

4. Ibid., p. 2.
5. Edward R. Lorens, "Requirements of Highway Department Appraisals," *J. Amer. Soc. of Farm Mgrs. and Rural Appraisers*, Apr. 1961, p. 61.
6. Gensurowsky and Smith, pp. 38, 44.

costs are to be minimized, highway planners should make every effort to avoid building complexes.

Adjustments:

. . . farm operators in the study area adjusted to any circuitous travel to severed parcels, usually by changing the land use or by selling the separated tract. They adjusted to the possibility of a permanent reduction in farm size by purchasing or renting additional land or by discontinuing farming.

Over-all effect:

The interviews pointed up areas of disagreement and confusion among occupants of the land. Whether they were fronting or astride the path of new highway construction, they held differing views of the extent of benefits and damages that result from restricted-access roads. Changes in the existing order help some people while disrupting the lives of others.

The last sentence above, pointing out that changes help some and hurt others, sums up the toughness of the damage problem in condemnation appraisals. The appraiser's task is to come as close as he can to a reasonable answer, keeping in mind that what is reasonable is determined by court cases but that these court cases themselves are sometimes erratic and also are subject to change as more information becomes available.

PIPELINES AND POWER LINES

When compared with highway takings, appraisals for pipeline and power line condemnations have some differences but many similarities. They are different particularly in that they are easements; an easement being a right to a certain specified use of real estate but only that use. An easement is not a sale; the owner who gives an easement retains title and full use except the use he has given up in the easement. A power line easement to a utility company gives the company the right to construct and maintain a power transmission line across the farmer's land, but the farmer still owns the land subject to the rights of the utility company. After the easement is granted the farmer can use the land as he sees fit as long as it does not conflict with the easement right he has granted the power company. The same situation holds for an easement granted a company to lay a pipline in the ground across a farm. Subject to the rights given the pipeline company, the farmer can use the land as he chooses.

In agreeing to an easement the owner should consider all the unfavorable developments such an easement might cause, some of which can only be estimated. Future use of land under a power transmission line should be considered as well as present use, with limitations on crops grown and disadvantages in use of farm machinery caused by poles or towers. And full consideration should be given to the attitude of buyers towards farms with and farms without these power lines. Pipelines may have an adverse effect on the land not only in the installation when the land is compacted by heavy machinery but also by the disruption of tile drainage. Also what reduction now or in the future is likely in the sale value of a farm because it has a pipeline across it?

On the other hand, the valuation of pipeline and power line easements are similar in many respects to highway takings. The main objective is to find the value of the farm before and after the easement, just as the before and after value for the highway taking. As with the highway, the difficult problem of the appraiser is making the valuation after the easement. The damages are so intangible it is difficult to pin them down to a money value. The appraiser does have a before value as a bench mark. He also usually has the advantages of comparable sales and of examples from farm owners and utility companies of other pipeline or power line easements from which he can obtain evidence that has a bearing on the appraisal being made.

MORE LOGIC NEEDED

More convincing evidence and more logic are needed in appraisal.[7] Appraisal systems and formulas are attractive but they do not take the place of thinking one's way through a valuation or of explaining the assumptions behind the reasoning. In too many appraisal reports the final values may stand the test of the market but the supporting evidence does not stand the rigorous test of logic. In brief the appraiser may have the right answer but is short on an explanation of how he obtained it. Unfortunately there also are cases where there is neither the right answer nor good logic.

This unsatisfactory situation has been brought into the open by appraisals presented in condemnation cases. Two examples will be used to show both the unfavorable and favorable sides. The first was a condemnation in which 14 acres were being taken out of a 126-acre farm, leaving the 112-acre farm in two separate pieces. At the trial, five appraisals were presented as follows:[8]

---

7. See William G. Murray, "Challenge Facing Real Estate Appraisers Today," *J. Amer. Soc. of Farm Mgrs. and Rural Appraisers*, Apr. 1961, pp. 55–58.

8. Appraisal Review Committee Reports, American Institute of Real Estate Appraisers, ARC No. 55–6, 1956.

Appraiser	Value of Whole Farm before Taking	Value of Remainder after Taking	Value of Taking and Severance Damages
A	$270,600	$169,600	$101,000
B	289,200	182,700	106,500
C	110,000	87,500	22,500
D	105,000	80,000	25,000
E	100,000	76,500	23,500
Award			$ 40,000

It is not difficult to identify from these figures the two appraisers presented by the owner and the three presented by the condemning authority. The width of the gap between them is cause for concern. Expressed in per acre value, the gap is from $7,600 to $1,600. The award of $2,850 per acre indicates the damage was severe. This represents approximately twice the value set by the lowest appraiser and two-fifths of the value of the highest. If appraisers claim professional standing, give accredited titles based on rigorous examinations, and subscribe to a code of ethics, what is the public likely to think when a list of appraisals like the five above are presented for the same 14 acres of farmland? They could likely assume that both sets of appraisers were more influenced by prejudice than logic.

The attack on a wide range of appraisals for the same property should start with the reasoning. Reasoning in this instance includes the logic which ties the appraisal together and makes the final valuation an understandable conclusion based on what precedes it. It also is involved in the selection of the facts and evidence appropriate for the valuation. This approach if followed consistently will narrow the gap and enhance the professional standing of appraisers.

The second example, a type which is fortunately becoming more common, represents a narrow spread between appraisals for opposing sides in a court condemnation. This case, which has been reported in detail, covered a 307-acre farm in Illinois through which a new highway was projected, taking 20 acres and dividing the remainder into two separate tracts. The seven appraisals presented in 1958 follow:[9]

Appraiser	Value of Whole Farm before Taking	Value of Remainder after Taking	Value of Taking and Severance Damages
1	$184,320	$147,700	$ 36,620
2	202,320	167,054	35,266
3	192,000	172,250	19,750
4	200,000	167,550	32,430
5	192,999	160,000	32,000
6	213,100	166,540	46,560
7	210,693	161,483	49,210
Jury Award			$ 39,815

9. Walter R. Kuehnle, *Partial Taking Midwest Farm*, American Institute of Real Estate Appraisers, Chicago, Ill., Case Study No. 5, Schedule 15.

In this group it is not nearly as easy to pick the three appraisers presented by the owner and the four presented by the condemning authority. Although not at all obvious, appraisers 5, 6, and 7 made appraisals for the owner. One unfavorable item, though a small one, was the false accuracy indicated in the appraisal before the taking by No. 7 whose figure was $210,693; this should have been rounded off. But this was of slight importance compared to the favorable narrow spread in the appraisals. Finally, in considering this narrow spread as favorable, one has to assume that the jury award was fair and that a competent appraiser, independent of the owner or of the condemning authority, would come up with a figure in the $35,000–45,000 range.

THE APPRAISER IN COURT

Some suggestions on court procedure will be presented for appraisers who have never presented a valuation before a judge or jury, or have never been cross-examined, but who may have this experience sooner or later. There are very few appraisers who did not feel the least bit nervous presenting their first appraisal in court. Most appraisers recognize the need for thorough preparation for a presentation in court and also recognize the ease with which damaging errors can be made in the courtroom, especially in cross-examination.

Among the suggestions which rate high on the list for appraisers in court are the following:

1. Stick to the facts and to opinions and judgments based on these facts.
2. Understand and be able to explain clearly the reasoning which connects the appraisal evidence with the final valuation.
3. Know the property "inside and out."
4. Show proper respect to questions; others in court may have a hazy idea of appraisal procedures.
5. Keep answers short and to the point; parading your knowledge only burdens the record and may provide the "loophole" the opposition has been hoping would appear.
6. If you make a mistake or do not know the answer to a question, admit the error and say No to the question, and get on to the next point.

This all adds up to professional competence expected of an expert appraiser, good manners, and ordinary common sense. It may appear elementary but the appraiser who fulfills these requirements need not worry even if it is his first court appraisal.

On cross-examination he may earn the grudging respect of opposing counsel in a manner which was especially well phrased by Judge Hale of Australia as follows:

> . . . if you really know your case no lawyer has a hope of shaking you. An advocate who finds himself dealing with a professional witness who really knows what he is talking about will be well advised to sit down before he finds himself in water too deep for his own comfort. The more he cross-examines such a valuer the more he is likely to demonstrate that the witness is an expert in fact and not merely in name.[10]

QUESTIONS

1. Obtain if possible a condemnation appraisal of a farm made in connection with a highway taking, airport project, a pipeline or power line easement, or some similar case. Comment on this appraisal as it relates to the value approaches of market, income, and cost and to the difficult problem of measuring damages.
2. Interview several farmers who have had land taken from their farms for highways or other purposes, or have signed easements for pipelines or power lines. How have these takings or easements affected farming operations and the attractiveness of the farms to potential buyers?
3. From a local court record obtain if available the details of a condemnation trial involving farmland. Discuss the issues raised in relation to the subject matter of this chapter.

REFERENCES

Consult current issues of the *Appraisal Journal* and the *Journal of the American Society of Farm Managers and Rural Appraisers.* Other references include those appearing as footnotes in this chapter.

Encyclopedia of Real Estate Appraising, Edith J. Friedman, ed., Prentice-Hall, 1959.

Schmutz, George L., *Condemnation Appraisal Handbook,* revised and enlarged by Edwin M. Rams, Prentice-Hall, Englewood Cliffs, N.J., 1963.

10. Justice John Hale, "The Valuer as a Witness," *Appraisal J.* Apr. 1961, p. 243.

27

LOAN APPRAISALS

ONE OF THE BEST WAYS to study the loan appraisal is to compare some of the forms used in the field. Before we consider a series of these forms, however, it is appropriate to refer to the section in Chapter 1 dealing with loan appraisal. In that section emphasis was placed on the long-range view needed in making a good loan appraisal. This long view, it will be recalled, was divided into three parts: a long-range forecast of farm productivity, an analysis of risks from price declines and natural catastrophes, and an estimate of community morale. In this chapter we will see how these long-range views, fortified by specific details and discussion in the chapters which followed Chapter 1, find expression in actual appraisal forms used by lending agencies.

LOAN APPRAISAL EXAMPLES

There are many different types of loan appraisals. In this chapter six have been chosen which illustrate a wide range in type and method. They come from three different lenders: the Farmers Home Administration, an insurance company, and four federal land banks.

APPRAISAL BY THE FARMERS HOME ADMINISTRATION

The FHA farm appraisal shown in Figure 27.1 gives three different values—comparable sale value in Part 2, income value in Part 5, and summation value in Part 6.

Form FHA 422–1
(Rev. 1–17–68)

UNITED STATES DEPARTMENT OF AGRICULTURE
FARMERS HOME ADMINISTRATION

FORM APPROVED.
BUDGET BUREAU NO. 40—R1064.10

APPRAISAL REPORT
(FARM TRACT)

APPLICANT'S NAME ..... John R. Jones

STATE ..... Iowa

ADDRESS ..... Fayette, Iowa

COUNTY ..... Fayette

PART 1. GENERAL INFORMATION

A. LOCATION OF FARM:

HOW MANY MILES FROM THE NEAREST TOWN? 3

IN WHAT DIRECTION? ..... NW

NAME OF TOWN ..... Podunk

HOW MANY MILES FROM THE NEAREST SCHOOL? 3

HOW MANY MILES FROM THE NEAREST CHURCH? 3

B. ROADS AND COMMUNITY:

IN WHAT CONDITION ARE ROADS IN COMMUNITY? ..... Good

ON WHAT KIND OF ROAD IS FARM? ..... Gravel
(DIRT, GRADED DIRT, GRAVEL, HARD-SURFACE)

NUMBER OF MILES TO NEAREST GRAVEL OR HARD-SURFACE ROAD ..... --

HAS APPLICANT A RIGHT-OF-WAY TO FARM? ..... yes

THE FARMING DISTRICT IS (E, G, F, OR P) ..... Good

WHAT TYPE OF FARMING IS IN COMMUNITY? ..... General

C. FARM SERVICES: (CHECK)

[X] R. F. D.   [X] PHONE   [X] POWER LINE   [X] MILK ROUTE   [X] SCHOOL BUS

INDICATE AVAILABILITY OF ELEVATORS, GINS, MARKETING FACILITIES: .....
3 miles to Podunk

D. COMMUNITY FACTORS:

THIS FARM AS COMPARED WITH THE AVERAGE COMMUNITY
WITH RESPECT TO LOCATION IS [X] AVERAGE [ ] BETTER [ ] POORER

" DESIRABILITY IS [X]   [ ]   "   [ ]   "

" SALABILITY IS [X]   [ ]   "   [ ]   "

" RENTABILITY IS [X]   [ ]   "   [ ]   "

E. CONDITION OF LAND:

HOW IS THE APPARENT GENERAL FERTILITY? (E, G, F, OR P) ..... good

NOXIOUS WEEDS (KIND AND PERCENT OF INFESTATION) .....
None to effect crop yields. A few Canada
thistles in pasture

CONDITION AND SUITABILITY OF FENCES ..... Fair - need 80 rds.
pasture fence and outside fence repair.

ACRES: TERRACED ..... none ..... ; LEVELED ..... none

CONDITION OF TERRACES ..... none

ACRES NEEDING: TERRACING ..... 40 ..... ; LEVELING ..... none ..... ; CONTOURING 205

DRAINAGE: (ACRES) TILE ..... ; OPEN DITCH .....

NATURAL ..... all ..... ; NEEDING DRAINAGE .....

NAME OF DRAINAGE OR RECLAMATION DISTRICT .....
..... none

IRRIGATION: ACRES IRRIGATED ..... -- ..... ; NAME OF IRRIGATION DISTRICT
--

(IF FARM IS TO BE IRRIGATED OR IS LOCATED IN DRAINAGE, OR LEVEE DISTRICT, ATTACH SPECIAL REPORT, FHA 422-2.)

F. NATURAL RESOURCES: (DISCUSS MINERALS, TIMBER, OIL, GAS, GRAVEL, ETC.)
..... none known

(IF MINERAL, TIMBER, OR OTHER RIGHTS ARE LEASED, RESERVED OR EXCEPTED, ATTACH SPECIAL REPORT, FHA 422-2.)

G. DOMESTIC WATER SUPPLY:

FARMSTEAD

SOURCE	well	ADEQUACY	good
DEPTH OF WELL	170'	DEPENDABILITY	good
CONVENIENCE	good	QUALITY	good

PASTURE

| SOURCE | well | DEPENDABILITY | good |
| ADEQUACY | good | DISTRIBUTION | fair |

DISCUSS ANY SPECIAL PROBLEMS: well at bldgs. supplies
water to stock in pasture. A pond in
pasture would be valuable development.

H. FARMING HAZARDS: (CHECK IF PRESENT)

[ ] OVERFLOW [ ] HAIL [ ] UNTIMELY FREEZES [ ] CROP DISEASES

[ ] HARDPAN [ ] ALKALI [ ] WIND EROSION [ ] DROUGHT

[X] WATER EROSION [ ] INSECT PESTS [ ] GRAVEL (DROUGHT) SUBSOIL

OTHER (SPECIFY) .....

DISCUSS HAZARDS CHECKED; INDICATE EXTENT, FREQUENCY: water erosion
can be controlled by contouring & limited
terracing.

I. OWNERSHIP HISTORY:
(WHEN MORE THAN ONE TRACT IS INVOLVED, LIST SEPARATELY)

YEAR ACQ'D BY PRESENT OWNER	57	58	63
ACRES ACQUIRED	80	51	127
HOW ACQUIRED	contract	contract	W. Deed
PURCHASE PRICE	$18,800	$3,000	$22,000
OWNERSHIP CHANGES DURING PAST TEN YEARS	0	0	1
EXPENDITURES FOR IMPROVEMENTS SINCE PURCHASE	$8,000	$ -	$ -

COMMENTS: All were open market sales with no
relationship between buyer and seller. New
house built on 80 in 1957.

J. RENTAL TERMS:

ON WHAT BASIS IS ANNUAL RENT? (SHARE, CASH, STANDING) ..... Share & cash

AMOUNT OF ANNUAL RENT (INDICATE AMOUNT OF CASH, SHARES, AND/OR QUANTITY OF PRODUCT) ..... ½ grain, $15/A meadow $6/A perm.past.

ESTIMATED CASH RENT AT CURRENT PRICES, $ ..... --

K. TAXES:

ASSESSED VALUE, $ 13,458

ASSESSED ON WHAT PERCENT OF VALUE ..... 27 ..... %

REAL ESTATE TAXES FOR 19 68 $ 1316.56

MILL LEVY 97.828

SPECIAL IMPROVEMENT DISTRICT TAXES, $ ..... none

DELINQUENCIES, IF ANY ..... none

FIG. 27.1. *Farmers Home Administration appraisal of an Iowa farm.*
(Courtesy Farmers Home Administration.)

456

PART 2. SALES DATA FOR COMPARABLE PROPERTIES

DATE OF SALE	IDENTIFICATION OF PROPERTY	PRICE		ACRES			COMPARED WITH PROPERTY BEING APPRAISED*		
		TOTAL	PER ACRE	TOTAL	CROP	LON	P	E	B
(1)	(2)	(3)	(4)	(5)	(6)	(7)	P	E	B
1968	A. Davis _Sec. 3, T92-R7	$42900	$ 275	156	148	256			X
1968	B. Smith _Sec. 6, T92-R7	56200	190	296	210	176	X		
1967	C. Jones _Sec.10, T92-R7	55200	230	240	210	225		X	
	D.								
	E.								

* P= POORER; E= EQUAL; B= BETTER

Compare above properties with property being appraised and describe in detail any differences.

A - This farm is located two miles south of subject farm. It is a smaller farm with a higher percentage of cropland most of which is Kenyon - Floyd soil types. There is no sand. Buildings are considered equal.

B - The Smith farm is located 3 miles west. This farm is predominately Fayette soil type with rather large scattered areas of Dickenson. Pasture land is very rough with areas heavily timbered. Fences are very poor. Buildings are similar except dwelling is older and less desirable.

C - This farm is one mile east and is very similar type land and about same percentage of crop land with possibly less of Dickenson soil type. Dwelling is much older but recently completely remodeled. Barn is poor and is in need of major repair. Location on a paved road is an advantage to this farm. Other service buildings are old, poorly maintained and will require major repair and maintenance within next several years. Although location on a paved road is an advantage to this farm, the age and condition of building offsets this advantage and the two are considered equal.

PART 3. LAND AND BUILDINGS

Comments on adequacy and appropriateness of buildings taking into consideration plans for building construction or repair as shown on Form FHA 424-1:

Buildings are modest but adequate for the general livestock program typical to the area. Will need to repair approximately 80 rods of pasture fence.

Comments on condition of land taking into consideration plans for land improvements as shown on Form FHA 424-1.

Land is in good state of fertility. Some pasture renovation will increase carrying capacity.

FIG. 27.1. *Continued.*

PART 4. OPPORTUNITIES FOR OTHER INCOME

(Evaluate and describe off-farm employment, rented land, etc.)

Very limited

PART 5. USE OF LAND RESOURCES AND IMPROVEMENTS

CROPS (1)	ACRES (2)	YIELD PER ACRE (3)	TOTAL PRODUCTION			CASH AND/OR RENTAL SHARE	
			AMOUNT (4)	PRICE PER UNIT (5)	GROSS VALUE (6)	RATE OR PERCENT (7)	VALUE (8)
Corn	82	90	7380	1.05	$ 7749	50	$ 3875
Oats	41	50	2050	.65	1332	50	666
Meadow	82			$ 12/A	984	100	984
PERMANENT PASTURE	43	ANIMAL UNITS	@	$ 5/A	215	100	215
WOODLAND							
FARMSTEAD, ROADS, ETC.	10		.				
CASH RENT							$
TOTALS	258				$10280		$ 5740

DEDUCTIONS

Real Estate taxes and assessments	$ 1450
Insurance on buildings,	107
Maintenance: buildings, fences, water supply, tile	300
Operating and maintenance costs for irrigation and drainage	-
Annual installments on bonded debts	-
Seed	323
Fertilizer, lime, spray material	800
Harvesting and marketing expenses	200
Total Deductions	$ 3180

NET FARM INCOME $ 2560

CAPITALIZATION RATE FOR AREA _____ 5 _ % = $51,200

FIG. 27.1. *Continued.*

PART 6. SUMMATION VALUE OF FARM

USE OF LAND (1)	ACRES (2)	SOIL DESCRIPTION (3)	DEPTH OF TOPSOIL (4)	KIND OF SUBSOIL (5)	TOPOGRAPHY (6)	VALUE PER ACRE (7)	VALUE TOTAL (8)
CROPLAND						$	$
	95	198 Floyd	6"-10"	Clay	Gent.Rolling	300	28,500
	70	163 Fayette	2"-6"	Clay	Rolling	250	17,500
	40	175 Dickenson	4"-12"	Sand-gravel	Rolling	125	5,000
TOTAL CROPLAND	205						
PERMANENT PASTURE	43				Rolling	60	2,580
WOODLAND, FARMSTEAD, ROADS, ETC.	10						
TOTALS	258				NORMAL MARKET VALUE OF LAND		$ 53,580

VALUE OF ESSENTIAL BUILDINGS _____ $ 6,000

PART 7.

RECOMMENDED NORMAL VALUE ---------------------------------- $ 56,000

PART 8. COMMENTS

This is a small family type unit that is suitable for crop and general livestock production that is prevelant in the area.

Although split by a county road, livestock facilities and water are available on each side. This is not considered to detract seriously from the value of the farm.

Date: February 24 _____, 19 69   (Signed) _____

County Supervisor
(Title)

GPO 806-923

FIG. 27.1. *Continued.*

USDA—FHA
Form FHA 426-1
(Rev. 8—22—67)

VALUATION OF BUILDINGS

FULL NAME OF BORROWER(S)							MAILING ADDRESS			

John R. Jones   Fayette, Iowa

KIND OF BUILDING (a)	YEAR BUILT (b)	CON-STRUC-TION (c)	KIND OF ROOF (d)	KIND OF FOUN-DATION (e)	SIZE DIMENSIONS OR AREA OF GROUND FLOOR (f)	STORIES (g)	CONDITION OF BUILDING (h)	DEPRECIATED REPLACEMENT VALUE AS IS (i)	DEPRECIATED REPLACEMENT VALUE AS IMPROVED (j)
1. Dwelling	57	Fr	AS	SM	24 x 48	1	good	$ 10,000	$ 10,000
2. Barn	18	Fr	AS	SM	30 x 40 / 16 x 32		good	4,000	4,000
3. Hog house	48	Fr	AS	SM	26 x 48		good	1,500	1,500
4. Crib	36	Fr	AS	SM	24 x 36		good	3,000	3,000
5. Cattle shed	48	Fr	steel	pole	24 x 52		good	1,000	1,000
6.									
7.									
8.									
9.									
10.									
11.									
TOTALS	xxxx	xxxx	xxxx	xxxx	xxx xxxxxx xxxxx	xxxxx	xxxxxx x	$ 19,500	$ 19,500

The following information to be furnished in a State or Territory where credits are given or charges are made customarily in the Board Rate for the items listed below:

ALL BUILDINGS (Show kind of building)

Approved lightning rods ....

Certified lightning rods* ....

Exterior painted within last 5 years   X

Electric or gas lighting throughout   X

Plumbing throughout   X

TOBACCO BARNS

Air-cured or when firing is done by salaman-ders using coke for fuel .... ____Yes .... ____No

Flue- or fire-cured (equipped with flue whether fire used or not) .... ____Yes .... ____No

*Show master label number.

DWELLINGS

Number of rooms ....   six

Basement .... __X__Full _____Partial _____None

Central heating plant .... __X__Yes .... ____No

All rooms plastered on lath or sheathed with plaster, wallboard, tongue and groove, or wood panel .... __X__Yes .... ____No

Approved spark arrester when roof is wooden † ....

Heated by electricity .... ____Yes .... __X__No

Chimney .... __X__Yes .... ____No

Standard chimney .... __X__Yes .... ____No

Number of stovepipe connections .... furnace only

Telephone .... __X__Yes .... ____No

†Show name of manufacturer.

LOAN SECURED BY

[X] 1st   [ ] 2d   [ ] 3d

MORTGAGE

GPO 804-832

SIGNED _____

County Supervisor   2-24-69
(Title)   (Date)

FHA 426-1 (Rev. 8-22-67)

FIG. 27.1. *Continued.*

In the comparable sales table in Part 2 the third heading under "Acres" is "LON" which means "Level of Normal." For example, the "256" for the first sale in this table indicates that the normal value of the farm is $256 an acre instead of the 1968 sale price of $275. For the third sale which is rated as equal to the farm being appraised the normal value is given as $225 compared to the 1967 sale price of $230 an acre. The 258 acres in the appraised farm multiplied by $225 gives a total normal value of $58,050, which is slightly over but close to the final normal value of $56,000 stated in the appraisal.

The capitalized income value comes out at $51,200 using a 5% capitalization rate. This value gives no allowance for other factors besides income such as the relatively new dwelling built in 1957. A separate statement on building values shows the depreciated value of the buildings at $19,000, but in this summation statement essential building value is listed at $6,000.

APPRAISAL BY AN INSURANCE COMPANY

The second example is a ranch appraisal by an insurance company. Insurance companies are admirably situated to make long-term farm mortgage loans because their reserves, based on life expectancies, are not subject to heavy short-term demands. As a consequence, insurance companies are heavy farm mortgage lenders, accounting for approximately $5 billion of outstanding farm mortgage loans throughout the country. In the making of these loans the companies have given thought and effort to the development of appraisal forms which call for the kind of information and judgments needed for intelligent and successful lending. And it goes almost without saying that appraisal forms by themselves are of little value if the persons who make the appraisals in the field are not highly skilled, intelligent appraisers.

The ranch appraisal shown in Figure 27.2 rests mainly on income with little information on sales. The information on sales appears under Nos. 18 and 20 near the end of the appraisal. Although insurance companies along with other lenders are chiefly concerned with the income value of the farm used as security for a loan, it would be appropriate to have more information on comparable sales and market value data to be used by the appraiser in making his sale value estimate. A more extended discussion of ranch appraisals is included in the next chapter.

Value in this appraisal example is determined by capitalization of the $1,111 net income at 5% to give a total figure of

RANCH APPRAISAL

No. _____ Name(s) _____ Known as _____ Ranch _____ Area.

Acres _1006.27_____ Deeded _____ Unpatented __240_____ Leased _Gunnison_ County, State _Colorado_

1. SOIL—Include parent material, topography, erosion, sub-soil, drainage, productivity, plant vigor, range conditions, trends.

All the soils are very deep black medium textured soils. Very few rock outcroppings and practically no erosion due to the excellent cover. The ranch is gently rolling to rolling. There are just a few clumps of aspen, balance large open parks. Hay could be put on many areas on the pasture. Arizona Fescue was 8 to 12' high when the ranch was inspected. The meadows would put up over $1\frac{1}{2}$ ton if they were not pastured as late as they are, however, they normally buy or lease some hay land just as they are leasing the _____ Ranch now.

March, May

2. CLIMATE—Precipitation __24 - 28___ inches. Principal months _July thru Sept._ Snowfall _167-175_____ inches. Growing Season _41-65___ days. Elevation _8867-9500_ feet. Average length of grazing season _6-7_____ months Effect of climate and elevation on supplemental feed required and type of feed produced. Drought frequency and duration. Severe storms. Disaster losses. This is a heavy feeding area where 2 ton is normally fed, however this past winter they only fed $1\frac{1}{2}$ ton. This ranch does not have enough hay, however they have always bought hay or preferably leased meadow land nearby. They could put up more hay on this ranch by not pasturing it so late, however this depends upon their lease setup. Now they have the _____ ranch leased and will put up plenty of hay to carry them through. This hay is very strong and feeds out exceptionally well. Not subject to drought, even this year they have excellent moisture conditions, the irrigation water does not hold up much past the middle of July normally, however these meadows sub naturally. Not subject to severe storms, do have heavy spring snows, but there are no losses from this to speak of.

3. Operation—Livestock adaptability. Convenience of operation. Balance of unit. Dependency on leases and permits. Operation expenses. % increase, livestock normally sold. Selling weight, average death loss, stability of area. Show effects on earnings.

This is a good cow and calf outfit very conveniently operated. Just a short distance from the headquarters in _____ to the main ranch and the _____ ranch just a couple miles below _____ The State Lease and Private Forest Allotment join the ranch and are very accessible. The new purchase joins in and really goes with this ranch. This ranch needs twice as much hay as they have, this is normally acquired by leasing land nearby or buying hay. The 120 acres of timbered land lying up _____ Gulch to the NW is not used by this ranch and they have no abstract on it, however it is to be included in the trust deed for what it is worth. Operation expenses are normal for this area. They normally have close to 90% calf crop and sell caves weighing 400 to 430 lbs. Average death loss is 2%. This is a very stable area, however I am making a $.25 reduction in the net income attributable to land because of their having to lease additional meadow land. However, this has been done for years right in their neighborhood.

A. DETRIMENTS—Poisonous weeds, insects, diseases, predatory animals, rustling, nuisances, encroaching brush and other undesirable vegetation. Show effect on income.

No detriments of any consequence.

FIG. 27.2. *Ranch appraisal form used by an insurance company for mortgage loans on ranches.*

4. WATER SUPPLY—Natural, artificial, depth of wells and capacity, quality, dependability of all supplies. **Include irrigation.**

Excellent supplies of stock water all over the ranch. Domestic and stock water from the town supply at

Ditch, being Ditch No. 222, Pri. No. 249 for 6 SF.

Ditch, being Ditch No. 227, Pri. Nos. 256 & 258 for 5 SF.

All supplies very dependable, irrigation water normally runs out about middle of July.

5. NATURAL RESOURCES: Oil, Gas, Hunting, Recreation, etc. This year it will be about normal.

Big game hunting, deer, elk, bear right on the ranch and nearby. Excellent fishing reported nearby. They have all the min. on the main 320 where the meadow lies. Some of the other

6. FORAGE AND VEGETATION: land has patent reservations and the min. are all reserved at the Hdqs.

Number (from plat)	Range Type	Acres	Density %	Prin. Forage Plants	Comp. %	Palat. H, M or L	Livestock Adaptation
1.	Meadow	156	95-100	Fescues, redtop	50	H	C or S
				Sedges	30	H	
				Timothy, Clover, M.	20	H	
2.	Irr. Past.	50	80-95	Fescues	65	H	C or S
				Wheat Grass & Brome	20	H	
				Blue Grass, Clover	15	H	
3.	Open Park Aspen Range	680	65-85	Aspen, few conifers	3	L	C or S
				Arizona Fescue	30	H	
				Wheat Grasses	20	H	
				Mt. Brome, Misc.	27	H	
4.	Timbered Range	120	50-80	Conifers, Aspen, Br.	30	L	C
				Fescues, Wht. Gr. Mis.	70	H	

7. CARRYING CAPACITY AND INCOME—

Actual: 19 60 : Cattle 190 Sheep Goats Horses 15 Total A.U. 205

Typical Cattle 150 Sheep Goats Horses 6 Total A.U. 156

HAY AND GRAIN PRODUCED AND FED TO LIVESTOCK ON SECURITY

Crop	Acres	Yield Per A.	Total Yield	Feed Units Per Ton	Total Feed Units	Animal Units	Income To Land
Hay	156	1.	156	23	3588	45	
The 6 A. of hay at		puts	up 15 to 18 ton according to				

GRAZING LAND—

Kind	Season Used	Range Condition Class	P'c'p't.	Acres	Acres /A.U.	Animal Units	Income To Land
Meadow	Sp, F	Good	25	156	6	26	
(This meadow is pastured late)							
Irr. Past.	Sp, Su, F	Good	25	50	3	17	
Aspen Park Range	Sp, Su, F	Fair	28	680	10	68	
Timbered Range*	Su, F	Fair	26	120			
(*This is not used by app. no credit given)							

Some good grazing along Gulch, mostly too heavy timbered for grazing.

TOTAL ANIMAL UNITS— 156 @ 12.00 $ 1872

CROP SALES

Crop	Acres	Yield	Total Yield	Price	Value	Share	Income To Land

TOTAL INCOME $ 1872

FIG. 27.2. *Continued.* (*Plat not shown.*)

463

8. EXPENSE—

Taxes 19 59 . $ 495.77 _____ Assessed Val. $6,560 _____ Est. Future Tax $ ___500___

Special Tax $ ___None___ .

Insurance: Amount Recommended $ ___5,800___ @ ___70___ rate $ ___41___

Maintenance: Buildings $ 145 _____ Fences $ 55 _____ $ ___200___

Miscellaneous: (Seed, Comb., Bailing, Fertilizer, Ditch Maintenance) $ ___20___

Total Expense	$ 761
Net Income	$ 1,111

9. EARNING VALUE— Capitalized at 5% $ 22,220

10. OPERATIONAL FACTORS—Net effect of facts discussed under Nos. 3 & 3A $ .25 Cap. at 5% $ -780

11. LOCATION—Markets, Community, Nationality, Utilities, Comparison with neighboring ranches, trucking and freight rates, accessibility.

Ranch is on gravel road 8 miles from paved road, and _____ miles _____of

_____ Pop. 500 , and _____ miles N of Gunnison , Pop 3,000

This ranch has an above average location, will be paved right to this summer. Main meadow unit is just 3 miles from headquarters NE of . REA and Tel service is installed. Stock is trucked to Salida or Denver, $.59 and $.86, respectively.

$ 2,000

12. LEASES AND PERMITS—Stability, probable reductions, etc.

Kind	Period of Use	Annual Cost	Expira-tion	Acres	Acres per A.U.		Animal Units	
F. P.	7/1--10/15	176	Term	80	head	Allot.	23	Private fenced Allotment
State	Sp, Su, F	84	7/14/66	240	12		20	
(Has immunity clause, Lease No. S)								
			Adjustment Per Animal Unit $ 60		43			$ 2,580

13. IMPROVEMENTS—Note presence of any unusual value. Include fences, wells, hdqts., corrals, drift fences, loading chutes, etc.

Good barn at headquarters with 2 corrals and two barns on the meadow unit, also loading chute, good cross fences, good water development. $ 1,000

14. HOME FEATURES—Church and school, educational and religious advantages.

Very comfortable home, one of the best in , has REA and with water, Tel, but is not modern. Churches and schools at Also, in Gunnison and College at Gunnison, Western State. $ 1,400

TOTAL NET ADJUSTMENTS . . . . $ ' 7,000

15. BASIC VALUE 1006 acres @ $28.25 per acres. $ 28,420

16. BUILDING APPRAISAL: Present value is replacement cost less observed depreciation and obsolescence.

Structure	W	L	H	Age	Material	Roof	F'd'n.	Paint	General condition	Present value	Practical repl. value	Remarks
House	28	50	1½	28	F	WS	C	F	G	3,000		New asbestos siding
Barn	24	50	20	42	L	WS	R		G	1,000		2 yrs. ago
House	15	40	1S	35	L	WS	R		FG	800		
Barn	30	60	16	36	L	M	R		G	500		
Barn	40	56	14	30	L	WS	R		G	500		
Garage	18	22	7	30	L	M	R					
										5,800		

FIG. 27.2. Continued.

17. OCCUPANCY: Effect on security, any change contemplated
            and his wife live on the small tract in
            and his family rent a house in   No change contemplated.

Tenant: Name   Age ....   Spouse   Age ....

        Address   Relation to owner

        Improvements owned by tenant:

18. APPLICANT: 52   46

        Name   Age 45   Spouse   Age 41

        Address   Nationality   Citizenship

        Health Good   Polish-American   U.S.

        Family Cooperation (List age and sex of children)

                does not have any children.   has two boys ages 6 and 3.

Both wives are from local ranching families.
Ranch experience, occupation or other interests:

Both men have ranched their lifetimes right here. They operate this property and
lease the   Ranch just below   on which we have a loan. No other interests.

Your conclusion as to applicant, future prospects and operation in event of change:

Both men are very conservative hard working old time ranchers. I have met them
previously around Gunnison and at the sale in Salida. They want to add more hay
land to their holdings when they can buy something at their price.

Financial statement—progress, condition of stock and equipment, chattel mortgages:

They show a satisfactory statement. They have always been quite conservative. Have
very good cattle. This year they are going to have a number of June calves since their
herd as well as many others experienced difficulty in getting their cows settled the first
time they were bred last summer. Their equipment is in good repair. This loan will
Details of the acquisition of the ranch, purpose of loan: reduce their chattel to a conservative $4,500.
Inherited most of the ranch in 1939 from their father. Purchased the 160 acres in
Sec. 14 this spring for $6,000 of which $500 has been paid down. They spent $1,400
two years ago putting on new asbestos siding and improving the house.

19. OPINION OF LOCALITY AS LENDING TERRITORY: Mention tax hazards if any:
This is at the upper end of one of our very best lending territories. We have a number
of securities just below this outfit, and are in the process of closing up several more.
Taxes are higher on this outfit than they should be since the main improvements and a
20. SALABILITY: Previous sales of this security: small acreage lies in   and adjoining it.
This will always be very salable property. Many ranchers below would like to buy this
outfit and pasture the whole thing.   whom I am sending in a ranch loan for also
21. RECOMMENDATION: Do you recommend both the applicant and security as worthy of a loan:
                      wanted the 160 A. pasture that   just bought.
Yes. Both boys are conservative hard working ranchers and well recommended by
everybody. They were working with   , branding, dehorning and castrating calves
when I was up inspecting these ranches. This is good security, however they should
have more hay land.

22. CONDITIONS FOR CLOSING LOAN: Co-signers or special conditions:

Amount $ 16,000   Rate 5½ % Term 20 yrs. (1) Assignment all water rights

Interest dates January 1   Prin. payments $500 ann. (2) Waiver 80 head FP on Gunnison Nat.

Closing date at once   Date of 1st Payt. 1-1-62   Forest.

Option No. 1 --- 1/5   Commission None   (3) Assign. State Lease No.   with

Agent   Immunity clause on 240 A.

Address Gunnison, Colorado   (4) Pay all title and closing expense.

*This land just to be included on security. (5) No abstract on W½NW¼ Sec. & NE¼NE¼ Sec. *
23. This is to certify that on May 25, 19 61, I appraised the above-described property and the statements contained
in the foregoing report are the results of careful personal investigation and are to the best of my knowledge correct, and the
opinions offered are based on full and fair consideration of all available facts.

Date July 7, 1961   , Appraiser.

FIG. 27.2. *Continued.*

$22,220. To this figure are added the values of certain intangibles—location, leases and permits, improvements, and home features. In this case the grazing permit and lease add $2,580 to the value.

The final part of the appraisal, sections 17–23, is related to the making of a loan on the property. While the preceding parts were devoted to the data which formed the basis for the $28,420 value, this last part provides the data which supports the loan recommendation of $16,000 at 5½% for twenty years.

APPRAISALS BY FEDERAL LAND BANKS

The twelve federal land banks in the United States have as their exclusive business, as provided by Congress, the making of first-mortgage farm loans. Their outstanding loans, made through local cooperatives called Federal Land Bank Associations, amount to around $6 billion. Federal land banks, like the insurance companies, have devoted major attention to appraisal forms, policy, and personnel. In fact they were pioneers in stressing income information in their appraisals, although they have never adopted the income-capitalization method of arriving at value.

Appraisal of a Corn Belt farm shown in Figure 27.3 comes from the Federal Land Bank of Omaha, Nebraska. Of special interest in this first example are the appraisal check sheets presented in the last part. The first part of the appraisal indicates a typical operator net return of $27,104. This net income is analyzed in detail in the appraisal checks, along with a breakdown of the receipts and expenses. The appraised market value of this farm in November 1968 was $700 an acre or $336,000. The normal value of the farm on the same date was $665 an acre or $319,000. The large investment represented in a farm like this is an important reason for the detailed analysis of income and expense in the appraisal checks.

Appraisal of a 300-acre timber tract in North Carolina, the second example, was provided by the Federal Land Bank of Columbia, South Carolina (Fig. 27.4). It shows the valuation of 100 acres of saw timber and pulpwood with present market value of the stumpage at $10,050. The other 200 acres is a tract of 10-year-old slash pine. On a per acre basis the 100 acres of saw timber and pulpwood is valued at $125 an acre or $12,500, and the slash pine is valued at $65 an acre or a total of $13,000.

A citrus grove appraisal provided by the Federal Land Bank of Berkeley, California, is the third example. This appraisal,

FIG. 27.3. *Federal Land Bank appraisal of a Corn Belt farm. (Courtesy Federal Land Bank of Omaha, Neb.)*

| THE FEDERAL LAND BANK OF OMAHA **|**

APPRAISER'S REPORT

FLBA Office at ___Hawkeye_____   Application No. _____

I have made a personal examination of and identified the property described in the above application made by _____
___John Doe___ of _____ for a loan of $_____ and report as follows:

A. LOCATION, TYPE, AND QUALITY OF FARM

1. State_____ County_____ Twp._____ Range_____ Total acres ___480___
2. The farm is located __10__ miles ___N___ from **Pleasant Ridge** ; _____ miles from _____
   (Direction)   (Nearest Town)   (Nearest Shipping Point)
3. Farm is on (kind) **gravel** Co._____ Public Road. If not on public road, describe right of way required on Line 6, Section F.
4. Type of farming in community is **grain-livestock** Conveniences available are (X)____**all available**_____
                                                                      (R.F.D., School Bus, Power Line, Telephone)
5. A comparison of this farm with the average in the community is: Location __**average**__ ; quality of soils__**good average**__
6. Value of improvements__**hi average**__ general desirability and salability___**high average**_____
7. Describe domestic water supply ___**2 bored wells and pressure system**_____
8. Adequacy of drainage (Explain) __**adequate**_____
9. Hazards (Explain) _____**some wind erosion in the fall**_____
                       (Frost, Hail, Frequency Damaging Drouth, Flood, Drainage, etc.)
10. _____
11. _____
12. The general condition of farm is ___**good**___ ; Bldgs. __**maintained**__ Land __**maintained**__
                                  (Excellent-Good-Fair-Poor)   (Being maintained, improving, or depreciating)
13. This farm __**is not**__ irrigated.(If irrigated, prepare supplemental report) Land __**is not**__ in a mineral area.(If a factor, discuss)
          (is-is not)   (is-is not)
14. _____
15. The farm is in a class __1__ area. It is a class __A__ farm _____
                         (1-2-3-4-5)   (A-B-C-D-E)

B. PURCHASE PRICE OF FARM

1. The information in the application concerning purchase price of the farm and capital improvements added _____ correct.
                                                                                                        (is-is not)
2. (If incorrect, complete lines 3 through 8.)(List acquirements of past 10 years.)

	Acres	Date Acquired Month - Year	Cash	Contract	Mortgage	Inherited Interest	Trade	Total Purchase Price
	160	1948	$	$	$	$ **all**	$	$
4.	160	1959						64,000
5.	160	1966	25,000	75,000				100,000

6. Capital improvements added (Itemize) _____   Total..........$
7. ___Tiling - $6000. Hog house, grain storage $10,000.___
8. Remarks_____
9. _____

C. DESCRIPTION OF BUILDINGS (Present cost of replacement less depreciation, including obsolescence)

Kind	Size	Material	Condition	Describe adaptability, and other features which affect desirability, including obsolescence and conformity with community standards. Indicate if dwelling has furnace, running water, bath, lights, etc.	Present Value Insurable
1. Dwelling 2. 3.	30x38x8/16	Frame	Good	8 rooms, 1 and 1/2 baths, full basement, central heat - story and a half built 1946	$ 16,000
4. Barn	52x52x10	Frame	Fair	Old barn remodeled for hogs	6,000
5. Hog Hse. 6.	24x36x8	Frame	Good	Farrowing house with outside feeding floor	4,000
7. M. Shed	30x90x12	Frame	Good	Comp. roof, conc. floor	6,500
8. 3 Bins	20,000 bu. capacity - one equipped with drier				12,000
9.					
10. Garage	16x24x8	Frame	Fair	Comp. shingle, conc. floor	600
11. 2 Cribs	30x40x14	Frame	Fair	Conc. fdn. and floor - shingle roof	4,000
12.					
13. Hen house and some old cribs - not used...					
14. Totals					$ 49,100

D. EARNING POWER OF FARM (Typical operator - Usual conditions)   INCOME AND EXPENSE

Crop	Acres Typical Operator	Average Yield Per Acre	Average Total Production	Unit Value (Normal Prices)	Average Gross Value	Average Normal Sales	Value Landlord's Share		Typical Operator	Rental
1. Corn	270	125	33750	$ 1.00	$ 33750	$ 28150	$ 16875	Sale of crops............ $ 45790		xxxxxx
2. Soybeans	180	40	7200	2.45	17640	17640	8820	Sale of livestock........ 15124		xxxxxx
3.								Dairy, poultry, wool, etc...		xxxxxx
4.								Total gross income....... $ 60914		xxxxxx
5.								Rental income...........	xxxxxx $ 26695	
6.								Cash operating expense... $ 9900		$
7.								Seed, fertilizer, lime... 11962		6235
8. Rot. past.	10			50.00	500			Upkeep-buildings......... 800		800
9. Pasture		___head for___ ___Mo. @___						Upkeep-fences, wells, etc.. 150		150
10. Bldgs., Lots	8							Repair & replmt. machy... 7200		xxxxxx
11. Roads Waste	12	Cash rental for pasture & imps.					1000	Insurance............... 300		300
12. Totals	480				$ 51890	$ 45790	$ 26695	RE tax 19**68**............ 3298		3360
13. Rentability of this farm is ___good___ . The usual rental terms are			(Good-Fair-Poor)					Personal property tax.... 200		xxxxxx
14.								O & M (inc. pumping)..... -		
15. Estimated annual outside income of typical operator $ __none__ .								B & I (for_____years)...		
16. Explain sources and dependability of outside income under usual conditions								Total Expense........... $ 33810	$ 10845	
17. ___Full-time family farm___								Net income............. $ 27104	$ 15850	

The legal description given in the application __is__ correct. Indicate corrections, if any, on plat and in remarks.
                                            (is-is not)

Any shortage in the acreage of this farm that may be developed upon examination of the title and verification of the plat which does not
exceed____2____percent of the total acreage shown in my report and which is not caused by the exclusion of a specified part of the
property will not affect my estimate of the normal value.

E. DESCRIPTION OF LAND

Present Use	Acres	Describe type and quality of soil, topography, fertility condition, durability, noxious weeds, drainage, erosion, crop adaptability, etc.
1. Crop and		
2. farmsted	468	Webster silt and silty clay loam, some Nicollet.
3.		Level to very gentle slope, not exceeding 3 percent.
4.		Good surface and internal drainage. This is a "duck back"
5.		farm - taking practically no water from other land.
6.		
7.		
8.		
9.		
10.		
11.		
12. Roads-Waste	12	
13. Total	480	

F. VALUATIONS AND ACCEPTABILITY

1. Normal Value $ __319,200__ Per Acre $ __665__   Present Market Value $ __336,000__ Per Acre $ __700__

2. Consideration has been given to O&M & B&I chgs. of $ __-__ per A. and project debt liab. of $ _____ on this land.

3. In my opinion this prop. __is__ satisfactory security for an FLB loan scheduled to be repd. in not to exceed ____ yrs.
   (is-is not)

4. (If limited, give reasons.) _____

5. The foregoing values assigned to this farm and opinion as to its acceptability for a loan are subject to the following

6. requirements being fulfilled: __no special requirements__

7. _____

8. Normal Value of Farm "as is" $ _____ and it _____ satisfactory security for a loan not to exceed _____ yrs.
   (is-is not)

9. subject to the following requirements being fulfilled: _____

G. PLAT   Use standard legend.   Show section centers.

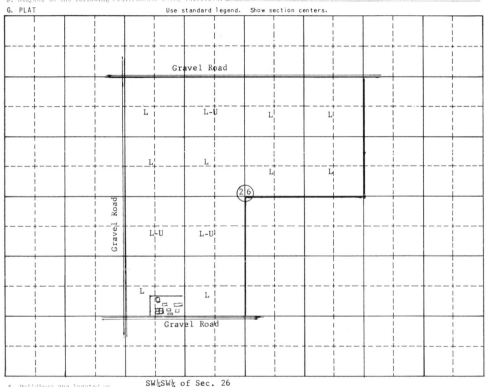

1. Buildings are located on __SW¼SW¼ of Sec. 26__
                                      (Legal Description)

H. REMARKS (Explain favorable, unfavorable & unusual features. If not satisfactory security for loan or if term of yrs. is restricted, give reasons.)

This is a very good farm, desirable to an owner-operator as well as an investor. It is located in a strong area of similar type farms. Land is top quality for the area, buildings are adequate and provide good living conditions, adequate machinery and grain storage and fair livestock facilities.

Date appraised __11-28-68__. The above is true and correct to the best of my knowledge and belief.

                                                    Land Bank Appraiser ————— Appraiser

For association use only: Maximum loanable including capital stock $ _____

FIG. 27.3.   *Continued.*   468

Name   John Doe   FLBA   Hawkeye

Feed Requirements

I. LIVESTOCK PROGRAM	Corn Equiv. Bushels	Silage Tons	Hay Tons	Supplement Tons	Pasture Acres
_____ Cows (Include feed for necessary replacements.)					
400 Hogs fed to 220 #	5,600			26 T	10
For seed ___ _____					
Total feed used	5,600			26 T	10
Total feed raised					
Feed to purchase					
Feed for sale					

II. SALES

1. Livestock Sales: _____ 398 hogs @ $38.00 _____ $ 15,124
   $ _____
   $ _____
           TOTAL LIVESTOCK SALES . . . . $ 15,124

2. Milk from _____ cows @ $_____ per cow . . . . $ _____
   Poultry income from _____ hens @ $_____ per hen . . . . $ _____
         TOTAL DAIRY AND POULTRY INCOME . . . $ _____

3. Cash sale of crops . . . . $ _____

III. CASH OPERATING EXPENSE

Labor   4   months @ $ 400 per month. . . . . $ 1,600
Custom Work: Shelling-grinding 5600 bu. @ 5 ¢; Silo filling $ _____ $ 280
   Combining _____ acres @ $_____ ; Picking _____ acres @ $_____ $ _____
   Topping _____ tons @ $_____ ; Weed control of table beans $ _____ $ _____
Fuel & Oil 480 acres @ $ 2.75 per acre . . . . $ 1,320
Livestock Expense:
   400 Hogs @ $ 2.00 per head (Veterinary, vaccine, boar) . . . $ 800
   _____ Cattle @ $_____ per head (Salt, mineral, vaccine & vet.). $ _____
Miscellaneous livestock expense – bull service, etc. . . . . $ _____
Feed Purchased: _____ bushels @ $_____ . . . . $ _____
     26 tons supplement @ $ 125 . . . . . $ 3,250
     _____ tons supplement @ $_____ . . . . . $ _____
Pasture rental . . . . $ _____
Farm car expense 10,000 miles @ 10 ¢. . . . . $ 1,000
Trucking expense . . . . $ _____
Hauling milk _____ cwt. @_____ ¢. . . . . $ _____
Dairy Expense: Cow testing, association fee, supplies, repairs, and replacement of dairy equipment. . . . . $ _____
Electricity and telephone. . . . . $ 300
Interest on operating capital. . . . . $ 1,000
Miscellaneous – pickup and truck license, liability insurance, etc.. $ 350

       TOTAL CASH OPERATING EXPENSES . . . $ 9,900

IV. SEED, FERTILIZER, AND LIME

				Operator	Rental
Corn seed 270 acres @ $ 6.00 per acre . . .				$ 1,620	$ 810
Legume seed 5 acres @ $ 6.00 per acre . . .				$ 30	$ 30
Bean seed 180 acres @ $ 4.25 per acre . . .				$ 765	$ 382
Beet seed _____ acres @ $_____ per acre . . .				$ _____	$ _____
_____ seed _____ acres @ $_____ per acre . . .				$ _____	$ _____
_____ seed _____ acres @ $_____ per acre . . .				$ _____	$ _____
Chemicals (Weed control 270 acres @ $ 3.50 per acre . . .				$ 945	$ 472
(Insect control 270 acres @ $ 4.00 per acre . . .				$ 1,080	$ 540
Baling cost _____ acres @ $_____ per acre . . .				$ _____	$ _____
Corn drying 33,750 bu. acres @ $ .03 per bu. . . .				$ 1,012	$ 506
Fertilizer 270 acres @ $ 20.00 per acre . . .				$ 5,400	$ 2,700
Bean herbicide 180 acres @ $ 3.50 per acre . . .				$ 630	$ 315
_____ acres @ $_____ per acre . . .				$ _____	$ _____
_____ acres @ $_____ per acre . . .				$ _____	$ _____
Lime 480 acres @ $ 1.00 per acre . . .				$ 480	$ 480
TOTAL COST.				$ 11,962	$ 6,235

FIG. 27.3. *Continued.*   469

APPRAISAL CHECKS

TYPICAL BASIS

*(For use on regular appraisals. Include income on
security offered plus typical outside income.)*

I.   DEBT REPAYMENT ANALYSIS

Net income from farm . . . . $ _____27,104_____

Other net income . . . . $ _____-_____

TOTAL NET INCOME . . . . $ _____27,104_____

Estimated Income Tax and Social Security . . . . . . $ _____2,188_____
Normal living costs. . . . . _____6,000_____
Installments on maximum FLB loan . . . . _____16,798_____

TOTAL DEDUCTIONS FROM INCOME . . . . $ _____24,986_____

BALANCE. . . . . $ _____2,118_____

This balance should ordinarily be sufficient to take care of required payment on
the primary debt on chattels and leave some for emergencies.

II. INCOME AND EXPENSE CHECKS

A. Conversion:
    (1) Farm~~-~~raised~~-~~feed. . . . . _____-_____ %
    (2) Total feed including purchased grain and supplement . . . . . _____162_____ %

B. Ratio of cash operating expense to gross income. . . . . _____16_____ %

C. Ratio of net income to gross income. . . . . _____44_____ %

III. RETURN ON INVESTMENT

		Operator	Rental
A.	Percent return on NV of $ __319,200__	__-__ %	__4.7__ %
B.	Percent Return on Total Capital Investment:		
	(1) Present Basis	__-__ %	__-__ %
	(2) Normal Basis.	__-__ %	__-__ %

C. Capital Investment

	Present Basis	Normal Basis
Farm . . . . $	__336,000__	$ __319,200__
Machinery. used value . . . . $	__30,000__	$ __30,000__
Livestock. $	__2,000__	$ __2,000__
_____ . . . . $		$
TOTAL CAPITAL INVESTMENT . . . . $	__368,000 (1)__	$ __351,200 (2)__

D. To Obtain Return on Capital Investment
Net farm income – this farm. . . . . $ _____27,104_____
Typical outside income . . . . $ _____-_____

TOTAL NET INCOME . . . . $ _____27,104_____

Estimated Income Tax – Federal – State . . . . . . $ _____2,188_____
Operator's labor . . . . $ _____6,000_____
Total deductions from income . . . . $ _____8,188_____
Balance for interest on capital. . . . . $ _____18,916 (3)_____

Return on Capital Investment:

Normal Basis (3) Balance for interest on capital . . $ _____18,916_____   _____5.4_____ %
    (2) Total capital – Normal Basis . . . . $ _____351,200_____

Present Basis (3) Balance for interest on capital . . $ _____18,916_____   _____5.1_____ %
    (1) Total capital – Present Basis . . . $ _____368,000_____

FIG. 27.3. *Continued.*

FIG. 27.4. *Federal Land Bank appraisal of a North Carolina timber
tract. (Courtesy Federal Land Bank of Columbia, S.C.)*

Approved Bureau of the Budget
Form No. 40 R 924
Form No. 1322, Rev. 8-4-67

APPRAISER'S REPORT ON SECURITY

TO THE F. L. B. A. OF___Clinston___AND FEDERAL LAND BANK OF COLUMBIA

Acres this tract 305

Application No. 123456   Name   John H. Doe $ 15,000   Total Acres 305

I have made a personal examination of and identified the property described in above numbered application and report as follows:
I did (X) did not ( ) interview the applicant (if not explain under Remarks)

A. LOCATION, TYPE AND QUALITY OF FARM

1. State N. C.   County Collier   8   SW   From Mooresboro
   (Miles)   (Direction)   (Town—Name)

2. 4   West   From Highway No. 901   --   The farm is located on State Paved
   (Miles) (Direction)   (Federal) (State)   (Kind and Type Road)

3. Road #1528   The Applicant has   a suitable right-of-way to the farm. (If not, discuss in remarks):
   (Has—Has Not)

4. It is not in a special improvement district or incorporated town. Name of town or district_____
   (Is—Is Not)   (Prepare Supplement If Necessary)

5. The farm is in a class 3 area. It is a class C farm (this tract). It is a class --- farm (all tracts).
   (1-2-3-4-5)   (A-B-C-D E)   (A-B-C-D-E)

6. It is a timber farm. Conveniences available are (√) X X   X   X
   (General, Dairy, Livestock, Poultry, Timber, Truck, etc.)   (RFD, School Bus, Milk Route, Power Line, Telephone)

7. Describe water supply Creek, pond and smaller streams   The general condition of farm is
   (Excellent, Good, Fair, Poor)

8. As a farm unit it is Fair   It is being maintained
   (Excellent, Good, Fair, Poor)   (Being Improved) (Being Maintained) (Depreciating)

9. Describe drainage condition, also overflow and erosion hazard adequate natural and artificial drainage;
10. no severe overflow or flood hazard; erosion controlled by vegetation.
11. Land is not in a mineral area. (Discuss if a factor)
    (is—is not)

12. Oil, gas, coal, mineral, timber, or other leases on this farm. Yes____ No X____ If yes, describe under remarks.
13. Approximate amount and species of timber and normal value 250 mbf of pine sawtimber and 300 cords of
14. pine pulpwood, predominately loblolly, with a normal value of $9,000; present market
15. value of $10,050. (See Supplemental Timber Valuation Sheet).

B. EARNING POWER OF FARM—TYPICAL OPERATOR—USUAL CONDITIONS

Crops	Acres Typ. Operator All Tracts	Average Yield Per Acre	Average Total Prod.	Unit Val. (Normal Prices)	Total Value	Typical Owner's Cash Sales
				$	$	
Forest Prod.	300	Current Annual Income			1153	1153
Forest Prod.	(300)	Average Annual Income			(3157)	(3157)
Pasture	-	For	Head of		for	months
Totals	300				$ 1153	$ 1153

Rentability of this farm is not customarily rented

C. OUTSIDE INCOME

Normal net outside income $ None   Describe sources and dependability
under normal conditions. The typical owner would live elsewhere
and have other sources of income.

D. PURCHASE PRICE

Farm acquired 1955 by purchased from T. B. Doe (uncle)
_____ price without personal property $ 12,000
Subsequent improvements Normal
--- Cost $ ---
Subsequent depreciations Normal - cut small
volume of pulpwood _____ Value $ 1,500

INCOME

Cash sales crops	$ 1153
Cattle No. _____	. -
Hogs No. _____	. -
Dairy Prod. Av. No. Cows____	. -
Volume of _____	. -
Volume of _____	. -
Poultry No. _____	. -
Eggs Doz. _____	. -
Cash Rentals _____	. -
Miscellaneous	. -
Gross Income	$ 1153

EXPENSES

Labor & Management . . $	60
Seeds, Plants, Trees	. -
Insecticides and Poisons . . .	. -
Fertilizer and Lime	. -
Gas and Oil	. -
Misc. and Custom Work . . .	. -
Taxes, Real and Personal . .	. 75
Ins. on Bldgs. and Per. Prop. .	. -
O & M Cost Incl. Irrigation . .	. -
Annual B & I Pay. for____Yrs..	. -
Upkeep on Improvements . .	. -
Repairs & Replacement Equip. .	. -
Int. on Short-Term Credit . .	. -
Other Sales Commission	. 57
Total Expenses	$ 192
Net Return	$ 961
*Expense chargeable to dwg. . .	$ -
*Adjusted net return	$ -

*These spaces to be filled out only where the net return shown above is a negative amount and the property is a part-time farm.

E. DESCRIPTION AND VALUATION OF LAND:(Designate separate parcels valued below as A, B, C, etc.)

Present Use	Type and Quality of Soil; Topography; Crop Adaptability; Ease and Economy of Operation; Any other important features that enchance or reduce desirability	Acres	Acre Value	Total Value
Planted Pine	Good stand of 10 yr. old planted pines. Soils are	200	$ 65	$ 13,000
timber	of Norfolk and Goldsboro series; over 2/3 of	100	125	12,500
	woodland well drained and suitable for timber			
	growth. 1/3 of woodland is fair quality mixed			
	soils, with scattered patches of mixed hardwoods			
	below merchantable size, and fair stand of young			
	pine timber.			
Roads and Waste		5	-	-

Total acres and total normal value of land 305   X X $ 25,500
Consideration has been given to O & M and B & I charges of $ -0- per acre and project liability of
$ -0- against this land.

F. BUILDINGS: Present cost of replacement less depreciation.
If new construction or improvements are involved carry all values forward to last column.

Description of Buildings	Map No.	Occupied By	No. Stories	Rms. No.	Length and Width	Material	Type of Roof	Condition	Value Insurable Bldgs.	
									"As Is"	After Improvements or New Const.
									$	$
				NO BUILDINGS						
							Totals		$	$

Discuss adequacy of farm improvements: Adequate farm roads for normal logging and management. No buildings or other improvements.

G. VALUATION AND ACCEPTABILITY: (This appraisal covers____1____tracts)
In my opinion the property offered in this application____is____ satisfactory
(Is—Is Not)
security for a land bank loan scheduled to be repaid in not to exceed ____40____ yrs.
If term of years limited, give reasons ____----____

NV this tract	$ 25,500
PMV this tract . . .	$ 30,000
NV all tracts . . .	$ --
PMV all tracts . . .	$ --
Amt. loanable . . .	$ 17,400

The foregoing values assigned to this farm and opinion as to its acceptability for a loan are subject to the following requirements being fulfilled.
NONE

"AS IS" VALUES AND INCOME: (All tracts included in application)' Not Applicable
Without the above requirements being fulfilled this property is a class_____ farm. It (is—is not)_____
satisfactory security for a land bank loan scheduled to be paid in not to exceed_____years. NV $_____;
PMV $_____ The net farm income is $_____ Net normal outside income $_____

H. REMARKS: Sections ____--____ THector's Creek ____--____
A desirable woodland tract with good potential from existing young pine plantation and timber stand. Property offers no immediate alternate uses or development potential.

(map area)
John Smith
1" = 1320'
Second Growth Loblolly Pine
State Paved Road No. 1528
10 Year Old Slash Pine
Roger Quay
Bill Smith

The above map was the basis for my appraisal and was prepared from surveyor's plat____X____, legal description_____, aerial map _____. Are boundaries well established? Yes__X__ No_____. Do you recommend a survey? Yes_____ No__X__. If answer is yes, give reasons under remarks. Any general acreage shortage of not more than ____5____% which does not result from exclusion of any specific parcel will not affect my estimate of value.

I certify that the foregoing is true and correct to the best of my knowledge. (Signed)_____R. B. Price_____
Date application received_____, 19_____
Date appraised_____, 19_____   _____Land Bank_____Appraiser
Date transmitted to F. L. B. A._____, 19_____
Date received by F. L. B. A._____, 19_____

Do not write in this space	
Reviewed by: _____	_____
	Initials Date
Reviewed by: _____	_____
	Initials Date

FIG. 27.4. *Continued.* 472

Approved Bureau of the Budget
Form No. 108-R019.1
Form No. 1322B

APPRAISER'S SUPPLEMENTAL TIMBER VALUATION SHEET

Tract Name: __Silver Creek__

Application No. __123456__ Name __John H. Doe__ Acres __305__ Tract No. __1__

State __N. C.__ County __Collier__ Located __8__ __SW__ From __Mooresboro__

Miles   Direction   Name of Nearest Town

__4__ __West__ From a Public __U. S. 901__ Road. It is a class __3__ Area: a class __C__ Farm.

Miles   Direction   Type   (1-2-3-4-5)   (A-B-C-D-E)

A. Condition of Property: Young timber below merchantable size with an adequate natural seed source; Acres: Good __300__; Fair __0__; Poor __0__. There are __0__ acres of young timber below merchantable size **without** an adequate natural seed source. Discuss species, plan of management (even-aged or all-aged), density, growth rate, improvement measures and hazards.

   __Mostly loblolly pine averaging 2500 BF and 3 cords per acre over approximately 100 acres; 200 acres of 10 year old slash planted pines with adequate spacing. The growth rates on the sawtimber averages 11.35% currently, while the planted stand currently averages 20.15%. There is adequate fire protection and no other serious hazards.__

B. Naval Stores:

   Worked by __Not Applicable__ ;

                   (Describe lease terms or share agreement)

   Present number of working faces on tract _____ Year of Working _____ ;

   Future cupping: Year _____ ; No. of cups _____ ; Year _____ ; No. of cups _____ :

   Current annual naval stores production: _____ barrels @ $ _____ per barrel.

   Average annual naval stores production: _____ barrels @ $ _____ per barrel.

C. Annual production and income from all forest products:

   Current annual normal income from timber and pulpwood .... $ __1153__

   Current annual normal income from naval stores operation .... $ __--__

   Total current annual normal income from all forest products .... $ __1153__

   Average annual normal income from timber and pulpwood .... $ __2004__

   Average annual normal income from naval stores operation .... $ __--__

   Total average annual normal income from all forest products .... $ __3157__

D. Cruise System:

   __5__ % cruise of __100__ acres by __1/4 acre plots__ . Log Rule—Pine __Scribner__

   __5__ % cruise of __200__ acres by __1/20 acre plots__ . Hdwd. __Doyle__

E. VOLUME AND VALUE OF COMMERCIAL TIMBER AND PRODUCTS: (Marketable under normal conditions)

PRODUCT (sawtimber pulpwood, etc.)	SPECIES	QUALITY	SIZE of trees— stump, B. H., or log	VOLUME & UNIT	NORMAL UNIT VALUE	TOTAL VALUE
Sawtimber	Pine	Fair	10" to 16"	250 mbf	$ 30	$ 7500
Pulpwood	Pine	Fair	6" to 8"	300 cds.	5	1500
Present Market Value, All Timber Products $ __10,050__					TOTAL $ 9000	

* * * * * * * *

(THE FOLLOWING SECTIONS TO BE COMPLETED ONLY WHEN USED AS A SEPARATE VALUATION SHEET)

F. Farm acquired 19____ by (purchase, inh., etc.) .... $ _____

   Subsequent improvements .... $ _____

   Subsequent depreciations .... $ _____

G. DESCRIPTION AND VALUATION OF LAND: (Designate separate parcels valued below as A, B, C, etc.)

Present Use	Type and Quality of Soil; Topography; Crop Adaptability; Ease and Economy of Operation; Any other important features that enhance or reduce desirability	Acres	Acre Value	Total Value
			$	$
Roads and Waste				

   Total acres and total normal agricultural value of land .... X X $ _____

   Consideration has been given to O & M and B & I charges of $ _____ per acre and project liability of $ _____ against this land.

FIG. 27.4. *Continued.*   473

H. BUILDINGS: Normal cost of replacement less depreciation including obsolescence.
If new construction or improvements are involved carry all values forward to last column.

Description of Buildings	Map No.	Occupied By	No. Stories	Rms. No.	Length and Width	Material	Type of Roof	Condition	Normal Value Insurable Bldgs.	
									"As is"	After Improvements or New Const.
Dwelling	(1)								$	$
	(2)									
	(3)									
	(4)									
	(5)									
	(6)									
	(7)									
	(8)									
							Totals		$	$

Discuss adequacy of farm improvements:_____
_____
_____
_____

I. FARM VALUATION: **PMV $**_____   **NMV $**_____   **NAV $**_____

J. LAND PLAT: Sections_____   Township_____   Range_____

[large blank land plat grid]

The above map was the basis for my appraisal and was prepared from surveyor's plat_____, legal description_____, aerial map_____.
Are boundaries well established? Yes_____ No_____. Do you recommend a survey? Yes_____ No_____. If answer is yes, give reasons under remarks. Any general acreage shortage of not more than_____% which does not result from exclusion of any specific parcel will not affect my estimate of value.

K. REMARKS:_____
_____
_____
_____
_____
_____
_____
_____
_____
_____

I certify that the foregoing is true and correct to the best of my knowledge.
Date appraised_____, 19_____
_____
_____Appraiser

FIG. 27.4. *Continued.*

shown in Figure 27.5, provides a large amount of factual data about the 40-acre citrus grove located in Tulare County in the San Joaquin Valley in California. Normal agricultural value is $3,250 an acre, and present market value is $5,500 an acre. Income data are provided, but in line with federal land bank policy the income is not capitalized into value. In this case no comparable sales are listed. Note the long production record provided covering 13 years. The land was purchased in 1948, improved and planted to orange trees in 1949.

A dairy farm appraisal provided by the Federal Land Bank of Springfield, Massachusetts, is the fourth and last appraisal example, shown in Figure 27.6. This appraisal of a farm in New York State is relatively simple with factual information on buildings, soils, income and expenses, and a plat of the land and the farmstead buildings. The farm, which has a capacity of 75 cows, was purchased for $28,000 in 1959; ten years later it was appraised at $60,000.

LOAN EXPERIENCE

Appraisers may study to advantage the experience of previous loans. The depression years furnished such an unmistakably clear record that even the busiest loan executive should take time to study it. Study is needed because the mere occurrence of more foreclosures in one area than in another does not of itself prove anything unless several other conditions are known. It is reasonable to assume that, with depression prices, the foreclosures and deeds in lieu of foreclosure would be about the same in two areas where outstanding loans were approximately equal. If, however, there were twice as many in one area as in the other, an error in appraisal may be suspected, but other probable reasons should be checked before definite conclusions are drawn. Natural factors (drouth, excessive rain, or grasshoppers) or economic factors such as low prices for certain products might have struck with more force in the one area than the other. The natural and economic causes which might apply should be analyzed, and then the degree to which the appraiser misjudged the land in the area of heavy losses can be determined. These postmortem surveys may not be pleasant, yet most appraisers welcome them because of the help they give in preventing recurrence of similar mistakes in the future.

A great many loan experience studies were conducted in the 1930–39 period, some by state agricultural experiment stations and government agencies and some by lending agencies. The appraiser has at his disposal, in the form of bulletins and

FLBA of... **Visalia** _____   Application _____

APPRAISER'S REPORT ON SECURITY TO THE FEDERAL LAND BANK OF BERKELEY

I have made a personal examination of and identified the property described in the above-numbered application.

Applicant.... **J. Doe** _____ Address... _____
Amount requested $ _____ - _____ Total acres _____ **40** County _____ **Tulare** _____ State _____ **California** _____

A   LOCATION, TYPE, AND QUALITY OF FARM

1   Farm is situated **1** miles **W** and **2** miles **N** from **Orosi** on **NE corner Ave. 432 and Road 120** road.
2   Land **is not** in a mineral area. Discuss if a factor. **-**
3   Comparison this farm with average in community: location **average**; quality of soils **average**;
4   value of improvements **above average**; general desirability – salability **above average**
5   Condition of farm **good** Farm is (being maintained—improved—depreciating) **being maintained**
6   Class of area and farm **3 - B** Type of farming in community **citrus, vineyard, cotton and alfalfa**
7   For coding purposes: Type of farm **citrus** Principal crop **oranges - Navels**

B   EARNING POWER OF FARM (Typical Operator—Usual Conditions)   **INCOME AND EXPENSE**

Crop	Age of Trees, Vines	Normal Acres	Average Yield Per Acre	Total Production	Unit Value	Total Gross Value		Owner Operator	Landlord
1 Navels	16	21	500 FB	10,500 FB	$1.85	$19,425	Crop sales	$32,786	$
2 Navels	10-11	13.5	525 FB	7,087 "	1.85	13,111	Livestock sales		
3 Limes	5-9	2.5	100 FB	250 "	1.00	250	Livestock product sales		
4									
5							TOTAL INCOME	$32,786	$
6							Hired labor	$	$
7							Production costs	9,022	
8							Misc. farm exp.	600	
9							Taxes—real estate	1,666	
10							Taxes—personal	100	
11 Past _____ acs. _____ hd. _____ for _____ mo. @ _____ per hd. mo.					$		Irrig. & drainage–O&M	455	
12 Size of livestock or poultry operation _____ Total					$32,786		Irrig. & drainage–B&I	-	
13							Building upkeep—Insurance	800	
14 Rentability _____ - _____ Usual rental terms **Would not be rented.**							Rep. & replac., pers. prop.	200	
15							Int. on short-term credit.	425	
16 Normal net outside income $ **5,000** Sources and dependability **This farm**							TOTAL EXPENSE	$13,268	$
17 **generally would be operated as a headquarters farm with**							NET RETURN	$19,518	$

other farm income. Normal living costs $7,000.

C   VALUATION OF BUILDINGS

Parcel No.	Kind	Size	Sq. Ft.	Material	Type of Foundation	Type of Roof	Age	Condition	Normal Value Insurable Buildings
1	House	Various	1980	Stucco	Concrete	Comp.	8)	Good	$18,000
2	Attached garage	24 x 28	672	Stucco	Concrete	Comp.	4)		
3									
4	Storage shed	30 x 40	1200	Alum.	Concrete	Alum.	15)		
5	" "	40 x 45	1800	Steel	Concrete	G. I.	4)	Good	8,000
6	Loading dock	5 x 40	200		Concrete		4)		
7									
8									

9   Total normal value uninsurable buildings $ _____   Total $ **26,000**
10   Adequacy and appropriateness of buildings—conformity to community standards **Buildings are fully adequate and**
11   **appropriate; but with swimming pool, etc., consider total building layout somewhat above community standards.**

D   VALUATION OF LAND

Parcel No.	Present Use	Type and quality of soil; drainage, hardpan, alkali; topography and irrigation; adaptability; climatic hazards; noxious weeds; any other important features	Acres	NAV Per Acre	NAV Total
	Navels	Soil is San Joaquin loam and clay loam from 24" to 40" in depth underlain with hardpan. Soil has been ripped to a depth of 40". Soil has good	21	3,000	63,000
	Navels	fertility and durability and is well suited to citrus. There is a swale or natural drain which meanders across the middle of farm. There is a	13.5	3,250	43,875
	Limes	natural slope to the swale; balance of the rows are graded satisfactorily for irrigation. There is adequate pipeline to irrigate satisfactorily as well	2.5	800	2,000
	Farmstead	as a return line with a sump pump.	1	3,000	3,000
	Road		2	-	-

     NAV per acre - $3,250   PMV per acre - $5,500

Total cultivable acres **37** Total irrigated acres **37**   Totals **40** | xxx | $ **111,875**
B & I charges (remaining term **-** years) and O & M charges are $ **12.30** per acre. Debt liability $ **-**

E   VALUATIONS OF FARM—ACCEPTABILITY FOR LOAN

1   NORMAL AGRICULTURAL VALUE $ **130,000** NORMAL MARKET VALUE $ **130,000** PRESENT MARKET VALUE $ **220,000**
2   This farm **is** satisfactory security for a Land Bank loan for a term not to exceed **20** years.
3   The foregoing values are subject to the following conditions: **-**
4 _____
5   NAV (as is) $ **-** and (as is) **-** satisfactory security for a Land Bank loan for a term not to exceed _____ years.

F   PURCHASE PRICE AND IMPROVEMENTS

1   Farm acquired: month **December** year **1948** by **purchase** for $ **5,000** (without personal property).
2   Purchase terms **were** confirmed. Improvements since purchase $ **81,000** Describe **Built house & garage**
3   **$17,700; tool shed $14,000; pipelines $7,500; well and pump $3,000; sump pump $300;**

Form 306 (Rev. 3-53) FLB Berkeley   **domestic well and pump $1,000; leveling $5,000; 4 wind machines $12,500;**
planted trees $10,000; patio, swimming pool dressing rooms and fences $10,000.

FIG. 27.5. *Federal Land Bank appraisal of a California citrus grove. (Courtesy Federal Land Bank of Berkeley, Calif.)*

G   **WATER RIGHTS** (The water and carrier rights described below are reflected in this appraisal as part of the security.)

1   ....shares in...; ....shares in..

2   ...40... acres in...Orange..Cove..Irrig. District or Project; ....acre-feet in...Project.

3   Other..

4   ....

5   Right of way for conveyance of water to farm...District..pipeline..borders..farm..on..the..west..side.........................

6   Water duty.......2....... acre-feet per acre; water pumped.........-.............acre-feet per acre; acres subirrigated.............................

7   If applicant owns less than 100 per cent interest in any well(s) or pumping plant(s), explain details of ownership—adequacy and desirability.

8   ....

9   Water supply is class....I....for...37...acres; class..................for..................acres; class..................for..................acres.

H   **IRRIGATION AND DRAINAGE DATA AND COSTS**

PUMPING PLANTS	ACRES SERVED	Well Depth	PUMP			PUMPING CONDITIONS					Total Pumping Cost	Gravity Supply		Other District	
			Size Setting	HP	GPM	Static Level	Draw-down	Boost	Reces-sion	Total Head		Acres	Total Cost	Acres	Total Cost
1 No..........		231	126	7½	275						$	37	$ 430		$
2 No..........															
3 No..........															
4 No...Power..and..depreciation..on..1..HP..tailwater..pump										25					

I   **REMARKS** (Indicate, on plat or in remarks, water right applying to each parcel. If a factor, comment on drainage, flood and erosion hazard, and adequacy of pumps and wells.)

There is a pump on the security, but water costs are based on a total use of district water.

There is a drainage system which uses 3600 feet of concrete pipeline for carrying water either to the return pump or it is transported across the road into a drainage ditch.

J   **PLAT AND LEGAL LOCATION OF FARM**

1   Section...S½..-..1....Township...15..-..S....Range...25..-..E....Meridian...MD...............Scale...1"..=..660'

2   The legal description.............is............correct. The property.....has............a right of way. If not, discuss in remarks below.

K   **REMARKS** (Explain favorable and unfavorable features of security; if not satisfactory security for a loan or term of years is less than usual limits, fully state reasons. No reference should be made to personal factors about applicant.)

This is a desirable citrus grove. It has a good home and a large storage shed equipped for grading and packing limes and provides living quarters for hired help. Also, the storage shed is used as headquarters for custom spraying and grove care operations. The original 21-acre block of Navels was planted in 1949, the year when the temperature dropped to 19°. There was no frost protection at that time, and many trees were killed to the bud union. This has resulted in some replants and topworking which makes this not as uniform as the 13.5-acre block. There is a low swale which goes across the entire 40 acres. Approximately 2.5 acres of the lowest part is planted to limes. The limes were severely damaged in the 1962-1963 freeze, but are now back in production.

Over-all, this is an above average grove for the area.

Any general acreage shortage up to........2............per cent will not affect my values, unless caused by excluding a specified part of property.

Date appraised..................19...... Date written.................19....... The above is true and correct to the best of my knowledge and belief.

Date copy mailed.................19......to FLBA of.. ....Study Group.............................
                                                                Appraiser

FIG. 27.5. *Continued.*

Approved
Bureau of the Budget
Form No. 40-R925

Applicant ___J. Doe___   Application _____-_____

Average Cash Operating Expenses

	Per Acre	TYPICAL OPERATOR
Cultivating: Non-tillage Weed oil and Karmex	$ 18	666
Hired Labor_____ Contract_____		
Pruning - Brush Disposal:		
Hired Labor_____ Contract	18	666
Thinning:		
Hired Labor_____ Contract_____		
* **Spraying:** Material and application		
Hired Labor_____ Contract	90	3,330
Material_____		
Dusting:		
Hired Labor_____ Contract_____		
Material_____		
Fertilizing:		
Hired Labor	3	111
Manure_____ Com.	15	555
Cover Crop_____		
Frost Protection: Including depreciation		
Heaters - Hired Labor_____ Fuel	20+	751
Wind Machines - Power or Fuel	47+	1,757
Hoeing_____		
Irrigation - Hired Labor	8	298
Propping - Hired Labor	12	444
Harvesting:		
Hired Labor_____ Contract_____		
Material_____		
Other_____		
Hauling_____		
Auto - Truck Expense_____		
Gas - Oil - Grease_____	2	74
Tree or Vine Replacements_____	5	185
Compensation Insurance_____	3	111
Misc. or Other Expense_____		
Pipeline repairs	2	74
TOTAL CASH OPERATING EXPENSE	$ 9,022	

Crop Record

YEAR	VARIETY	UNIT	TOTAL UNITS	INCOME
1952-53	Navels	FB	164	$
1953-54			525	
1954-55			1,982	
1955-56			3,360	
1956-57			4,848	
1957-58			6,848	
1958-59			11,593	
1959-60			9,576	
1960-61			9,370	
1961-62			10,888	
1962-63			11,031 - Packed	
			7,209 - Juice	
1963-64			21,796	
1964-65		Est.	23,000	
1960-61	Limes	FB	40	
1961-62			200	
1962-63		Frozen		
1963-64			6,120 lbs	$897.90

Average weight of unit   50 pounds

* Spraying costs include some foliar nutrient sprays which supplement the fertilizer program.
Comment on following:

Frost hazard (average hours heating and/or wind machine operation per season) Estimate 100 hours of wind machine operation on each machine plus an average of 10 hours of heating.
Wind hazard None

Condition of plantings (number diseased - number budded over) Navel trees are in a thrifty condition.

Root Stock 15-year-old trees on Rough lemon - 10- to 11-year-old trees on Trifoliate.

Plantings are set _18 x 24_ feet apart. ☐ Diagonal ☒ Square   Number per acre 100

Desirability of varieties Washington and Frost Nucellar Navels

Quality of fruit - (Compare size and grade with community or industry standards.) Size and quality have been above the average in the area because trees are young and capable of producing good quality fruit.

Source of information as to production figures used in Appraiser's Report on Security Grower's records and estimate for the 1964-65 crop.

Future trend of production Believe this grove will produce 500 plus boxes for many years to come.
Remaining economic life of plantings - size considering age Trees are generally good size considering age and have 25 years plus remaining life under normal conditions. The 16-year-old trees were planted in 1949, the year there was 19° weather and there was no protection for the young trees. Many froze and were replanted so this part of the grove is not completely uniform. There are approximately 25 trees which grew two or three trunks after the frost damage. Others froze to the ground and later a Rough lemon sucker came up. They were budded high and have not made good trees. There are about 100 of these and they gradually are being replaced. The limes suffered severe damage in the 1962-63 freeze but appear to be coming out of it.

Date_____

Study Group_____
Appraiser

Form 190 (Rev. 2-62) FLB Berkeley

FIG. 27.5. *Continued.*   478

SUPPLEMENTAL APPRAISAL DATA

WIND MACHINES AND HEATING

OPERATING COSTS OF WIND MACHINES

1 - 30 HP electric
1 - 35 HP electric
2 - Machines operated by power take-off shaft on tractor - capacity of these machines is 45 HP at the propeller.

Demand charge - 2 electrics	$ 437	
Power on 2 electric machines - 100 hrs.	97	
Fuel and oil for tractors operating machines	130	
Repairs and maintenance of electric machines	50	
Repairs and maintenance of tractor-powered machines	200	
Depreciation costs on four machines - total cost $13,000 - 15 yrs. life	867	$1,757

OPERATING COSTS OF HEATERS

Operating of heaters for 37 acres
10 hours average heating - 450 heaters used
Average of 300 heaters fired at one time

Filling heaters	$ 55	
Moving heaters	26	
Cleaning heaters	6	
Firing heaters	37	
Oil - 2,500 gallons fuel @ 15¢	375	
Torch fuel	5	504

DEPRECIATION OF HEATING EQUIPMENT

	Total Cost	Depreciation	
450 heaters @ $7.50	$3,375.00	$ 225.00	
6 thermometers @ $6.00 - 10 yrs. life	36.00	3.60	
4 torches @ $6.00 - 15 yrs. life	24.00	1.60	
3,000-gal. storage tank - 25 yrs. life	425.00	17.00	247

WATER DUTY

2 AF x 37 acres = 74 AF

ORANGE COVE IRRIGATION DISTRICT

74 AF x $5.00	$ 370
O & M - $1.50 x 40	60
Power and depreciation on return pump	25
	$ 455

FIG. 27.5. *Continued.*

APPRAISER'S REPORT

Approved Bureau of the Budget
Form No. 108-R019.1

To the FLBA and The Federal Land Bank of Springfield on Security Offered in Connection with the Loan Application of:

John Doe .... State **New York** County **Caledonia** Town **Rutland**

A. Location and
Quality: Farm is on .... **Rt. 12, good macadam** .... , **2** miles **N** from **Rutland**
(Road — kind, number and condition)   (direction)   (nearest town)

Land is .... **well maintained** .... Buildings are **well maintained** General condition **very good**
(improved — being maintained — depreciating)   (excellent, good, fair, poor)

It is a class .... **B** .... farm in a class .... **2** .... area.   Elevation .... **900'**
(A,B,C,D)   (1,2,3,4)

B. PURCHASE DATA OF FARM

Applicant acquired farm in ....**1959**.... by ....**purchase**....   Improvements since purchase: Describe
With personal $ ..**29,750**.... without $ ..**28,000**....   Added pole barn. Cleared some pasture for
Cash payment $ ..**13,550**.....   use as tillage and tillable pasture. Many
Contract or mortgage $ ..**16,200**.....   minor improvements to all buildings.
Trade $................

C. DESCRIPTION OF BUILDINGS AND LAND (Refer to plat dated ....**Feb. 4, 1969**........)

Kind	Size	Mat.	Description—Condition—Improvements—Adaptability
Dwelling Rooms 9	30x36	Brick	Two story, bath, hot air furnace, concrete cellar, closed porches, good condition. Good style farm home in very good condition.
Ells	24x36 24x55 20x24	Fr	Composition siding, fairly good condition, used for car, trucks, and tractors.
Barn	36x120	Fr	Gambrel roof type barn with modern stable with 66 chain ties, gutter cleaner. Good hay storage overhead. Very good general condition.
Milkroom	12x24	Fr	Very good building with new bulk milk tank (not considered real estate).
Pole barn	52x100	Pole	Metal roof, very good, loose housing for dry and young cattle.
Silo	20x50	Conc.	Good. Metal roof.
Machine Shed	44x80	Fr	Clear span truss roof, aluminum siding and roof. Very good machine storage. Farm shop in one end.

Acres	Present Use	Type and Quality of Soil — Topography, Fertility, Ease and Economy of Operation, Drainage
158	Tillage	Mapped as Ontario loam, Palmyra and Hilton gravelly silt loam. Lies gently rolling in large fields convenient to buildings. Adapted to all dairy farm crops and also suitable for some cash crops.
117	Pasture & Woods	A small amount of mixed 2nd growth, some fair feed for young cattle. About 15 acres have been worked over for pasture improvement.
275	Total Acres	Water supply: Farmstead ....**springs & pond, good**........ Pasture ....**springs, ponds, small brook, good**....

Any general acreage shortage of not more than ....**8**.... per cent revealed by title examination which does not result from exclusion of any specific parcel will not affect my estimates of value.

FIG. 27.6. *Federal Land Bank appraisal of a New York dairy farm. (Courtesy Federal Land Bank of Springfield, Mass.)*

D. EARNING POWER OF FARM (Typical Operator — Usual Conditions)

Capacity of farm ......75..... cows, ... hens, ..., ...

Crop	Typical Acres Oper.	Aver. Yield Per Acre	Average Total Prod.	Unit Value Normal Price	Aver. Normal Sales
Hay	100	4	400 tons)		
Ensilage Corn	20	20	400 tons)	Feed on farm	
Oats	20	70	1400 bu.)		
Balance of tillage used as rotated pasture					

Sq. ft. Poultry Buildings ....

Dependable sources of outside income ......limited to small mills and plants in the Rutland area.

INCOME AND EXPENSES (Typical Operator)

Sale of crops .... $..........................
Sale of livestock .... $.....4,000.....
Sale of livestock products .... $...50,115.....
13,000# @ 5.14 .... $...........................
Total Gross Income .... $...54,115...
Real estate taxes.................... $..........636......
Personal property tax .... $..........450......
Ins. bldgs. and pers. .... $..........750......
Building repairs .... $.........800......
Repair, replace. equip. .... $...4,300......
Hired labor .... $...4,800......
Other cash expenses ~~Interest on capital loans not included~~ $...27,379......
$..........................
Total Expenses .... $...39,115...
Net Return .... $...15,000...

Normal net outside income $ ...None...........

E. VALUATIONS—(Acceptability, Requirements)

Present Market Value | $ 60,000 | Normal Value | $ 60,000 | Loanable | $ 41,000

This property ...is... satisfactory security for a loan not to exceed ...40........ years subject to the following requirements being
(is—is not)

fulfilled: ...none...

....

....

This report has been developed assuming that the foregoing requirements would be completed. Without ....
(list requirements that

...., the normal agricultural value of this
may be omitted)

property is $ ...., it is a class .... farm, and it .... satisfactory for a loan not to exceed .... years.
(A,B,C,D)   (is—is not)

F. REMARKS (Comment on particularly good or bad features, special hazards, history of farm, etc.):

This is a very good, productive farm of economical capacity located in a very good area. The buildings and land are being improved continually. Farm would be readily salable at all times.

....

....

....

....

....

....

I have made a personal examination of this property and to the best of my knowledge and belief this report is true and correct.

Date appraised ......February 4,.........19 69

Wm. Smith
Appraiser

FIG. 27.6. *Continued.*

PLAT OF LAND AND BUILDINGS REFERRED TO IN SECURITY REPORT
MADE IN CONNECTION WITH APPLICATION OF:

Applicant	Date Appraised			Title
John Doe	Feb. 4, 1969	by	Wm. Smith	Appraiser
		by		
		by		

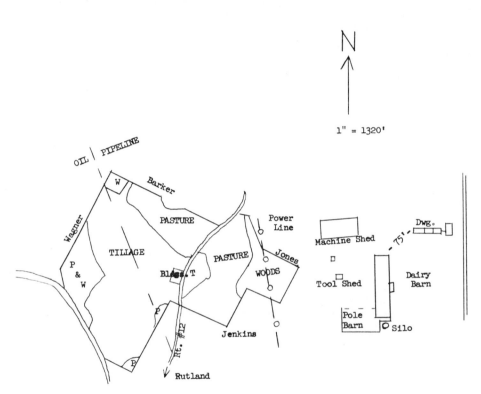

N

1" = 1320'

Record title .... acres
Releases: .... acres dated ....   leaving .... acres
....   leaving .... acres
....   leaving .... acres

FIG. 27.6. *Continued.*

special reports, the experience of the past with all its mistakes set out in bold letters and figures.[1]

The most prevalent mistake in loan appraisal has been overvaluation of poor land. The exceptions which have occurred are usually explained by the fact that the land is so obviously poor that no loans or only small ones were made. The appraisers have tended to hold too close to the average in a given area, putting the better land too low and the poorer land too high. This mistake is not difficult to explain because, as noted in Chapter 20, many of the factors determining farm productivity (such as depth of surface soil, character of subsoil, and drainage) are not evident on a superficial examination. Emphasis on such factors as these may lead to more satisfactory appraisals preventing overvaluation of the relatively low producing land in the future. Another contributing condition is the tendency for men reviewing loans in an office distant from the farm to accept readily those appraisals which are close to the average and to hold back those which are either high or low for the territory. This is the natural reaction of the reviewer who does not know the actual farm, but it works against accurate appraisals by encouraging a concentration around the average.

A factor contributing to overvaluation of poor land has been the effective limitation of loans on the best land. During the years 1919–29, many loan agencies set $100 an acre as their maximum loan. This policy resulted in too many maximum loans and too many heavy loans on poor land. When prices dropped in the early thirties, the mortgages on the best land survived the test while the poor land did not produce enough to pay the required interest and eventually came into the hands of the mortgage lenders. It is unfortunate that the poor land was overvalued, because this is the type of land which suffers most from erosion and hard treatment before and during a foreclosure. If

1. Phil E. Eckert and Orlo H. Maughan, *Farm Mortgage Loan Experience in Central Montana*, Mont. Agr. Exp. Sta., Bull. 372, 1939.

F. F. Hill, *An Analysis of the Loaning Operations of the Federal Land Bank of Springfield from its Organization in March, 1917, to May 31, 1929*, Cornell Univ. Agr. Exp. Sta., Bull. 549, 1932.

E. C. Johnson, *Farm Mortgage Foreclosures in Minnesota*, Minn. Agr. Exp. Sta., Bull. 293, 1932.

J. B. Kohlmeyer, J. W. Van Hoy, and S. O. Kessler, *The School Fund Mortgage Loan Situation in Indiana with Special Reference to Land Use in Martin County*, Purdue Univ. Agr. Exp. Sta., Bull. 422, 1937.

E. H. Mereness, *Farm Mortgage Loan Experience in Southeast Alabama*, Ala. Agr. Exp. Sta., Bull. 242, 1935.

William G. Murray, *Farm Mortgage Foreclosures in Southern Iowa, 1915–36*, Iowa Agr. Exp. Sta., Res. Bull. 248, 1938.

Stanley W. Warren, *Results of Farm-Mortgage Financing in Eleven Counties in New York State*, Cornell Univ. Agr. Exp. Sta., Bull. 726, 1939.

lenders have to make the mistake of overvaluation, how much better it would be if they made this mistake on the best land which is relatively easy to manage and resell.

EXPERIENCE BY AREAS

Not all low-value land was overvalued. In some areas like southern Missouri the land was correctly appraised with comparatively few mortgage foreclosures as a result. An indication of the special areas of distress and of the areas that had few foreclosures is provided in Figure 27.7.

Three areas of distress stand out above the others: the northern Great Plains including North and South Dakota and northeast Nebraska; a portion of the Corn Belt, especially northern and southern Iowa, western Minnesota, and northern Missouri; and portions of the Cotton Belt. An evaluation of these three areas is given by Lawrence A. Jones:

> *Great Plains* — In the Great Plains the heavy debt burden of farmers resulted largely from failure to recognize the production limitations imposed by the climate. Also, the decline in wheat prices to levels not anticipated by many was important. In the eastern Great Plains during the 1920's debt distress was also great but, because of better crop yields, many farmers were able to continue fairly successful operations until the 1930's. There then occurred several years of drought which, combined with very low prices, caused the failure of many indebted farmers who had managed to struggle through the previous decade. The variability of yields and the bunching of bad years in this region constituted a problem in meeting fixed-charge obligations.
>
> *Corn Belt* — The weather risk was not so important a cause of debt distress in the Corn Belt States. In that region the heavy rate of foreclosures was founded on the great inflation of the dollar values of land and of debt which took place in the World War I period. The increase in land values between 1910 and 1920 ranged from 65 percent in Ohio to 137 percent in Iowa. Expansion of farm-mortgage debt during this decade was even more significant. In Iowa it amounted to as much as 176 percent. Overoptimism resulting in widespread land speculation brought the debt load of many farm owners to a level in 1920 that could have been carried only if highly favorable farm incomes had continued.
>
> Variability in the productivity of farm land also influenced the degree of farm-loan difficulty in the Corn Belt. In this region, where so much of the farm land is highly productive, the tendency was to overlook or not properly recognize variations in soil fertility, drainage conditions, and topography, all of which have an important effect on the profitability of farming. As a result, land in the rougher, poorer areas such as southern Iowa, northern Missouri, and parts of southern Illinois and Indiana was overvalued to a much greater extent than was true in other areas. Debt based on these values was particularly excessive

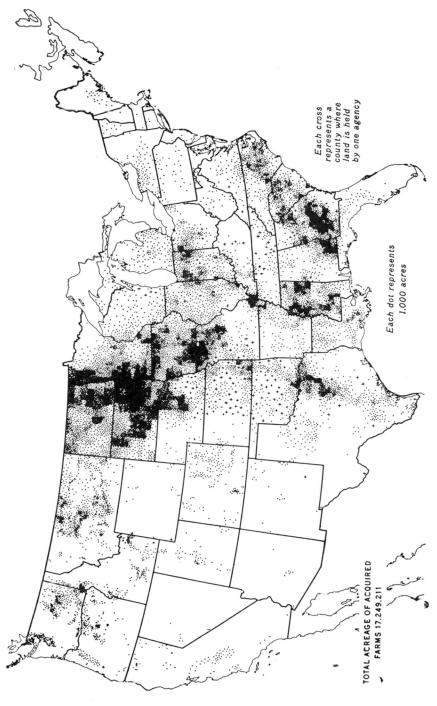

Each cross represents a county where land is held by one agency

Each dot represents 1.000 acres

TOTAL ACREAGE OF ACQUIRED FARMS 17.249,211

FIG. 27.7. Total acreage of acquired farms held by a group of leading lending agencies, Dec. 31, 1934. (Courtesy USDA.)

and, as a result, the farm-loan experience in those areas was the worst in the Corn Belt.

Failure to evaluate adequately the productivity of these "fringe" areas was common in many sections of the country. However, the worst loan experience often did not occur in the poorest farming regions. Where productive limitations were sufficiently obvious or had become generally known after years of experience, farmers and lenders were cautious in their use and extension of credit. Under such conditions the proportion of all farms indebted ordinarily is relatively low and the average size of loans small. In the rough, poor sections of southern Missouri, southeastern Ohio, and the Appalachian region, cases of debt distress were fewer than in some of the better farming areas. Any financial trouble and adjustment in these areas probably occurred at a much earlier time, before their agricultural limitations had become generally recognized.

Eastern Cotton Belt — One of the worst farm-debt trouble spots of the country was in the eastern Cotton Belt and covered most of South Carolina, Georgia, and the southeastern part of Alabama. In this section, which includes the old plantation Piedmont, cotton was very profitable during most of the World War I period and, between 1910 and 1920, dollar values of land increased in South Carolina and Georgia 166 and 152 percent, respectively. The boll weevil, which had been spreading from the West, did little damage in this part of the Cotton Belt before 1920. Thus, there was less restraint on the use of credit here than in other sections of the South where the weevil already had lowered production. The expansion of farm mortgage debt between 1910 and the peak on January 1, 1923, for South Carolina and Georgia was 377 and 474 percent, respectively. In Alabama, where the boll weevil hit earlier, this debt increased only 215 percent.

Following the decline of cotton prices in 1920 the boll weevil damage in the eastern Cotton Belt became very serious, reducing yields by half in many localties. In view of the greatly inflated debt and the dependency of the whole economy on cotton, the fall in prices and yields was disastrous. Farmers had had little experience with other types of farming; unskilled labor hindered a sudden shift; and the heavy soils in many localities were not well adapted to cash crops such as peanuts and tobacco. Those who attempted to continue to produce cotton ran more deeply into debt. Soon farms were neglected, and erosion became more serious. Numerous farms were abandoned. This further reduced earning capacity and land values; losses by both farmers and lenders became great. In the decade from 1919 to 1929 the acreage of cotton harvested in Georgia and South Carolina declined about 27 percent, whereas for the Cotton Belt as a whole it actually increased 28 percent. Thus, low prices of cotton were not the only cause of debt distress in the eastern Cotton Belt. Low production and inflexible farm organization also contributed to the distress.[2]

---

2. Lawrence A. Jones, "Potentialities of Farm Debt Distress," *Agr. Fin. Rev.*, USDA, Washington, D.C., Nov. 1949, pp. 2–4. For a more complete study of this whole subject of loan experience see Lawrence A. Jones and David Durand, *Mortgage Lending Experience in Agriculture*, National Bureau of Economic Research, Princeton Univ. Press, Princeton, N.J., 1954.

TABLE 27.1.   Loan Experience Data of Federal Land Bank in Four County Regions of North Central Iowa

	Area I	Area II	Area III
Number of loans			
1917–32	435	170	92
1933–47	1,307	517	285
Foreclosures			
Proportion of loans	7.8%	6.5%	22.8%
Gain or loss per			
$100 loaned	$ .98 gain	$ .69 gain	$ 3.50 loss
Average loan per acre			
1917–27	$77	$78	$73
1933–36	63	58	50
1937–41	67	59	48
1947–48	85	72	58

Source: A. G. Nelson, *Experience of the Federal Land Bank with Loans in Four North Central Iowa Counties, 1917–47,* unpublished Ph.D. thesis, Iowa State University Library, Ames, 1949, pp. 46, 50, and 61. The four counties are Franklin, Hamilton, Hardin, and Wright.

SPECIAL STUDIES

Detailed studies of loan experience reveal appraisal mistakes. One such study made by A. G. Nelson of federal land bank loans in four north central Iowa counties is a good example.[3] It covers not only predepression and depression years but also postdepression years which show the effort made to correct the earlier mistakes (see Table 27.1). The territory covered in the study was classified into the following three areas: Area I made up mainly of A farms, Area II of B farms, and Area III of C farms. A major portion of the territory fell in Area I, and two-thirds of the federal land bank loans made in the territory from 1917 through 1947 were in Area I with approximately one-fourth in Area II and the remainder in Area III. A map showing one of the four counties classified into the different areas appears as Figure 24.1 in Chapter 24.

Practically all the mortgage foreclosures were federal land bank loans made in the 1917–32 period. In fact up to 1948 only four loans made after 1932 were foreclosed in the region studied.

Loan experience was favorable in Areas I and II but highly unfavorable in Area III. Slightly more foreclosures percentagewise occurred in Area I than in Area II, but when the acquired farms were sold after foreclosure, a somewhat higher gain was realized on the farms in Area I. On the other hand almost one out of every four loans in Area III was foreclosed, and losses

---

3.   A. G. Nelson, *Experience of the Federal Land Bank with Loans in Four North Central Iowa Counties, 1917–47,* unpublished Ph.D. thesis, Iowa State University Library, Ames, 1949.

TABLE 27.2. Federal Land Bank Loan Experience in Western Washington, 1917–32

Net Income Area	Number of Loans	Percentage Acquired	Loss per $100 Loaned
1	187	2	$ 0.30
2	806	5	1.20
3	3,427	9	2.70
4	2,463	16	7.60
5	769	23	11.20
	Total 7,652	Average 12	Average $ 4.20

Source: K. O. Hanson, *Federal Land Bank Loan Operations in Western Washington, 1917–49*, unpublished Ph.D. thesis, Iowa State University Library, Ames, 1950, p. 29.

averaged $3.50 on every $100 of mortgage credit extended in this area.

Back of the foreclosures in Area III was a serious mistake in appraisal. Farms in this area were appraised too high relative to farms in Areas I and II. When the new loan policy was established in 1933, and many loans were made on this basis, a much greater spread was evident between the appraisals of Area I and Area II land on the one hand, and Area III land on the other. This spread widened in the 1937–41 period, providing evidence of the successful educational campaign to place more emphasis on productivity in farm appraisals. The increase in loan per acre in the 1947–48 period was the result in part of the increase in federal land bank loan limit in 1945 from 50 to 65 percent of the normal agricultural value of a farm.

A study of federal land bank loan experience in western Washington by K. O. Hanson (Table 27.2) reveals results similar to those in the Nelson study in Iowa.[4] The same classification system as in Iowa was used with five areas instead of three and with No. 1 the best and No. 5 the poorest. Classification was based on desirability of the areas as a place to farm, to live, and to make money. The results of the Washington study, as shown in Table 27.2, indicate that the higher the net income, the better the loan experience.

The same tendency of overvaluation of low-income land that was evident in Iowa can be seen in this western Washington study. Losses on loans in Area 5 land were especially heavy in comparison with those on the better grades. From this and other studies of a similar nature it is apparent that a major weakness in appraisal has been insufficient spread between the values of good and poor land. Fortunately there is evidence to show that this weakness has been corrected, at least in part.

---

4. K. O. Hanson, *Federal Land Bank Loan Operations in Western Washington, 1917–49*, unpublished Ph.D. thesis, Iowa State University Library, Ames, 1950.

28

SPECIAL ENTERPRISE APPRAISALS

MANY SPECIALIZED FARM VALUATIONS are recognized as farm appraisals even though they may not include any cropland. Important among these special farm types are livestock ranches, irrigated farms, vineyards, orchards, citrus groves, and timber tracts. In addition there are government program allotments for such crops as cotton and tobacco which frequently have a decided effect on land value.

There is no separate appraisal theory or body of principles which applies to these special enterprise appraisals. The three methods—comparable sales, income capitalization, and cost—are used in varying degree in valuing these tracts just as they are in appraising a grain or dairy farm. But there are features of these types which present unique problems in the application of appraisal principles and methods. This chapter indicates how the general principles can be used in appraising certain specialized farm types.

LIVESTOCK RANCHES

The characteristics of a typical ranch appraisal which set it apart from the usual farm appraisal are carrying capacity of range pasture, grazing permits, and water rights for irrigated lands. Carrying capacity is generally measured in terms of acres per animal unit. Grazing permits are the means of obtaining leases or rights to graze livestock on federal lands. These federal lands are managed ordinarily by the Bureau of Land Management or the Forest Service. There are also permits to graze livestock on cooperative range units. Water rights for irrigated lands are especially important because water may be scarce and legal rights exist by which each individual owner can obtain certain amounts of water from a river, stream, or irrigation ditch.

Productivity is the key word in describing carrying capacity in terms of animal units. How many cattle or sheep can the ranch carry? This is a critical question because the answer gives an immediate and clear-cut estimate of ranch size. If the normal operation of Ranch *A* is 300 head of cattle and of Ranch *B* 1,200 head, the appraiser has in these figures not only basic evidence indicating size of the ranches but helpful information in reaching an appraised value. Crop and livestock farms are valued in dollars per acre. Livestock ranches, however, do not lend themselves readily to the dollar per acre figures because the typical ranch is made up of different kinds of land—irrigated hay land, range lands of different carrying capacity, and possibly grazing permits on forest lands. A more meaningful measure of ranch value is dollars per head. For example, three comparable ranch sales were described in an appraisal at $625, $615, and $630 per animal unit of carrying capacity.

Important in carrying-capacity evaluation is accounting for year-round feed for the livestock. This is referred to as a balanced operation. A ranch might have grazing capacity for 1,000 head of cattle to midsummer but only for 750 head from midsummer to fall. Or the grazing capacity might be 1,000 for the grazing season but hay land and wintering facilities might be insufficient to care for the 400 usually kept during the winter after the sales in the fall.

The real test on carrying capacity for the appraiser is his ability to estimate the amount of forage produced by the range. This forage amount is the basis for the appraisal of the number of animals the ranch will support. The same kind of appraisal know-how is needed as the cropland appraiser has who estimates corn and soybean yields. An example of carrying capacity is provided by a Colorado ranch appraisal made by Vern A. Englehorn of the Western Farm Management Co. of Phoenix, Arizona. The ranch in question had fee-owned lands, grazing permits on forest lands, and an interest in a stock growers' cooperative grazing unit. This last unit was an area of range owned by a group of ranchers with each rancher entitled to graze a certain number of livestock in proportion to his holdings in the cooperative unit.

A statement in this appraisal covering carrying capacity and ranch balance follows:

> The major part of the outside leased lands in connection with this ranch is known as the Stock Growers Grazing Unit. This is a block of land composed of Bureau of Land Management land and 6,070 acres of scattered parcels of fee-owned land. There are twelve ranchers who own the 6,070 acres and who use this unit on a ten-year permit basis. The current permit runs from July 1, 1961, to June 30, 1971.
>
> The ownership in this association is determined by the number of head of cattle that can be run by each operator and not by shares of stock. Last year this ranch grazed 780 head in this area, which was 24.1 percent of the cattle carried by the entire association. The ranch paid the Bureau of Land Management a grazing fee of $587.10.
>
> The carrying capacity for the range country of this ranch as set by the Bureau of Land Management on the federal range is as follows:

No. & Class	Period From	To	% Fed. Range	% Fee Owned	AUMs[a]
10 Horses	5–11	5–15	81	19	2
394 Cattle	5–11	5–15	81	19	53
20 Horses	5–16	10–21	81	19	89
788 Cattle	5–16	6–30	81	19	957
616 Cattle	7–1	10–31	81	19	1,996
					3,097

[a] AUM = grazing for one animal unit for one month.

> The combined headquarters areas, deeded range land, and leased property has the following carrying capacity showing the amount credited to the commensurate deeded land and that allowed on the range (Bureau of Land Management acreage figures).

Area Ref. No.	Fee Acres	Fee Land AUMs	Fed. Range AUMs
1	640	2,427	2,837
2	540	70	69
3	600	48	79
4	632	58	69
5	160	27	—
1D	640	76	60
2D	591	71	54
3D	241	11	18
4D	581	1,016	904
6	40	5	—
	149	15	—
		3,924	4,090

This ranch also has the right to use certain forest lands. This is known as the South Beaver Allotment and this is a permit to graze 172 head of cattle from July 1 to October 31 for an additional 688 AUM's. Last year the grazing fee was $165.12.

The analysis of all the resources of this ranch including the rating on the leased land results in a carrying capacity estimate of 725 head of animal units for 12 months. A summary follows:

Fee-owned land	3,924
B.L.M. acres and Stock Growers area	4,090
Forest permit	688
Total AUM's	8,702
Animal units on a year basis:	
8,702	725
12	

An animal unit is usually defined as a mature cow, a 1,000-pound steer, or a mature horse. Five sheep are generally considered equivalent to one animal unit. Other kinds of livestock such as calves are figured according to their equivalent of one cow or a 1,000-pound steer. Carrying capacity then is the number of acres or of feed required to provide for one animal unit expressed either in terms of one month or one year.

Grazing permits are important because of the fees charged and the permanence of the right to graze a certain number of animals at the rate charged. For example the Colorado ranch appraised by Englehorn had a specific forest grazing allotment which permitted the grazing of 172 cattle from July 1 to October 31 for a specified fee. If the charges for grazing were based on competitive rates for grazing lands there would be no difficulty in using these grazing permits. Under competitive conditions the rental fees paid for grazing permits would be similar to the rental rates which exist for cropland. But this is not the case with the federal grazing permits. The Bureau of Land Management and the Forest Service set the fees and the conditions for grazing. As it works out the fees set are usually less than competitive with a value attaching to the grazing permits. Thus if a ranch has a permit to graze a specified number of cattle on federal lands for a definite period of time, this right is likely to have a value in itself because the fees charged are less than what the competitive price would be for this amount of grazing. Since the value of the grazing permits will vary, it is highly important that the appraiser determine the value of any grazing permits that exist.

Water rights, often a major problem on ranches, are essential in the operation of irrigated land if there is a scarcity of water and more than one landowner is concerned. The critical nature of these rights stems from the limited amount of water available at the height of the growing season when everyone wants to use irrigation water. And it is at just such times that the crop can be made or lost depending on the availability of water. Consequently the question of who has the right to the limited amount of water is highly important. Water rights go with the land and not with the landowner, and the value of an irrigated tract depends to a great extent on the priority and amount of the rights. Priority can be significant. Ranch *A* may be the first in line to take a certain amount of water from a stream while Ranch *F* may be the last in the line. Ranch *F* may get no water at the height of the season because the year is a dry one with insufficient water to supply all the legal rights.

An example of water rights is provided by the following list from the Colorado appraisal by Englehorn:

Ditch Number	Priority Number	Priority Date	Amount of Water (cu ft/sec)
31	31	9–1–1882	4.69
31	186	4–2–1925[a]	13.31
31	274	4–10–1909[a]	0.80
11	11	5–1–1878	1.57
11	174	4–2–1925[a]	4.43
3	3	5–20–1876	1.57
3	169	4–2–1925[a]	4.43
10	10	4–15–1878	0.53
10	173	4–2–1925[a]	1.47
139	149	5–2–1906	2.60
139	261	4–2–1925[a]	7.40
48	48	5–1–1884	2.87
48	195	4–2–1925[a]	8.13
262	407	8–20–1951	1 to 15

[a] Indicates supplemental rights in same ditch with a historical right as first given in ditch but a decreed right as of date shown.

Comparable sales and income capitalization can be used in ranch valuation similar to their use in other farm appraisals. In the case of the Colorado ranch, Englehorn used both comparable sales and income capitalization. He arrived at an appraised value of $510,000 or $700 per animal unit for the carrying capacity of 725 units. This value was based on five comparable sales and on a capitalization at 2.7% of an estimated net annual income of $13,900.

VINEYARDS, ORCHARDS, AND CITRUS GROVES

The unique features of fruit and similar specialty appraisals are the heavy cost of land development, the waiting period, and the depreciation of the trees as they age and their production declines. The development cost and waiting period before annual gross income rises above annual expense introduces an entirely new aspect not present in the usual appraisal of a farm or ranch.

During development and waiting the cost method of valuation can be used as a good approximation of estimated value along with any comparable sales that are available. For example the cost of development in the second year of a citrus grove might be something like this:

	Dollars per Acre
Value of bare land as determined by comparable sales	1,200
Cost of land leveling	300
Cost of trees and tree planting	400
Taxes, insurance, and miscellaneous	300
Interest on the above from date of expense to date	250
Total	2,450

Most of these costs were incurred one and two years earlier; the purchase of the bare land was over two years earlier which accounts for a large part of the accumulated interest cost. In the listing of costs it should be noted that the $1,200 for the bare land is not a cost but a market price based on comparable sales. The other items are strictly cost items. (Items which can be reproduced are considered costs while land which cannot in the usual sense be reproduced is not a cost item.) The use of this comparable sale-cost approach assumes that there has been no change in the outlook for citrus during the more than two years this grove has been under development. If prospects were rosier on the day of the appraisal than two years earlier, this value would be conservative since it would not reflect the improved situation which has occurred since the grove was started. But if the prospect were less favorable than two years earlier, the sale value would probably be lower than the $2,450 an acre which represents the original purchase price of the land and costs incurred to the date of the appraisal.

When the grove or orchard reaches the bearing stage the appropriate methods to use are market value and income with less or little use of cost. An example of the income method on both an owner-operator and a landlord basis is given below as presented by James C. Paddock.[1] The orchard, which is 10 years old, has a predicted life of 25 years. Paddock uses straight-line depreciation on the trees which means a depreciation or recapture rate of 4 percent added to the capitalization rate of 8 percent. The raw land value estimated at $500 an acre had a preparation cost of $100 which gives a $600 land cost for the orchard.

---

1. James C. Paddock, "California Orchard and Vineyard Appraisals," *Appraisal J.*, 36(4):576–88, Oct. 1968.

Owner-operator income approach:
Income: 500 units at $1.25 harvested, per acre .... $ 625
Culture and maintenance $225
Taxes 35
Management 25
Interest on operating capital 10

Expenditures .... −295

Net to entire property before recapture .... $ 330
Return of irrigation system investment ($150, 25 years)
   and heating system ($210, 25 years) plus return
   on average investment at 8% ($180 × .08) (rounded)   −29

Net to land and plantings .... $ 301
Less interest on land ($600 at 8%) .... −48

Net to plantings before recapture .... $ 253
Plant life interest 8%
Plant life recapture (return of 25 years) 4%

Capitalization rate (for plant life) 12%
Indicated value plant life (rounded) $2,100
Add value of land (with irrigation system) .... +690
Indicated value land and plant plantings (rounded) ... $2,790
Indicated value as going concern including
   heating equipment ($2,800 + $125) $2,925

Landlord income approach:
    Using the same property, the plant life residual technique
on the landlord share basis is brief, generally more conservative,
and will require lower interest rates to even approximate the
owner-operator basis.
Income cash rent or share, per acre ....$ 300
Real estate tax $35
Return on and of investment in heating and irrigation systems 29   −64

Net to land and plantings ....$ 236
Less 7% on $600 land .... −42

Net to plantings before recapture ....$ 194
Capitalization rate (7% interest and 4% recapture): 11%
Indicated value plant life $1,760
Add value of land with irrigation system .... +690

Indicated value land and plantings $2,450
Indicated value as going concern (with $125 for heating
   equipment) $2,575

    For a different approach to citrus grove appraisals the
reader is referred to the example of a California orange grove
presented in Chapter 27. The example does not include either
the comparable sale or the income capitalization method but

does provide a large amount of factual data which is used by the appraiser in arriving at his final value.

TIMBER VALUE

The valuation of timber tracts introduces a new and unique factor—the value of standing timber. Practically all farmland yields an annual return; that is, an annual crop is harvested such as grain, vegetables, fruit, or grass. With timber tracts the annual growth accumulates year after year until the timber crop is cut and sold. Either an area will be harvested all at once or individual trees wherever they are will be harvested when they are ready to cut.

The only way to put a timber tract on an annual basis is to harvest each year a portion of the timber equal to the amount of timber growth during the year. This annual growth concept is realistic and practical where the tract is on a sustained yield basis; that is, where the timber cut each year equals the growth. For example a timber crop is produced in 50 years and 2 percent of the area or 2 percent of the trees are harvested each year. In this example the whole tract is harvested every 50 years and the volume of standing timber remains approximately the same year after year. The only difficulty with this concept is that there are not many tracts where the annual timber harvest equals the annual growth.

If we have a sustained yield, with annual timber harvested equal to annual growth, the regular sale value and income valuation methods are appropriate and relatively easy to apply. Market value can be ascertained using comparable sales, and income value can be obtained by capitalizing annual net income which can be estimated from annual yield of timber which remains practically constant.

If the timber tract is not on a sustained yield basis the valuation becomes more complicated. In this situation the market value and income methods are still appropriate with an added stumpage or conversion value. Stumpage is the value of the standing timber. It is common to value the standing timber because lumber companies are often interested in buying the timber and not the land. In this case stumpage or standing timber bids by lumber companies are based on the value of the finished lumber less the estimated costs of processing into finished lumber.

In Chapter 27 the federal land bank appraisal of a North Carolina timber tract, Figure 27.4, gave a present market value (stumpage value) of $10,050 for the saw timber and pulpwood on 100 acres. The normal value for the timber on this same tract was $9,000.

Three methods of valuing timber tracts recommended in the California assessor's handbook are explained as follows:

TIMBER APPRAISAL METHODS

The appraisal procedures outlined will be based on (1) analysis of timber sales, (2) conversion value, and (3) capitalization of income.

1. *Sales Analysis*

The sales analysis approach to value gives an indication as to what timber properties are worth to people dealing in this class of property. The thought processes used by buyers and sellers of timber properties vary widely, and the reasons for the purchase price paid or received for a particular property may preclude the use of a transaction as a value indicator. Where effective sales data are available, an appraisal is strengthened by the existence of this type of information.

2. *Conversion Value*

Stated simply, the conversion value is the estimated worth of standing timber to an average operator in the business of converting timber to a saleable product. The selling price of the product minus the total conversion costs including a return for profit and risk, leaves a residual value which is considered a stumpage value. A large part of the timber sold for immediate cutting in the present market is appraised in this manner. Average costs and selling prices generally are used in deriving a conversion value.

3. *Capitalization of Income*

The net periodic income derived from timber property operations may be capitalized or converted into present worth and used as an indicator of market value. Capitalization of income is dependent upon an estimate by the appraiser of the regularity and duration of an income stream.

Income may be obtained from systematically liquidating a timber stand over a period of time, from harvesting an amount equal to the annual increment on a sustained yield basis, or a combination of the two. Benefits may be forthcoming from multiple use sources exclusive of timber production.

In the capitalization process, care should be taken to insure that incomes derived through different periods of time are capitalized separately. In combining values to arrive at an estimate of total worth, values must be added or subtracted at the same point in time.

In general, the appraisal methods outlined in this manual will be concerned with estimating unit market value for timber species by the sales analysis and conversion value approaches, and capitalizing anticipated future benefits to arrive at an estimate of present worth.[2]

Examples of the methods recommended in the California assessor's handbook are presented in Appendix D.

---

2. *The Appraisal of Timber Property*, Assessor's Handbook, AH551M, California State Board of Equalization, Sacramento, 1965, Sections 220.0–220.3.

CROP ALLOTMENTS

Acreage allotments for certain crops like tobacco and cotton represent a special problem in valuation of farms, because the allotment of a specified number of acres of a certain crop has a definite value over and above the value of the farm without this allotment. The major question is What is the value of the specific allotment to a given farm being appraised? Since tobacco is an especially good example of extra value attaching to the acreage allotments, the subject will be discussed from the standpoint of tobacco allotments. Since the tobacco allotment program in effect controls supply by limiting the number of acres in tobacco, the price of an allotment per acre is in one sense the payment to get into a government-regulated monopoly. A study of the tobacco allotments for the years 1934–62 by Seagraves and Manning found, by use of regression analysis, that in the period studied the value of the tobacco allotment over and above the value of the farmland increased substantially.[3] The authors concluded that the continuation of the allotment program had led to increased certainty that the program would become permanent. This has meant that the young man buying a farm with a tobacco allotment had to pay a price which assumed the program would continue indefinitely. In short, the value of the program had been capitalized into the value of the land.

Capitalized value per acre for tobacco allotments independent of the land, in the study of Seagraves and Manning, increased from $42 an acre in 1934 to $993 an acre in 1950 and to $3,281 in 1962. Some of this increase can be explained in terms of rising prices. However, allowing for this inflation by using constant 1957–59 dollars, the results in this same study showed an increase from $89 an acre in 1934 to $1,182 in 1950 and to $3,123 in 1962. From this it is obvious that the main question that is likely to be raised by the purchaser of a tobacco farm is What is the tobacco allotment? Some idea of how important this question is can be gained by noting how much a farmer will pay to rent an acre of tobacco allotment for just one year. In this same study annual rentals per acre in 1963 for tobacco allotments varied from $157 an acre in one county to $327 in another county.[4]

---

3. J. A. Seagraves and R. C. Manning, *Flue-Cured Tobacco Allotment Values and Uncertainty, 1934–1962,* Econ. Res. Rept. No. 2, N.C. State Univ., Raleigh, 1967.
4. Ibid., p. 11. These annual rental figures were obtained from an unpublished M.S. thesis by A. F. Bordeaux, Jr., N.C. State Univ., Raleigh, 1964.

APPENDIX

A

LEGAL DESCRIPTIONS

Descriptions currently used, many of them handed down from generation to generation, are not always in good form. Since appraisers should be acquainted with what is considered best practice, the following extracts from a government publication, *Specifications for Descriptions of Tracts of Land,*[1] are presented. Only the more pertinent statements from this bulletin are included; those interested in a more complete discussion are referred to the bulletin.

I. RECTANGULAR SURVEY

The first series of statements relates to the public domain included in the rectangular survey.

Description of lands within the scope of the public land rectangular surveys should conform to the accepted nomenclature of that system, citing the name of the proper reference meridian, the appropriate township and range numbers and, where necessary, the section and sectional subdivisions shown upon the official plats of survey. Each reference meridian has its own base line; and, therefore, the words "and base line" are usually omitted. The name of the reference meridian should be spelled in full. If the lands have not been surveyed, the description should conform to the legal subdivisions that will, when established, include the lands.

---

1. Prepared by the Committee on Cadastral Surveys, Board of Surveys and Maps of the Federal Government, Washington, D.C., rev. ed., 1942. Quotations taken from pages 11–21.

The township approximately six miles square, containing 36 sections, each one mile square, numbered from 1 to 36, is the unit of survey. The section lines are usually surveyed from south to north and from east to west, with any excess or deficiency placed against the north and west boundaries of the townships.

The section is subdivided into quarter sections by straight lines, connecting established quarter-section corners on opposite boundaries. This unit is usually designated by symbol in tabular descriptions (NW¼ sec. 10; SE¼ sec. 22). The 40-acre unit, resulting from the subdivision of quarter sections into quarter-quarter sections, is designated by symbol, as NW¼NE¼ sec. 10; SE¼NW¼ sec. 22. Occasionally the quarter-quarter section is further subdivided into its aliquot parts by mid-point subdivision. The resulting 10-acre unit is designated by symbol as NE¼NW¼SW¼, sec. 22.

Contiguous units may be combined. For example, if both NW¼ sec. 10 and SW¼ sec. 10 are included, the symbol W½ sec. 10 is used. Where NE¼NW¼ sec. 22 and SE¼NW¼ sec. 22 are included, the resulting 80-acre unit can be designated E½NW¼ sec. 22. In using symbols, the usual punctuation is omitted. Note that the period is omitted after N, NE, S, SE, etc., and that there is no comma and no space between symbols indicating a quarter-quarter section (NE¼SE¼ sec. 10).

The fractional units, usually resulting from the subdivision of the quarter sections in the northern tier and western range of sections or developed because of the existence of meandrable bodies of water or irregular boundaries of claims, are designated by lot numbers (lot 1, sec. 4; lot 1, sec. 15).

Abbreviations

The words "township" and "range" and the designations "north" or "south," "east" or "west" are sometimes written in full when used in the text, but the land description itself should be in tabular form and these terms abbreviated and capitalized where appropriate. The principal abbreviations are as follows:

Townships(s)	T., Tps.
Range(s)	R., Rs.
Section(s)	sec., secs.
North	N
Northeast	NE, etc.

Where two or more township units are to be grouped in the description, the plural abbreviation "Tps." should always be used, even though all the townships have the same number north or south of the base line. The term "range" is abbreviated in the singular or plural as the meaning may require, for example:

Tps. 3 S, Rs. 16 and 17 W
Tps. 4 and 5 N, R. 14 W
Tps. 1, 2, and 3 N, Rs. 6, 7, and 8 W

PREFERRED ORDER

The preferred order of listing is to begin with the lowest-numbered section in each township, giving first the lot numbers in order, then the subdivisions within each quarter section, in the order NE, then the NW, SW, and SE; if parts of the quarter sections are to be described, the same order is to be observed. If several townships are included, the primary order is determined by the range number, beginning with the lowest, and within each range by the township numbers also beginning with the lowest.

Where townships north and south of the base line or east and west of the reference meridian or both are involved, the order of listing is optional but usually follows the order given above; namely, first those north and east of the initial point, followed by those north and west, south and west, and south and east in the order named.

II. METES AND BOUNDS

The second series of statements relates to the approved forms for description by metes and bounds.

DEFINITION OF TERMS

The location and limits of a tract of land may be defined by describing its boundaries; by naming natural or artificial monuments to, from or along which they run; by stating the lengths and directions of the lines connecting successive monuments; or by giving the boundaries of abutting tracts of land.

A monument may consist of an object or mark which serves to identify the location of a line constituting a part of the boundary; it may be either natural such as a river, lake, ledge of rock, tree or ridge; or artificial such as a wall, fence, ditch, marked stone or post.

The type of metes-and-bounds description most commonly used in Executive orders and proclamations is based upon an actual survey of the tract of land involved. The lengths and directions of the lines forming the boundaries are ascertained by the survey and the record thereof describes the monuments marking the corners or angle points. The plat and field notes furnish the data for the description.

If the lines of an adjoining tract of land form a common boundary with the tract in question, the description should note this fact, identifying the adjoining tract by the name of the owner, survey designation, or other appropriate means.

DIRECTION OF LINES

The direction of a line in land surveying is generally expressed by giving the angle from the meridian[2] within one of the four quadrants, referred to either the north or the south point as may be appropriate. When so expressed (e.g. N. 70° 19′ E.; S. 24° 10′ W.), it is called the "bearing" of the line and unless otherwise stated, is to be interpreted as a "rhumb" line—that is, one that maintains a constant angle with the meridian throughout its length.

Occasionally the basic data have been developed in the execution of geodetic surveys and the direction of certain lines may be given by recording the angles which such lines make with the meridian measured clockwise from south. In such cases the lines are

2. As defined by the axis of the earth's rotation.

generally to be regarded as great circles rather than rhumb lines and the angles referred to are designated as azimuths. At any two points on a great circle, the forward and back azimuths of the line differ by 180° plus or minus the angle of convergence of the meridians passing through the points. Where either forward or back azimuths or both are given, they should be so designated.

LENGTHS OF LINES

In land surveying, horizontal distances are generally measured and recorded at the mean elevation of the ground. In some cases a general ground-level datum may be used for an entire survey or group of surveys. However, in geodetic surveys, the horizontal distances are adjusted to sea level.

The unit of measurement employed will usually depend upon the particular class of surveys upon which the description is based. The foot unit is used in many metes-and-bounds surveys and in town-site and city subdivisions; the chain is the linear unit in the public land surveys; and the meter is employed in the cadastral surveys of the Philippine Islands and surveys of similar character. Other units such as the vara and the arpent were employed in the surveys of the Spanish, Mexican and French land grants but slightly different values for these units are found in various localities. Consequently, in using data involving these units, it is necessary to ascertain definite equivalents in terms of the foot or chain units which are to be used in the descriptions. For this purpose, examination should be made of the early surveying records and court decisions.

CONVENTIONAL SYMBOLS AND ABBREVIATIONS

The conventional symbols for degrees (°), minutes ('), and seconds (") of arc should usually be employed in giving the direction lines.

The abbreviations for the units most frequently used are:

Chain(s) ....ch., chs.
Link(s) ....lk., lks.
Foot(feet) ....ft.

SEQUENCE AND CLOSURE

The bearings and distances of the courses connecting the turning points or corners of a tract are usually given in regular order around the perimeter thereof. Each course is written on a separate line and if any corner or course is coincident with a corner or course of another tract, notation should be made of this fact. The final course should note the return to the place or point of beginning.

An exception to the foregoing is found in the description of a tract of specified width on each side of a definitely described center line such as a right of way. In such a case the terminal point as well as the beginning point should be fully indentified.

POINT OF BEGINNING

The location of a tract of land may be defined by stating its position in relation to established monuments of known position or by stating its geographic position (latitude and longitude). In metes-and-bounds descriptions this is generally accomplished by a complete description of the point of beginning. The information furnished should be sufficient to enable a competent surveyor to locate and identify the initial point. Frequently a statement regarding nearby topographic or cultural features or objects is of great value. The general location (State, county, etc.) is usually given in the first part of the Executive order or proclamation and need not be repeated in the description.

If the point of beginning is an established corner of an official survey or is connected by survey to such a corner, the latter should be described by corner and survey number or other appropriate designation without detailed description of the monument itself. The latitude and longitude should be given unless the beginning point is a corner of the public-land rectangular surveys or connected by survey to such a corner.

B

Farm Real Estate Taxes: Levies per Acre by State, Average 1909–13 and Selected Years, 1930 to 1967 (in dollars)

Region	State	Average 1909–13	1930	1940	1950	1960	1967
Northeast	Maine	0.28	0.81	0.84	1.29	1.94	2.62
	New Hampshire	.31	.76	.88	1.39	2.13	3.21
	Vermont	.21	.58	.54	.86	1.42	2.53
	Massachusetts	.81	2.16	2.70	3.41	6.38	9.78
	Rhode Island	.46	1.35	1.70	2.46	6.10	8.91
	Connecticut	.48	1.63	1.86	3.20	5.91	8.90
	New York	.41	1.04	1.10	1.69	3.13	4.53
	New Jersey	.72	2.74	2.31	3.78	9.23	15.70
	Pennsylvania	.49	1.30	.98	1.38	2.39	3.65
	Delaware	.25	.50	.33	.58	1.07	2.00
	Maryland	.38	.93	.81	1.18	2.32	3.95
Lake States	Michigan	.43	1.34	.46	.80	2.36	3.89
	Wisconsin	.34	1.05	.78	1.57	2.50	3.71
	Minnesota	.23	.87	.66	1.30	2.09	2.76
Corn Belt	Ohio	.47	1.36	.69	1.08	2.21	3.11
	Indiana	.52	1.47	.76	1.36	2.42	4.23
	Illinois	.40	1.16	.98	2.08	4.03	5.67
	Iowa	.40	1.24	1.00	1.92	3.06	4.18
	Missouri	.14	.45	.32	.52	1.09	1.46
Northern Plains	North Dakota	.14	.38	.22	.43	.65	.82
	South Dakota	.13	.44	.28	.46	.69	.94
	Nebraska	.16	.44	.30	.64	1.11	1.53
	Kansas	.19	.55	.36	.72	1.16	1.74
Appalachian	Virginia	.11	.34	.27	.46	.83	1.35
	West Virginia	.12	.46	.16	.23	.31	.50
	North Carolina	.08	.59	.37	.51	1.00	1.51
	Kentucky	.15	.43	.32	.63	.74	1.26
	Tennessee	.14	.47	.38	.46	.66	.94
Southeast	South Carolina	.13	.40	.30	.38	.71	.97
	Georgia	.11	.30	.14	.32	.43	1.04
	Florida	.11	.70	.32	.54	1.42	2.84
	Alabama	.09	.25	.20	.25	.30	.43
Delta States	Mississippi	.14	.63	.34	.37	.42	.58
	Arkansas	.15	.32	.28	.32	.73	1.06
	Louisiana	.15	.57	.31	.40	.67	.87
Southern Plains	Oklahoma	.19	.47	.24	.36	.51	.73
	Texas	.06	.23	.14	.26	.47	.67
Mountain	Montana	.06	.14	.11	.21	.31	.48
	Idaho	.24	.64	.45	.85	1.21	1.54
	Wyoming	.03	.09	.06	.14	.19	.31
	Colorado	.11	.28	.20	.35	.59	.80
	New Mexico	.02	.07	.04	.09	.15	.17
	Arizona	.06	.22	.13	.36	.59	.77
	Utah	.15	.52	.30	.47	.59	.77
	Nevada	.06	.15	.15	.17	.26	.37
Pacific	Washington	.28	.71	.32	.62	1.15	2.10
	Oregon	.15	.40	.33	.80	1.54	2.39
	California	.35	1.14	.83	1.87	3.95	8.82
48 States		.21	.57	.39	.69	1.22	1.89
	Alaska	...	...	...	...	1.81	1.92
	Hawaii	...	...	...	...	1.73	2.58
United States		...	...	...	...	1.22	1.89

Source: USDA, *Farm Real Estate Taxes, Recent Trends and Developments*, RET-8, Dec. 1968.

C

Present Value of 1 at Interest Rates of 3 to 8% and from 1 to 25 Years

Years	Interest Rates					
	3%	4%	5%	6%	7%	8%
1	.9709	.9615	.9524	.9434	.9346	.9259
2	.9426	.9246	.9070	.8900	.8734	.8573
3	.9151	.8890	.8638	.8396	.8163	.7938
4	.8885	.8548	.8227	.7921	.7629	.7350
5	.8626	.8219	.7835	.7473	.7130	.6806
6	.8375	.7903	.7462	.7050	.6663	.6302
7	.8131	.7599	.7107	.6651	.6227	.5835
8	.7894	.7307	.6768	.6274	.5820	.5403
9	.7694	.7026	.6446	.5919	.5439	.5002
10	.7441	.6756	.6139	.5584	.5083	.4632
11	.7224	.6496	.5847	.5268	.4751	.4289
12	.7014	.6246	.5568	.4970	.4440	.3971
13	.6810	.6006	.5303	.4688	.4150	.3677
14	.6611	.5775	.5051	.4423	.3878	.3405
15	.6419	.5553	.4810	.4173	.3624	.3152
16	.6232	.5339	.4581	.3936	.3387	.2919
17	.6050	.5134	.4363	.3714	.3166	.2703
18	.5874	.4936	.4155	.3503	.2959	.2502
19	.5703	.4746	.3957	.3305	.2765	.2317
20	.5537	.4564	.3769	.3118	.2584	.2145
21	.5371	.4388	.3589	.2942	.2415	.1987
22	.5219	.4220	.3419	.2775	.2257	.1839
23	.5067	.4057	.3256	.2618	.2109	.1703
24	.4919	.3901	.3101	.2470	.1971	.1577
25	.4776	.3751	.2953	.2330	.1842	.1460

D

DATA FROM STATE ASSESSMENT MANUALS

Appraisal manuals containing a wide range of information on land and building valuation for assessment purposes have been published in recent years by state tax commissions. Although designed specifically for assessors, these manuals lend themselves readily to use by appraisers for other types of farm valuations. A few examples have been selected for this section to show the type of material contained in these valuation handbooks. Most of the manuals cover both urban and rural appraisals; and in the rural appraisal part, cover both land and building valuation. In rural land appraisal most manuals include the use of county soil survey reports and SCS use capability maps, and since this phase of appraisal was discussed in detail earlier it will be omitted from the examples presented.

I. RURAL APPRAISAL PROCEDURE:

A. From *Assessor's Manual, Real Estate, State of Arkansas*[1]

1. Obtain aerial photographs size 1" = 1320 ft., and stereoptic coverage 60% overlap.
2. Obtain pocket stereoscope for visual inspection of aerial photographs.

---

1. *Assessor's Manual, Real Estate, State of Arkansas*, published by Assessment Coordination Division, Arkansas Public Service Commission, Little Rock, 1956, pp. 43–44.

3. Obtain acreage-counting grid for computing acres in irregular tracts.

4. Any soil information or map that can be supplied by the Soil Conservation Service for transferring data to aerial photos.

5. Locate and mark on photos the section lines and mark numbers of sections in the exact center of section on photo.

6. Obtain proper farm record card and get all information pertaining to ownership and description on card, dividing each ownership into 40-acre tracts on the card.

7. Take the individual farm record card and show house lot in proper 40 as it shows on aerial photo, designated as one or two acres in house lot priced at the high value in the forty. Then take out acreage for road 1320 ft. $\times$ 33 ft. $= 1$ acre along the side of each forty on which the road occurs or cuts across. Next, show breakdown or Soil Capability Group Ratings as delineated on aerial photos. Upon completion of this, the card is ready for the field check and appraisal of rural buildings. After the land values have been established and the field work completed, the card may be returned to the office and the dollar per acre value put on each Capability Group Rating. It may be calculated, totaled, and any percentage due to accessibility may be applied. This completes the rural land appraisal procedure.

B. From *Iowa Real Property Appraisal Manual*[2]

1. Prepare adequate maps for the areas and tracts to be appraised.

2. Classify the land by soil type distribution, capability of use, or a combination of these methods to reflect the variations in values caused by physical features.

3. Make a market value study of the area by compiling and analyzing data relative to the sales of farmlands in the area. (Sales data covering a period of not less than 5 years is recommended.)

4. Make a study of the earning capacity of the area by compiling and analyzing long-term production records (10- to 15-year period), prices received from sale of crops, costs of production, and rate of capitalization. Establish base values from this study.

5. Calculate the values of individual parcels by (a) applying the appropriate base value to each kind of soil and land use, and (b) adjusting the base values as necessary to reflect the effects of benefits or hazards peculiar to the parcels.

6. Record the information on property record cards designed for the purpose of developing and maintaining an inventory of individual parcel values within a given area.

7. Determine special influences to individual tract values.

8. Value building by replacement cost-depreciated method. Add to land values to obtain total value of rural parcel.

---

2. *Iowa Real Property Appraisal Manual*, Property Tax Division, Iowa State Tax Commission, Des Moines, 1959, p. 36.

C. From *Real Property Appraisal Manual, Commonwealth of Kentucky*[3]

Establishing Base Values Using Sales Comparison Method

The method of establishing base values for the various land classes involves the use of information on land use, together with data on sales and trends as to land use. An effective way to obtain and apply such information is through contacts with the secretary of the Farm Loan Association, local land appraisers, farm managers, and others having thorough knowledge of land conditions and values.

In the mass appraisal of property, a local land value advisory committee composed of representatives of local farm and other related business groups is usually appointed to assist in the establishment of base values.

Farm sales are the key to the establishment of base land values for the various land classes. Sale data can be secured from the recorded deeds in the county clerk's office, real estate men, and various loan organizations. The sale price of each property should be verified through contacts with buyers, sellers, local real estate officers, and other qualified persons. The sales which are determined to be representative of the market are plotted on a county map to insure that all sections of the county are represented. By use of the manual, the reproduction cost less depreciation of the buildings is estimated and deducted from the total sale price to give a dollar value applicable to the vacant land, type of road, tobacco allotment, and location of the property.

II. USE OF AERIAL PHOTOGRAPHS:

A. From *Assessor's Manual, Arkansas*[4]

There is much to be said for aerial photo-maps as base maps for the development of tax maps and other maps. The very nature of a photograph is assurance of complete visual coverage of all detail. Refinement in photogrammetric equipment and techniques has now resulted in a fully acceptable accuracy for assessment purposes and practices. Several important aspects of aerial photo mapping may be enumerated and outlined as follows:

Aerial maps for base tax maps have a twofold advantage in that they are more rapidly and more economically produced. (Savings both in time and money of over 50% are usual.)

Flights are made so that the contact prints for the area flown overlap sufficiently so as to provide stereoscopic coverage of the entire area. Then, by viewing adjacent prints under a stereoscope, it is possible to ascertain relative elevations, to spot buildings or

---

3. *Real Property Appraisal Manual*, Department of Revenue, Commonwealth of Kentucky, 1962, pp. 73, 74.
4. Op. cit., p. 8.

structures not previously assessed, to determine swampy areas, to see wooded and rocky areas, and to observe areas that are almost inaccessible on foot.

There are two types or procedures that can be followed in aerial mapping. The first is to use direct enlargements. The second is to make a mosaic of a number of contact prints. The use of the direct enlargements will result in a lower cost. The process of making a mosaic requires cutting, fitting of contact prints (by controlled or built-up template methods) and rephotographing. The main advantages of the mosaic process lie in the increased accuracy and the more uniform sheet arrangement.

III. APPRAISAL OF FARM BUILDINGS:

A. From *Iowa Real Property Appraisal Manual*[5]

FARM BUILDING APPRAISAL AND ADJUSTMENTS

Iowa assessors are required by law to evaluate the land and improvements separately, but they are also charged with the responsibility of reflecting a true market value picture of the total. Since the buildings are sold as part of the farm, they then must be considered from the assessment standpoint and a value placed on them which will reflect the influence of the building value to the total market value of the farm.

One approach to the derivation of a realistic value on individual types of farm buildings refers principally to the obsolescence of the building as reflected by its use at the time of the appraisal.

After the replacement cost of the building has been determined and physical depreciation applied for its physical condition, consideration of the actual use of the building and its adaptability to that use will reveal its individual efficiency as a contributing factor in the operation of the farm.

The old-fashioned gable (or hip-roofed) horse and hay barn may be totally obsolete for the purpose for which it was originally built. Yet it still retains usefulness from the standpoint of a general livestock shelter building. Its physically depreciated replacement value would require an adjustment to reflect the change in its use. To determine the percentage of adjustment necessary to show this obsolescence, the difference between its depreciated replacement cost as it now stands and the cost of replacement with a modern pole type shelter building erected for the same purpose would provide a basis for arriving at this percentage adjustment.

It must be remembered that the cost of converting a certain type of building to one of another type could very well reflect the obsolescence of the building in its original state.

For example the old corn crib as we generally know it in Iowa is rapidly becoming obsolete due to the introduction of self-propelled picker-sheller type of corn harvesting machines and the advent of grain-drying equipment. However, many enterprising grain farmers are converting the old corn crib into a perfectly useful shelled corn or small grain storage unit, complete with drying ducts and augers

---

5. Op. cit., pp. 122–23.

for grain handling. This conversion cost is generally not as great as the cost of replacement of the corn crib with an all new grain storage bin of equal capacity equipped with new drying and grain-handling equipment. This cost of conversion may very well represent the percentage of obsolescence to be reflected in the depreciated reconstruction cost of the slatted corn crib.

FARM DWELLINGS

A problem encountered in recent years in the appraisal of farm dwellings is the appraisal of a new high-quality house erected on the farm, either as a replacement for the former house, or as a second house to serve as a home for the retiring owner who does not wish to leave the farm.

There is a strong tendency for most assessing officials to assume that the cost of such a house can never be recovered in the case of sale. Therefore it is appraised at a considerably lower value than a comparable house in town.

While it is true that the rural homeowner must install and maintain his own utilities and sanitary services, it is also true that a measure of this overhead is reduced when the adjustment is made on the total farm value for its locational position with reference to the social and market facilities of the community.

It is necessary for the assessor to study the sales trends in his jurisdiction to ascertain what the local market value of the rural residence is for either a replacement or an additional dwelling on the farm. This problem should also be studied from the area basis and can be accomplished by a study of the rural sales of several neighboring assessment districts carried out on a cooperative basis by the assessors of those districts.

In view of the modern ease of transportation, the trend toward rural residence, and the standard of living enjoyed by farmers, it appears more logical to accept the amount of investment involved in the replacement or construction of the house as a fair measure of its value and to assess accordingly, unless and until the market proves this assumption incorrect.

IV. APPRAISAL OF TIMBER:

A. From *The Appraisal of Timber Property*, California State Board of Equalization[6]

EXAMPLES OF THE APPLICATION
OF THE TIMBER PROPERTY APPRAISAL METHODS

To assist the assessor and his staff in reaching a better understanding of the application to actual properties of the timber property appraisal methods outlined in this manual, the following hypothetical cases for the various types of operations are presented:

Assume for purposes of these examples a present appraised stumpage value of $30 per MBM (thousand board feet) for sugar

---

6. *The Appraisal of Timber Property*, Assessors' Handbook AH551 M, California State Board of Equalization, Sacramento, 1965, Sec. 910.

pine, $25 per MBM for ponderosa and Jeffrey pine, $18 per MBM for redwood, $15 per MBM for Douglas fir, $8 per MBM for white and red fir, and $5 per MBM for cedar. A present appraised value of $10 per acre will be used for land. As an illustration, a safe interest rate of 5½% which includes an allowance for property taxes will be used with a factor of safety of .73 for capitalizing annual incomes.

EXAMPLE: *Long Term Liquidation*

A timber operation contains 25,000 acres of land and 600 MMBM of merchantable timber. The operator plans to cut 30 MMBM annually for 20 years. The timber stand consists of 60 percent redwood and 40 percent Douglas fir and will be cut in these proportions each year.

Present Worth of the Timber

Step 1 — Determination of total future benefits from timber.
          Annual cut
            18 MMBM redwood × $18 stumpage value/M = $324,000
            12 MMBM Douglas fir (40% of cut)
            30 MMBM — Total
          Value of annual cut
            18 MMBM Redwood × $18 stumpage value/M = $324,000
            12 MMBM Douglas fir × $15 stumpage value/M =   180,000
                                      Total value   = $504,000
          Total future benefits
            $504,000 value of annual cut × 20 years operating period
            = $10,080,000.
Step 2 — Determination of the present worth valuation factor (discount factor).
          Using compound interest tables for an annuity, select the factor for a safe rate return of 5½% (safe rate plus allowance for taxes) and a discount period of 20 years.
          Discount factor for 20 years @ 5½% = .5975
          Multiply safe rate return discount factor times the factor of safety to arrive at valuation factor. For this discussion we assume a factor of safety of .73.
          Valuation factor = .5975 × .73 = .4362
Step 3 — Determination of present worth of timber.
          Present worth = Total future benefits × valuation factor
                        = $10,080,000 × .4362 = $4,396,896
Land Value
          25,000 acres @ $10/a = $250,000 Total land value
Present Worth of Land and Timber
          Present worth, timber = $4,396,896
          Present worth, land   =   250,000
          Total present worth = $4,646,896

EXAMPLE: *Sustained Yield Operation*

A timber property consists of 25,000 acres of taxable young-growth timber. Calculations indicate an annual growth of 10.0 MMBM in perpetuity. The annual cut of 10.0 MMBM will consist of 50% ponderosa pine, 20% sugar pine, 25% white fir, and 5% incense cedar.

Step 1 — Value of annual cut.
            5.0 MMBM ponderosa pine @ $15 = $75,000
            2.0 MMBM sugar pine   @ 15 = 30,000

ASSESSOR'S COPY
RURAL REAL ESTATE TRANSFER

Twp, County or City _____

Date_____ Book_____ Page_____ Type of Transfer_____

Grantor_____Improved _____

Grantee_____Unimproved _____

Legal Description_____

_____

_____Acres_____

Revenue Stamps_____   Assessed

Indicated Consideration_____   Value_____
(Stamp value less $250)

Verified Consideration_____Sales/Assessment Ratio_____

(Use back of card for comments on transfer)

Sales Record Form (Sample No. 1)

REAL ESTATE TRANSFER

Deed Reference_____

Grantee_____County_____

Grantor_____Township_____

Location: (Street address, lot, block for town property - section number or other
approximate location for township property)_____

_____

_____

Size of Land Parcel: (dimensions or number of acres)_____

Vacant _____Improved_____Description of Improvements:_____

_____

_____Minerals included:_____

Federal
Date of Sale:_____Consideration:Stamps $_____

Name of Agent or
Attorney_____Remarks:_____

SALES DATA CARD

Sales Record Form (Sample No. 2)

FIG. D.1—*Sales record forms. (Courtesy Iowa State Tax Commission.)*

$$\begin{array}{llll}
2.5 \text{ MMBM white fir} & @ \$5 = & 12,500 \\
0.5 \text{ MMBM incense cedar} & @ \ \ 5 = & \underline{2,500} \\
\hline
10.0 \text{ MMBM} & & \$120,000
\end{array}$$

Step 2 — Present worth of standing timber property.

This step consists of capitalizing the annual income at the safe rate return of 5½%, then applying the factor of safety to give allowance for risk.

$$\begin{aligned}
\text{Present worth} &= \frac{\$120,000 \text{ annual income}}{.055 \text{ safe rate}} \times .73 \text{ factor of safety} \\
&= \$2,181,818 \times .73 \\
&= \$1,592,727
\end{aligned}$$

In the determination of the value of a sustained yield unit, no land value is computed directly. In this case as well as in above example, after a total value of a property has been computed, a land value per acre may be shown on the records to be consistent with the premise that bare land values are basic and other values are residual. In this manner land values remain constant throughout an operating period, and timber values are shown as a changing residual. This generally does not necessitate a large change as the land value is only a small part of the combined value for land and timber.

As an example of how the value of the above sustained yield unit may be divided between land and timber, assume, as before, a land value of $10 per acre.

$$\begin{aligned}
\text{Total present worth of land \& timber} &= \$1,592,727 \\
\text{Land @ \$10/a for 25,000 acres} &= \$ \ \ \ 250,000 \\
\text{Timber value} = \text{total value} - \text{land value} \\
&= \$1,592,727 - \$250,000 \\
&= \$1,342,727
\end{aligned}$$

Using this approach, land will show the same value each year as long as the $10 per acre figure is used. Timber too remains constant for a sustained yield unit but will vary each year for the other operations listed.

B. Further reference on timber assessment: *State Guides for Assessing Forest Land and Timber—1966,* Misc. Publ. 1061, compiled by Ellis T. Williams, Forest Service, USDA, Washington, D.C., 1967. This publication contains excerpts from and references to assessment guides covering forest land and timber for each of the 36 states which had issued such assessment guides.

V. MARKET VALUES AND ASSESSMENT-SALE RATIOS:

A. From *Assessor's Manual,* State Tax Commission of Missouri[7]

The assessor should be guided whenever possible by his knowledge of recent or current selling prices. While one sale does not necessarily establish fair value, a file containing numerous sales can

---

7. *Assessor's Manual,* State Tax Commission of Missouri, Jefferson City, 1957, p. 24.

be very useful to the assessor in appraising properties which can be compared with properties for which there are known sales.

Less recent sales can also be used if they are adjusted forward to present market levels. (The manual includes a table for assessors to use in adjusting sales of previous years to the current year.)

B. From *Iowa Real Property Appraisal Manual*[8]

It is important that a perpetual record of rural land sales be maintained by the assessor. It is this information that enables him to watch the trends of market values and the importance placed on certain areas of his jurisdiction by the market demand.

Since farm sales per year are seldom numerous enough in a given area to actually establish a stable basis for determining the market value of the land, it is necessary that sales for period of not less than five years be compiled, qualified concerning their bona fide status and analyzed to determine their usefulness in a reconciliation of market demand against the actual earning capacity of the land derived through capitalization of income.

The types of sales record cards designed for use in keeping a sales record file are illustrated in Figure D.1. Through the use of the information accumulated in this file, the assessor will be able to keep a current record of the market trends throughout his assessment districts.

---

8. Op. cit., pp. 65, 66.

E

NOTE ON REFERENCES

The United States Department of Agriculture is a gold mine of information for the appraiser. Of special importance are the following publications issued by this Department: *Statistical Yearbook*, containing useful series of yield and price data; *Agricultural Finance Review*, including, in addition to articles, statistics on taxes and related finance factors; and *Farm Real Estate Market Developments*. This last publication, appearing several times a year, should be a "must" in every appraiser's tool kit. It contains estimates of farm real estate values as of March 1 and November 1 for each year with comparisons for the same dates on previous years. Another important reference source is the state agricultural experiment station, which is generally located with the land-grant college or university in each state. An up-to-date list of publications available from this source should be kept on hand, and this should include a list of soil survey bulletins for the counties in which an appraiser is operating.

Among the government agencies which make appraisals or publish materials of interest to appraisers are the following: The Farm Credit Administration, Farmers Home Administration, and Soil Conservation Service, all of which can be reached through the USDA in Washington, D.C.

The International Association of Assessing Officers issues publications that contain helpful material, particularly for assessors but also of interest to appraisers in general.

The following periodicals frequently contain articles on valuation and farm appraisal:

Journal of the American Society of Farm Managers and Rural Appraisers.

The Appraisal Journal. American Institute of Real Estate Appraisers.

Appraisal Institute Magazine. Appraisal Institute of Canada.

Right of Way. American Right of Way Association.

Finally, the appraiser may obtain up-to-date lists of available materials, some of which apply to farms, by writing to:

American Institute of Real Estate Appraisers, 155 East Superior St., Chicago, Ill. 60611

American Right of Way Association, 500 S. Virgil Ave., Los Angeles, Calif. 90005

American Society of Appraisers, 1101 17th St. N.W., Washington, D.C. 20036

American Society of Farm Managers and Rural Appraisers, Box 295, DeKalb, Ill. 60115

Appraisal Institute of Canada, Suite 502, 177 Lombard Ave., Winnipeg 2, Manitoba, Canada.

International Association of Assessing Officers, 1313 E. 60th St., Chicago, Ill. 60637

Society of Real Estate Appraisers, 7 S. Dearborn St., Chicago, Ill. 60603

INDEX

523

203925

DATE DUE

GAY

PRINTED IN U.S.A.